Urodynamics

Upper and Lower Urinary Tract

Edited by

W. Lutzeyer · H. Melchior

Redaction Assistants

P. Rathert · K. Kraemer-Lutzeyer

With 209 Figures

Springer-Verlag
Berlin · Heidelberg · New York 1973

Professor Dr. W. LUTZEYER and Privatdozent Dr. H. MELCHIOR,
Abteilung Urologie der Medizinischen Fakultät an der Rhein.-
Westf. Techn. Hochschule Aachen, D 5100 Aachen, Goethestr.
27-29

ISBN-13: 978-3-642-65642-2 e-ISBN-13: 978-3-642-65640-8
DOI: 10.1007/978-3-642-65640-8

Softcover reprint of the hardcover 1st edition 1973

Preface

Once urology was established as a medical specialty, the initial anatomic-morphological point of view was soon supplemented by functional considerations and diagnostic procedures, thanks to the introduction of the cystoscope (NITZE, 1877) and retrograde pyelography (VOELCKER, JOSEPH and VON LICHTENBERG, 1903). Intravenous urography (VOLKMANN, 1924; VON LICHTENBERG, PFLAUMER and SWICK, 1929) lent roentgen examinations a more functional aspect.

The word "urodynamics" refers to the functional approach to the urinary tract – from the renal pelvis and calyces, the ureter and bladder down to the bladder neck and urethra. The embryologically defined topographic regions of the urogenital tract are peripheral relays, while the actual site of action of both normal and pathological urodynamics is to be found in the various hollow muscular organs. Despite extensive basic research and much knowledge acquired in this field, the active transport system between site of production and storage, i.e. the renal pelvis and ureter, is still regarded in clinical practice mainly from the anatomic-morphological viewpoint.

It is not easy to obtain adequate information for the clinical assessment of functional disturbances of the ureter. For instance, retrograde exploration methods may often be complicated by infection.

Clinically applicable methods for functional assessment of the upper urinary tract are (RUTISHAUSER): direct observation of the ureteral orifices, intravenous and retrograde visualization by roentgen contrast medium, sequential pyelography, roentgen kymography, and cinematography. Such methods allow good functional exploration by an experienced investigator but they yield hardly any objectively reproducible records.

Electromyography has not yet found any clinical application despite the investigations of TUCHIDA. It is still at an experimental stage and not easily interpreted by means of objective parameters. After technical refinement it may qualify as a supplementary method for use in combination with other parameters of flow and pressure.

Manometric methods permit refinement of urodynamic diagnosis by the introduction of dynamic pressure monitoring with electromanometric methods (KILL). The pressure transducer can be situated with the tip of the probe in the organ itself or extracorporally. The latest achievement of urodynamic diagnosis is represented by tiny ureteral probes designed for different measurements, for example, Aachen model (MELCHIOR, LUTZEYER, NAUMANN). Experimental and clinical progress is beginning to be seen.

The bioengineer is integrated in the urodynamic team in the Department of Urology at Aachen. The main targets of their combined efforts are the problem of incontinence, i.e. the bladder neck, the development of models of the ureter, and pressure problems in the storage and voiding organs of the urogenital tract. It proved possible to simulate the pressure profile in the ureter, to construct physical models of the ureter, and to derive functional parameters from them (SHAPIRO, WEINBERG and BOYARSKY).

During the first six years of the newly established medical faculty at Aachen team-work with the classical technical faculties has been very successful. In urology this team-work between physician and engineer is attested by the development of the cryoscalpel (LUTZEYER, LYMBEROPOULOS and WERNER), a combined pressure and flow probe for uroheomanometry in the ureter (MELCHIOR, LUTZEYER, NAUMANN and SIMHAN), and of the ultrasonic lithotriptor (LUTZEYER, TERHORST, POHLMANN and CICHOS).

Pharmacodynamics, or the study of the influence of drugs on the urinary tract, has been advanced by the application of new functional investigation methods (BOYARSKY, HINMAN, LUTZEYER, MELCHIOR, RUTISHAUSER).

It is clear that bladder and bladder neck no longer ought to be considered separately. Combined bladder tonometry, measuring sphincter resistance, urine flow, and maybe the pressure profile in the urethra and the activity of the pelvic floor, can supply information about normal and pathological processes of micturition which are specific to age, sex and certain diseases (BOYARSKY, HINMAN, MILLER, TANAGHO, TURNER-WARWICK, WHITESIDE, ZINNER).

Thus, when we speak of "urodynamics" we do not mean an isolated specialty directed to a specific region of the urogenital tract. It is a discipline that includes all dynamic processes from the renal calyces to the ureter, bladder and bladder neck, whether isolated or in combination. It concerns the relevant anatomy, physiology, instrumentation, pharmacodynamics and clinical projection.

Therefore the sphere of interest of urodynamics includes not only malformations like vesico-ureteral reflux but also clinical entities like hypertrophy of the prostate which influences urinary transport.

It is self-evident that the cardiologist uses the ECC not just routinely but continuously in an intensive-care unit, where it is computerized and always on call. It is self-evident, too, in gastroenterology and pulmology that a functional analysis is based on readings of different electronic probes. In urology, however, we tend to rely on optical control by cystoscopy, static pictures of intravenous urography, chromo-cystoscopy and some biochemical data in order to reach a functional interpretation that is based on this information alone.

From this awareness we derive our demand that the functional point of view be given a broad experimental and clinical basis. This will require not only improved cooperation between technical and medical disciplines, and worldwide exchange with interested groups, plus technical and instrumental refinements but also the projection of the urodynamic approach on to clinical practice.

Aachen, autumn 1973 WOLFGANG LUTZEYER

Contents

List of Contributors*

ALLEN, T. D., Department of Surgery, University of Texas, 5323 Harry Hines Boulevard, Dallas, TX 75235, USA (p. 137)

AUVERT, J., 78, Avenue de Suffren, 75 Paris 15e, France (p. 179)

BALOGH, F., Pécsi Orvostudományi Egyetem, Urológiai Klinikája, Munkácsy Mihály u. 2, Pécs, Hungary (p. 80)

BARBER, F. E., Departments of Urology, Chemical Engineering, Electrical Engineering and Center for Bioengineering, University of Washington, Seattle, WA, USA (p. 274)

BASTIAN, H.-P., Abteilung Urologie, Klinische Anstalten der RWTH Aachen, 5100 Aachen, Goethestr. 27–29, W. Germany (p. 270)

BISCHOFF, P. F., Urologische Klinik, Elisabeth-Krankenhaus, 2000 Hamburg 20, Heilwigstr. 28, W. Germany (p. 189)

BOEHNCKE, H., Kinderklinik Hochallee, 2000 Hamburg 20, W. Germany (p. 189)

BOTTACCINI, M. R., College of Engineering, Aerospace and Mechanical Engineering Department, The University of Arizona, Tucson, AZ 85721, USA (p. 301, 327)

BOYARSKI, S., Department of Surgery, Division of Genito-Urinary Surgery, Washington University, 4960 Audubon Avenue, St. Louis, MO 63110, USA (p. 1, 90)

BREINING, H., Abteilung Pathologie, Klinische Anstalten der RWTH Aachen, 5100 Aachen, Goethestr. 27–29, W. Germany (p. 123)

BURGHELE, TH., Strada Olga Bancic 4, Bukarest, Roumania (p. 319)

BYRNE, J. C., College of Engineering, Aerospace and Mechanical Engineering Department, The University of Arizona, Tucson, AZ 85721, USA (pp. 301, 327)

CHOW, D., Departments of Urology, Chemical Engineering, Electrical Engineering and Center for Bioengineering, University of Washington, Seattle, WA, USA (p. 274)

* The numbers following the addresses are the pages on which the authors' papers begin.

Court, B., 78, Avenue de Suffren, 75 Paris 15e, France (p. 179)

Debled, G., Clinique Urologique, Hôpital Universitaire Brugmann, Place Van Gehuchten, 1020 Bruxelles, Belgium (p. 160)

Duarte-Escalante, O., Clinique Urologique, Hôpital Universitaire Brugmann, Place Van Gehuchten, 1020 Bruxelles, Belgium (p. 90)

Frimodt-Möller, C., Urological Laboratory, Surgery Department H., Gentofte Hospital, 2900 Hellerup/Copenhagen, Denmark (p. 317)

Gessner, F. B., Departments of Urology, Chemical Engineering, Electrical Engineering and Center for Bioengineering, University of Washington, Seattle, WA, USA (p. 274)

Gil Vernet, S., Advenida Generalissimo Franco, 399, Barcelona 8, Spain (p. 171)

Gittes, R., F., Division of Urology, Department of Surgery, University of California, San Diego, CA, USA (p. 327)

Gleason, D. M., The University of Arizona, College of Engineering, Aerospace and Mechanical Engineering Department, Tucson, AZ 85 721, USA (pp. 301, 327)

Golenhofen, K., Physiologisches Institut der Philipps-Universität, 3550 Marburg/ Lahn, Deutschhausstraße 1–2, W. Germany (p. 46)

Graber, P., Kantonsspital Basel, Urologische Klinik des Departements für Chirurgie, 4000 Basel, Switzerland (pp. 199, 264)

Grasset, D., 10, Avenue du Professeur Grasset, 34 Montpellier, France (p. 185)

Gregoir, W., Clinique Urologique, Hôpital Universitaire Brugmann, Place Van Gehuchten, 1020 Bruxelles, Belgium (p. 160)

Hannappel, J., Physiologisches Institut der Philipps-Universität, 3550 Marburg/Lahn, Deutschhausstraße 1–2, W. Germany (p. 46)

Harris, J. H., Departments of Urology, Chemical Engineering, Electrical Engineering and Center for Bioengineering, University of Washington, Seattle, WA, USA (p. 274)

Haubensak, K., Urologische Universitätsklinik, Landeskrankenhaus, 6650 Homburg/Saar, W. Germany (pp. 26, 107)

Hedges, J., Departments of Urology, Chemical Engineering, Electrical Engineering and Center for Bioengineering, University of Washington, Seattle, WA, USA (p. 274)

Hinman, F., Jr., 1000 Francisco Street, San Francisco, CA 94109, USA (p. 14)

Ichim, V., Strada Batistei Nr. 1–3, Bukarest, Roumania (p. 323)

Kettner, A., Abteilung Urologie, Klinische Anstalten der RWTH Aachen, 5100 Aachen, Goethestraße 27–29, W. Germany (p. 112)

Kiil, F., Institute for Experimental Research, Ullevaal Hospital, Oslo, Norway (p. 57)

Klinkhammer, B., Abteilung Urologie, Knappschafts-Krankenhaus, 5124 Bardenberg, Dr.-Hans-Böckler-Platz, W. Germany (p. 123)

Kulov, N., Institut de Médecine, Chaire de Maladies Chirurgiques avec Orthopédie et Urologie, Varna, Bulgaria (p. 127)

Lassrich, A., Universitätskinderklinik Hamburg-Eppendorf, Radiologische Abteilung, 2000 Hamburg, W. Germany (p. 189)

Lutzeyer, W., Abteilung Urologie, Klinische Anstalten der RWTH Aachen, 5100 Aachen, Goethestraße 27–29, W. Germany (pp. 71, 337)

Lymberopoulos, S., Abteilung Urologie, Knappschafts-Krankenhaus, 5124 Bardenberg, Dr.-Hans-Böckler-Platz, W. Germany (p. 123)

Lynen, F. K., Abteilung Chirurgie, Klinische Anstalten der RWTH Aachen, 5100 Aachen, Goethestraße 27–29, W. Germany (p. 270)

Madersbacher, H., Lehrkanzel für Urologie an der Chirurgischen Universitätsklinik Innsbruck, Anichstraße 25, 6020 Innsbruck, Austria (p. 23)

Malatinsky, E., Urologische Universitäts-Klinik, Zabotova 14, Bratislava, CSSR (p. 102)

Marberger, H., Lehrkanzel für Urologie an der Chirurgischen Universitätsklinik Innsbruck, Anichstraße 35, 6020 Innsbruck, Austria (p. 23)

Mateu, A. T., Facultad de Medicina de Barcelona, Cátedra de Anatomía 2ª, Barcelona, Spain (p. 165)

May, P., Urologische Universitätsklinik, Landeskrankenhaus, 6650 Homburg/Saar, W. Germany (pp. 26, 107, 131, 195)

Melchior, H., Abteilung Urologie, Klinische Anstalten der RWTH Aachen, 5100 Aachen, Goethestraße 27–29, W. Germany (pp. 30, 71, 112, 149, 270)

Miller, E. R., Radiologic Research, School of Medicine, San Francisco Medical Center, San Francisco, CA 94122, USA (p. 204)

Moormann, J. G., Urologische Klinik, Universitätskliniken im Landeskrankenhaus, 6650 Homburg/Saar, W. Germany (p. 195)

Notley, R. G., Spindlewood, 59 Pewley Hill, Guildford, Surrey, Great Britain (p. 85)

Rathert, P., Abteilung Urologie, Klinische Anstalten der RWTH Aachen, 5100 Aachen, Goethestraße 27–29, W. Germany (p. 149)

Ratto, D., Facultad de Medicina de Barcelona, Cátedra de Anatomía 2ª, Barcelona, Spain (p. 165)

Reid, J. M., Departments of Urology, Chemical Engineering, Electrical Engineering and Center for Bioengineering, University of Washington, Seattle, WA, USA (p. 274)

Ritter, R. C., Department of Physics, University of Virginia, Charlottesville, VI, USA (p. 274)

Roberts, J. A., School of Medicine, Tulane University, 1430 Tulane Avenue Orleans, LA 70112, USA (pp. 98, 118)

Ruano Gil, D., Facultad de Medicina de Barcelona, Cátedra de Anatomía 2ª, Barcelona, Spain (p. 165)

Rutishauser, G., Urologische Abteilung der Chirurgischen Universitätsklinik, Bürgerspital, Rämistraße 100, 4000 Basel, Switzerland (p. 199)

Schindler, E., Urologische Universitätsklinik, Landeskrankenhaus, 6650 Homburg/Saar, W. Germany (p. 107)

Schreiter, F., Urologische Universitätsklinik, 6500 Mainz, Langenbeckstraße 1, W. Germany (p. 42)

Schulman, C. C., Clinique Urologique, Hôpital Universitaire Brugmann, Place Van Gehuchten, 1020 Bruxelles, Belgium (p. 90)

Schulte, K., Urologische Klinik, Städt. Krankenanstalten, 4600 Dortmund, Westfalendamm, W. Germany (p. 26)

Scott, F. B., St. Luke's Episcopal Hospital, 6720 Bertner Avenue, Houston, TX 77025, USA (p. 292)

SIMHAN, K. K., Aerodynamisches Institut Anstalten der RWTH Aachen, 5100 Aachen, Wüllnerstraße, W. Germany (p. 30)

SÖKELAND, J., Urologische Klinik, Städt. Krankenanstalten, 4600 Dortmund, Westfalendamm, W. Germany (p. 26)

STERLING, A. M., Afdeling Urologie, Academisch Ziekenhuis Leiden, Leiden, The Netherlands (p. 274)

SWOBODA, H. P., Lehrkanzel für Urologie an der Chirurgischen Universitätsklinik Innsbruck, Anichstraße 35, 6020 Innsbruck, Austria (p. 23)

SZOMOR, L., Pécsi Orvostudományi Egyetem, Urológiai Klinikája, Munkácsy Mihály u. 2, Pécs, Hungary (p. 80)

TAMMEN, H., Urologische Klinik rechts der Isar der TU München, 8000 München 80, Ismaninger Straße 22, W. Germany (p. 40)

TANAGHO, E. A., San Francisco Medical Center, University of California, San Francisco, CA 94122, USA (p. 215)

TERHORST, B., Abteilung Urologie, Klinische Anstalten der RWTH Aachen, 5100 Aachen, Goethestraße 27–29, W. Germany (p. 270)

TURNER-WARWICK, R. T., 61 Harley House, Marylebone Road, London, N.W. 1, Great Britain (p. 237)

TZEKOV, Tz., Institut de Médecine, Chaire de Maladies Chirurgiques avec Orthopédie et Urologie, Varna, Bulgaria (p. 127)

VELIKOV, K., Institut de Médecine, Chaire de Maladies Chirurgiques avec Orthopédie et Urologie, Varna, Bulgaria (p. 127)

WEINBERG, S., Department of Surgery, Washington University, 4960 Audubon Avenue, St. Louis, MO 63110, USA (p. 1)

WHITESIDE, C. G., X-Ray Diagnostic Department, The Middleesex Hospital, London, W. 1, Great Britain (p. 35)

ZIMSKIND, P. D., Department of Urology, Jefferson Medical College, 1025 Walnut Street, Philadelphia, PA 19107, USA (p. 154)

ZINNER, N. R., Department of Surgery (Urology), Charles R. Drew Postgraduate Medical School, Martin Luther King Jr. General Hospital, 1620 East 119th Street, Los Angeles, CA 90059, USA (p. 274)

Urodynamic Concepts

S. Boyarsky* and S. Weinberg**

The purpose of this talk is to give an overview of the salient features of urological physiology which must serve as assumptions for any urodynamic theory or measurement and to present one example of urodynamic modeling.

Kidney

As the unit of renal function, the nephron maintains accurate homeostasis of bodily composition. Its function is exquisitely sensitive to the pressures and flows of the filtered urine within its lumen. Its pressures and flows are in large part dependent upon capillary filtration, on plasma protein pressures, and in the final analysis, derive from cardiovascular pressures and forces. The mural tension of the epithelial walls plays a containing role. This subject would be best left for a future time, since concepts of active transport, bulk flow, diffusion and the interplay of physical and chemical factors need to be considered.

The renal pelvis plays a reservoir function and may possibly contain a pacemaker, or participate in pacemaker function.

Pressures, volume, geometric shapes, mural tension and mixing of its contents need to be considered in any discussion of renal pelvic urodynamics. The presence of neurologic control function is suggested by the nerves in its serosa, although the intrinsic innervation of the renal pelvis itself is less rich than that of the ureterovesical junction.

Ureter

The ureteropelvic junction has surgical importance. That disease should be more common at this location may merely reflect that this region of physiologic narrowing is also a critical transition point in urodynamic function and that disease produces a greater disturbance in function here than higher or lower.

The luminal caliber and shape changes abruptly between pelvis and ureter. This region has also been proposed to participate in pacemaker function though with less evidence. Urodynamically, the ureteropelvic junction and conus are a specialized segment of ureter; the bolus at this point does not influence the pressures, flows or volume above it appreciably, nor does it influence the nature of the bolus, contents or pressures below.

---

* Division of Genitourinary Surgery. Departement of Surgery. Washington University School of Medicine, St. Louis, Mo. 63110, USA

** Departement of Mechanical and Aerospace Engineering. Washington University School of Engineering, St. Louis, Mo. 63130, USA

Obstructions can be viewed as a narrowing of caliber, as regions of increased resistance, as areas where the range of excursion of the wall is restricted, or even regions where the mural elasticity is diminished.

The ureterovesical junction can be treated in terms of the length of the intramural ureter, its caliber, the shape of the ureteral orifice, the intramural tension of the bladder and ureteral walls, the bladder pressure, the compressibility, contractility and elasticity of the walls of the ureter.

Physiologically, the valvular action should be manifested as a differential in pressure between the ureteral lumen and the bladder lumen which is largely independent of bladder volume itself, more nearly dependent upon bladder pressure or possibly mural tension.

In this area there is little doubt that an intimate neurological control system exists although its exact mechanism has yet to be completely clarified.

Bladder

In the bladder, the variables of volume, pressure, and mural tension are paramount. The interplay with urethral function cannot be ignored except in a first simplifying approximation.

The extreme sensitivity of the bladder to stimulation, its rich dual innervation and the numerous reflexes arising from the bladder to impinge on targets in the heart, lung, uterus, rectum, blood vessels, and brain, warn us that urodynamic considerations of the bladder function in the absence of neurologic considerations may be incomplete. The urinary bladder fulfills two paradoxical functions with the same anatomy, filling and emptying; the nervous system makes this morphologic economy possible.

Bladder neck function can be considered in terms of cross-sectional area and geometry, longitudinal shape and geometry, and intramural tension. The role of the auxiliary muscles of the pelvic floor may or may not be important even though they are studied methodologically in attempts to assess ureteral resistance. The realignment of the histologic elements into varying functional combinations by the nervous system apply to the bladder neck or to the bladder.

The urethra can be considered in terms of caliber, geometric shape, length, elasticity, and flow state, whether laminar or turbulent. The limiting, smallest, and the effective caliber may be useful concepts.

The separation of the bladder neck sphincteric area from the urethra proper may be identified only by the urodynamicist who may tell us how much of the urethra is important in the function of continence and how much of it is merely a conducting passageway. There is experimental evidence that the length of the urethra is a factor in determining continence.

Overall urethral resistance, related to bladder function was a helpful first step in the development of our understanding of this area. It remains a useful diagnostic and physiologic concept, but less of a urodynamic concept.

Urethra

There is no specific anatomic entity such as the internal sphincter. The external sphincter, though it is identifiable anatomically, is episodic in function.

The external sphincter can compress the urethra during mid-stream even while the proximal urethra and vesical neck remain open, but the continent bladder shows no urine in the urethra below the level of the internal sphincter and our present understanding is that urethra tone and mural tension suffice to maintain continence under ordinary conditions.

Micturition must be studied in three phases initiation of voiding, maintenance of voiding, and termination of voiding. The first and third phases are transitional with the receptacle function, or continent state. The middle phase is idealized to be a steady state condition. How closely and for how long the steady state condition is sustained depends upon the refinements of our measuring techniques.

Urethral pressure can be viewed as a profile in time, according to the phase of voiding, and can also be studied as a profile in length, as the recording catheter is gradually drawn from the bladder in the outside.

The pressure profile can be analyzed by appropriate pharmacologic blocking experiments or nerve stimulation experiments to separate the contribution of the smooth muscles and the skeletal muscles.

The initiation of voiding can be summarized urodynamically as a gradual but progressive drop in urethral closure pressure below a concurrently rising intravesical pressure. Neurologically, initiation of voiding is due to the relaxation of the pelvic floor and external sphincter skeletal muscle, as manifested by diminished electromyographic activity from these regions. Geometrically and radiographically, voiding is initiated by the relaxation of the pelvic floor to lower the bladder base and facilitate funneling of the proximal urethra and bladder neck.

In different individuals, the pattern and relative importance of these factors vary. However, the detrusor muscle, the bladder neck, the proximal urethra, the pelvic floor, the abdominal wall and diaphragm (as the accessory muscles of micturition) and the lumen of the bladder and urethra influence the first phase of micturition.

The middle phase of micturition requires that the detrusor muscle maintain a steady state of contraction, the urethra maintain a steady state of resistance, and the flows and volumes remain within certain critical limits. Whether a steady state is ever attained depends upon the tolerances of data within which we define the steady state. Certainly the volume and mural tension are changing even though the pressure and flow may be constant.

Normally, voiding pressure averages 40 to 50 centimeters of water and the flow rate averages 25 cc per second. The physiologically effective diameter of the urethra in man is calculated to be 9 Fr even though the urethra calibrates instrumentally at 26 to 28 Fr.

The termination of voiding, Phase III, is assumed to be a reversal of the events initiating voiding, a rising pressure in the urethra and a falling pressure in the bladder, to tilt the balance between the two in the opposite direction.

Neuromuscular Tissue

The ureteral and bladder wall show increased resistance to distention described mathematically by the elastic modulus, which represents the restoring force within the tissue resisting stretch and distension. It is due to the viscosity and elasticity of collagenous fibrils and the macromolecular polymers of actin and myosin.

The elastic modulus is measured by the ratio of the changes in length to the distorting force. The most satisfactory model contains three elements, two elastic elements, like springs and one viscous element, a dashpot. The behavior of neuromuscular tissue is too complex to correspond to any single physical model or a single formula.
Histological spiral windings of smooth muscle need to be considered in terms of geometric transformation of linear tensions to three-dimensional stresses.
Detailed electrophysiological concepts remain outside the present discussion although they provide a theory which can explain or describe certain urodynamic events, particularly those of abrupt changes in caliber and tension due to muscle contraction or relaxation.
All overall concept of urodynamics should include simultaneous monitoring of the following parameters:

1. pressure variations
2. volumetric flow rates
3. geometric shape and wave length
4. wave speed
5. frequency
6. mural tension
7. interface reactions such as absorptive, secretory

An all-inclusive armamentarium for urodynamic study would include the following techniques and methods:

1. Gross description, anatomy, pathology, surgical observation
2. Histology, histochemistry and microdissection
3. Biochemical analysis, specific collagen, catecholamine
4. Radiographic visualizations
5. Geometric description and measurements; size, shape
6. Probing and calibration
7. Resistances and tissue elasticity determinations
8. Flow and volume measurements
9. Pressure measurements including determination of vesical, urethral, anal and rectal values.
10. Perfusion and sphincterometry
11. Plethysmography
12. Electrophysiology
13. Sound recordings

Modeling

The process of modeling requires one to develop descriptive summary analogies and then improve them to approach the true situation more closely. Analogies can be drawn physically, abstractly or conceptually on a theoretical plane. Modeling combines ideas, facts and characteritics which are thereby summarized. A model can develop a certain existence of its own and lead to further experiments, discussions, deduced rules, and as we have seen, even attack and defense. In a sense, every experiment, every operation, every patient is a model, if we learn so as to improve our handling of the next one.

The historical interpretation by KUHN of scientific revolutions also applies to models. The scientific community invests effort, equipment and publications in certain models, and then goes on to discard these for better ones – not without some internal struggle. Allied to the philosophic problem of modeling are those of definitions, assumptions and dialogue. It is important to make our assumptions explicit. Since many of us may use the same words in different senses, much confusion will be avoided by using consistent terminology when defining our terms. Misunderstandings due to ambiguity or unrecognized vagueness are more difficult to detect and overcome than these due to differences in language.

Our concepts are operationally restricted by our measurement techniques evenmore than we realize. Theory and methodology are intimately interdependent in the same way as anatomy and physiology. Improvements in, and combinations of techniques lead to widening and elaboration of knowledge and theory. Our ignorances are often reflections of the limitations of our technique.

Hydrodynamical Model of Ureteral Function*

Peristaltic transport has been studied theoretically by engineers for approximately seven years, and clinically by urologists for many decades. By combining both approaches, progress on the understanding of peristaltic transport can be greatly enhanced as is evident in the recent Hydrodynamics workshop of 1969. Original areas of disagreement between engineers and urologists and between engineers and other engineers have been clarified resulting in a much more comprehensive concept of ureteral urodynamics.

Objectives

Peristaltic waves generated by neuromuscular action, are the main mechanism of urine transport in the ureter. In one part of this talk the anatomical and physiological data available for the human ureter, in so far as they are known, are used to set up a simple hydrodynamic model. The results of the hydrodynamic calculations are compared with urometric observations in order to see if the hydrodynamic theory can account for the pressure-time variations and urine flows reported for the ureter.

Data for the Human Ureter

KIIL has made numerous pressure measurements in the ureter using catheters, ranging in size from 4F to 8F, connected to strain-gauge pressure transducers. Although the diameters of the catheters were equal to or larger than the lumen size, KIIL assumed that ureteral function was not perturbed because the catheters did not cause urine retention and the pressure-time recordings were not very sensitive to the catheter size in this range.

These recordings, of which Fig. 1 is typical, show how the pressure changes with time at the location of the recording. There is a succession of solitary waves, rather

* Taken from WEINBERG, S. L., JAFFRIN, M. Y. and SHAPIRO, A. H.: Hydrodynamic model of ureteral funktion Chapt. 17, Urodynamics, Hydrodynamics of the Ureter and renal pelvis. New York: Academic Press 1971.

than a continuous wave train, passing the measurement point. The pressure pulses are much the same at different measurement points. The peak pressure is typically equal to 25 mmHg, which corresponds to an amplitude of about 15 mmHg for the pulse. A most important observation is that there are uniform regions of resting pressure ahead of and behind each solitary wave and that the resting pressure is virtually the same before and after the wave passage. Thus, the net pressure rise due to the passage of a wave is zero. This finding is in agreement with the observation that the pressure in the renal pelvis is equal to the resting pressure in the ureter even when a wave is present.

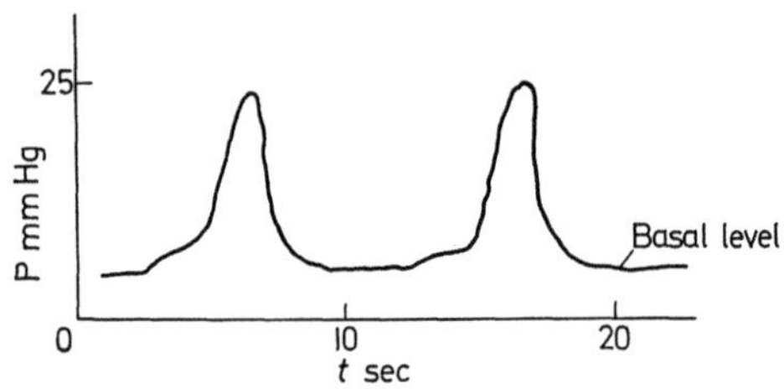

Fig. 1. Pressure recording within ureter (Ref. Kiil)

The resting pressure in the ureter is unaffected by the vesical pressure and remains equal to 5–10 mmHg if the vesical pressure does not exceed 25 mmHg. From this it appears that the ureterovesical junction acts as a check valve and that urine can be discharged into the bladder only if the pressure within the ureteral wave exceeds the vesical pressure. This conjecture is supported by the observations that urine flows into the bladder in spurts and, further, that each spurt corresponds to the arrival of a peristaltic wave at the ureterovesical junction. For these reasons, we postulate in the hydrodynamic model that there is no flow in the inactive parts of the ureter. The fact that there is no appreciable pressure gradient, either ahead of or behind the pulse wave, is additional evidence that this postulate is correct.

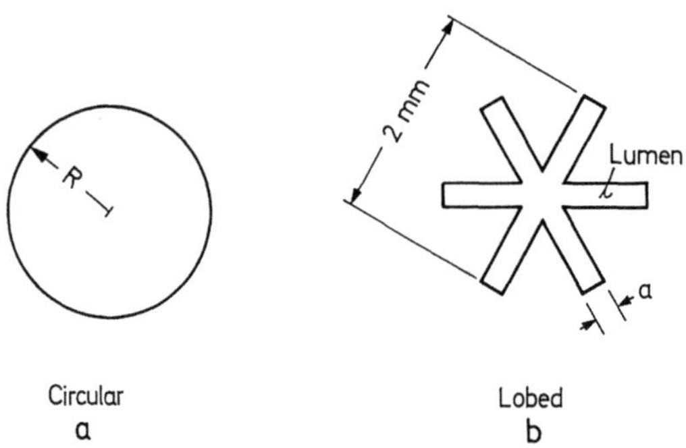

Fig. 2. a Axis-symmetric lumen. b Idealized lobed shape model

The cross section of the lumen of the ureter is almost circular when distended, but it folds when contracted. In fact, if the inner perimeter of the ureter remains approximately constant, a large change in cross-sectional area requires that the contracted lumen assume a lobe shape. Fig. 2 is a schematic representation of the lobed geometry which will be used in the theoretical model.

Radiographic pictures taken with radio-opaque fluid in the ureter show a dilated region (bolus) about 5 cm long and from 2 to 4 mm in diameter, preceded and followed by what looks like a very narrow lumen. The diameter of the constricted lumen, or indeed its shape cannot be easily evaluated because of the limited picture resolution.

Only one or two waves are present at a time in the ureter. The wave speed is about 3 cm/sec. Since the pulse width based on the zones of maximum gradients, is about 2–3 sec, the wave length is about 6–9 cm. The amount of urine volume in each spurt entering the bladder is, of course, highly variable and lies between 0.02 and 1 ml.

Calculational Model of the Ureter

Based on the foregoing discussion, we summarized here the quantitative information introduced into the hydrodynamic model of the ureter: a) There is no flow in the inactive sections of the ureter ahead of or behind the peristaltic wave. b) The net pressure change across each wave is zero. c) The pressure pulse is characterized by an amplitude of 15 mmHg and a width of 7 cm. d) The wave speed is $c = 3$ cm/sec. These values are taken as typical. Moderate changes in them would not alter the conclusions.

We believe that the choice of a particular wave shape is not critical to the elucidation of the fluid mechanics of ureteral function as long as the wave shape is of the correct general form, as dictated by the shape of the pressure pulse. Presumably, the wave shape varies from one wave to another, which accounts for the variations between

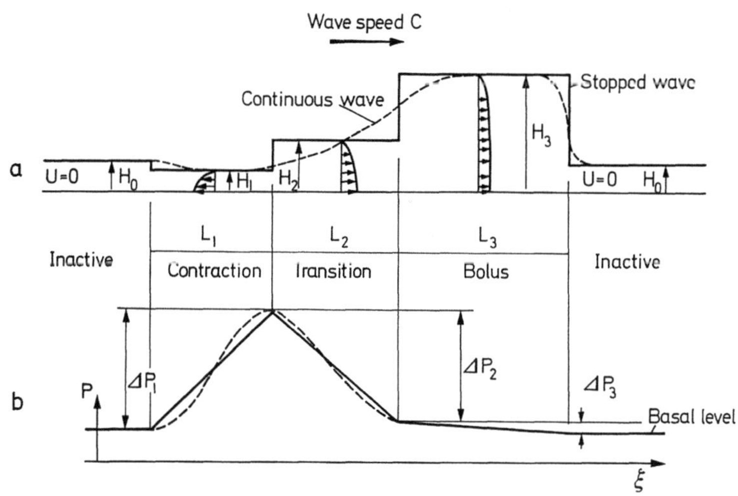

Fig. 3. a Stepped (*solid line*) and continuous (*dashed line*) wave geometry. b Corresponding pressure pulses

successive pressure pulses. Moreover, it varies among individuals. For these reasons, we choose here a stepped wave (solid curve of Fig. 3a) because it leads to especially simple calculations and a transparent understanding of ureteral fluid mechanices. Of course the stepped wave represents a rough approximation to the shape of the smoothly varying wave (dashed curve of Fig. 3a). For example, the pressure pulse of the stepped wave is triangular in shape (Fig. 3b) in comparison with the continuous wave shape of the ureter (Fig. 1).

The triangular pressure wave is also a reasonable approximation to the wave shape generated by the continuous wave model. Actually, any wave shape could be handled by using the procedure of (Shapiro *et al.*), but the calculations would lose in simplicity and clarity with little gain in accuracy because of the gaps in physiological data.

The elementary calculations for the stepped wave shape of Fig. 3 will become clear in the appendix. Fig. 3b shows p as a function of distance x at a fixed instant of time t in a horizontal ureter. A catheter transducer at a fixed location x would see the mirror image of the pressure pulse of Fig. 3 as the pressure pattern is swept to the right over the fixed transducer. Two simple rules may be summarized here. a) In the dilated section, the flow is in the direction of the wave, while in the contracted section, the flow is in the direction opposite to the wave. b) Consequently, the pressure in the ureter rises with time wherever the lumen radius exceeds the resting radius R_0, and decreases with time wherever it is below R_0. Thus, the only way to obtain a positive pressure pulse, with zero net pressure change, is to assume that a bulge precedes a contraction. Reversing the order would cause a dip in pressure instead of a rise. We note that it is impossible to satisfy the condition of no net pressure rise with the presence of a bulge alone or a contraction alone. However, it turns out that the diameter of the bolus necessary to account for the volume transported is much too large to account for the pressure rise of 15 mmHg. Therefore, we postulate that this pressure rise occurs during a transition between the bolus and the contraction. In this simplified model, as pictured in Fig. 3, the transition has a constant radius $R_2 > R_0$, while the contraction which follows has a radius $R_1 < R_0$. The pressure pulse reaches its peak when the lumen radius takes its resting value R_0 which occurs, here at the end of the transition. As seen in the next section, the transition radius R_2 is so much smaller than the radius of the bolus that the transition cannot be readily distinguished from the contraction on the radiographic pictures. A confirmation of this model may be found in the observation of Barry that the peak pressure follows what appears to be the beginning of the contractile event by a time lag of about 1.2 sec. Our model provides an explanation of this lag: it is the passage time of the transition over the point of recording, which is also approximately equal to the rise time of the pressure pulse. The corresponding length of the transition would then be about 3.5. cm.

The entrance of a wave into the ureter and its exit seem to have no effect on the pressure recordings. The reason may be that the contraction becomes strong only when the wave is entirely within the ureter. Accordingly, we ignore here the entrance and exit processes and consider only the fully formed wave within the ureter.

Now, if we knew the exact cross-sectional shape and dimensions of the lumen, we would proceed with the hydrodynamic theory and calculate the pulse amplitude and the volume of urine carried in each wave. The validity of the model would then be assessed by comparing these values with the reported pressure amplitude of 15 mmHg and the reported volume of between 0.02 and 1 ml per spurt. However, the shape and internal dimensions of the lumen in its active and inactive configurations are really not known at all. Thus, we are led to an inverse procedure: to see whether we can infer useful information regarding shape and size by combining the cited physiological data with the simple hydrodynamic theory. The criterion of success is a self-consistent result that agrees with the known physiological facts and also satisfies the laws of fluid mechanics.

Results

Here, the bolus section is assumed to be always cylindrical. However, in view of the uncertainty as to the lumen shape in its inactive ano contracted stages, we have carried out the calculations for two limiting configurations: either a) the lumen remains circular at all stages of activity (Fig. 2), or b) the lumen takes the lobed shape when inactive and also during the passage of both the contraction and the transition. This assumption is substantiated by the result that, in order to account for the observed pressure-pulse amplitude, the cross-sectional areas of the inactive, contracted, and transition stages must all be much smaller than the cross-sectional area of the bolus. Presumably this requires that the inner wall of the ureter be folded except when distended by the bolus. For the calculations presented here, we assume that the cross section is formed of six narrow lobes, each of about a 1mm span (see Fig. 2). The inner perimeter, assumed constant, is the same as that of a 4 mm diameter circle. The cross-sectional area is then proportional to the thickness a of the lobes.
The formulation of the fluid mechanics for these two geometries is developed in the appendix. In the calculations, the viscosity of urine at 38 °C is $\mu = 0.7$ cp while $L_1 = L_2 = 3.5$ cm.
We note that the pressure change in the bolus section is so small that it is negligible. Thus, the bolus may be thought of as essentially accounting for the volume carried by each wave, while the pressure pulse is generated by the transition-contraction regions.

Circular Lumen at All Sections

Dimensions

Due to the constraints imposed on the pressure pulse, the cross-sectional area S_0, S_1 and S_2 of the inactive, contracted and transitional sections. respectively, must satisfy

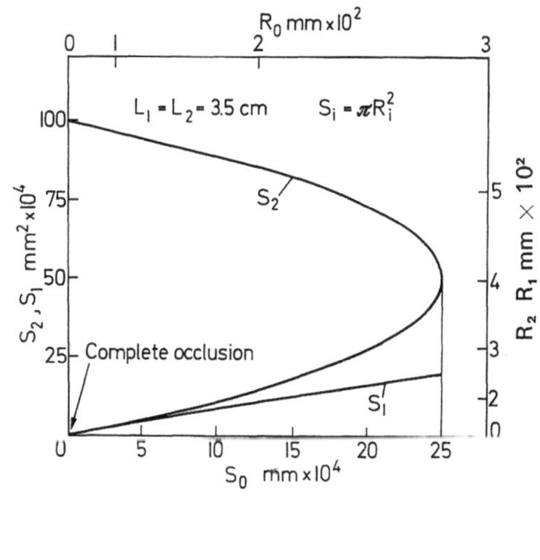

Fig. 4. Cross-section areas of the lumen during passage of a peristaltic wave, for a circular geometry

Eqs. 6 and 7[1]. It follows that a one-parameter family of dimensions is permissible for a given pressure pulse even though the wave shape has been severely restricted. Fig. 4 shows the values of S_1 and S_2 as a function of the inactive cross-sectional area S_0. Note that there are two possible values of S_2 for each value of S_0. The permissible ranges for the radii of each cross section are

$$0 < R_0 \leqslant 2.8 \times 10^{-2} \text{ mm}$$
$$0 < R_1 \leqslant 2.5 \times 10^{-2} \text{ mm}$$
$$0 < R_2 \leqslant 5.65 \times 10^{-2} \text{ mm}$$

These radii are much smaller than the bolus radius of 2 mm. According to this model, then, the ureter is practically empty between successive waves. Actually, it seems unlikely that the circular bolus radius of 2 mm could diminish to 0.06 mm without buckling or folding. It is worth noting that the observed urometric pressure pulses are compatible with a complete occlusion of the ureter, except of course in the bolus.

Volume per Spurt

The volume of fluid carried by a wave is given by

$$V = (S_3 - S_0)\, L_3 + (S_2 - S_0)\, L_2 + (S_1 - S_0)\, L_1$$

or since S_0, S_1, and S_2 are so much smaller than S_3, to a very good approximation,

$$V \cong S_3 L_3$$

The volume carried by the wave is nearly equal to the volume of the bolus and independent of the structure of the contraction. Any volume in the observed range can be accommodated by varying the cross section S_3 or the length L_3. If $R_3 = 2$ mm and $L_3 = 4$ cm, for instance, the volume is 0.5 ml per spurt, which is within the physiological range.

Effect of Gravity

When a person is standing up, the gravitational force acts on the urine in the longitudinal direction along the ureter. If the ureter were a passive duct, gravity would, of course, be the principal agent of flow. This is not so in the presence of peristaltic pumping, for then the flow of urine to the bladder depends mainly on the size of the bolus and the speed and frequency of the wave. The contribution of gravity is, at most, relatively small since the inactive part of the ureter is so narrow it offers a very high resistance to flow. Indeed, if the ureterovesical junction acts as a closed valve, gravity has no effect whatsoever on the flow except indirectly, if it changes the wave shape. The main effect of gravity is to superpose the well-known hydrostatic pressure distribution on the pressure pulse of Fig. 3. This means that the resting pressure varies along the length of the ureter.

Lobed Lumen in Inactive, Transition, and Contracted Sections

The procedure is the same as before but yields quantitatively different results. The dimensions of the lumen are obtained by solving Eq. 1 by numerical or graphical methods.

---

[1] The numbered equations appear in the appendix

The thickness of the lobes a_1 and a_2 in the contracted and transitional stages are plotted versus the inactive thickness a_0 in Fig. 5.
The dimensions are smaller but of the same order of magnitude as in the circular model. The permissible range of the thickness of the lobes in each stage of activity is

$$0 < a_0 \leqslant 2.56 \times 10^{-2}\ \text{mm}$$
$$0 < a_1 \leqslant 2.25 \times 10^{-2}\ \text{mm}$$
$$0 < a_2 \leqslant 6.6 \times 10^{-2}\ \text{mm}$$

Here again the volume per spurt is determined by the bolus size given by the earlier formular and is essentially unaffected by the geometry of the contraction.

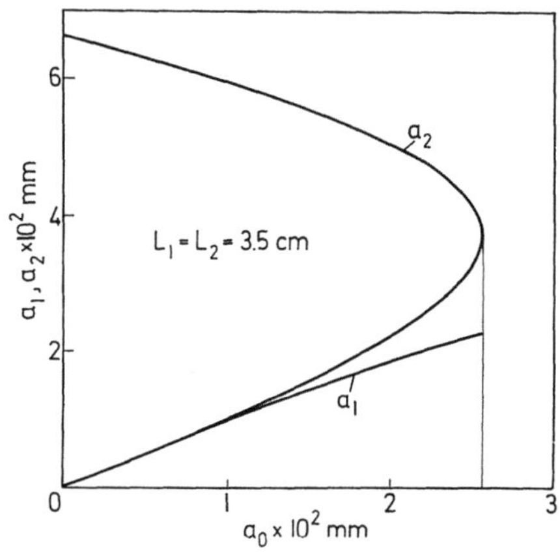

Fig. 5. Thickness of the lobes of the lumen during passage of a wave for the configuration of Fig. 2b

Conclusions

It has been shown that the range of urometric data available could be accounted for by a simple model of peristaltic transport. Also, the time lag of the apparent beginning of the contraction relative to the peak pressure has been interpreted. The model, in its present form, does not allow one to determine unambiguously the internal dimensions of the ureter, but it sets upper limits for these dimensions.
An important conclusion of this analysis is that, if the pressure measured is the actual pressure in the ureter and not an artifact resulting from obstruction by the catheter, the transverse dimensions of the contracted and inactive lumen must be extremely small-less than 1/10 mm. Therefore the ureter may be considered as nearly occluded and empty between waves. In other words, the ureter "milks" itself and behaves as a positive displacement pump.
This analysis suggests several remarks: First, peristaltic waves in a relatively large duct, such as a ureter dilated by an obstruction or the renal pelvis will produce insignificant pressure fluctuations (less than 1/10 mmHg). Thus, the failure to record large pressure variations in the renal pelvis does not rule out peristaltic activity there. Second, large pressure amplitudes are not at all necessary to transport urine in the ureter. The same quantity of urine could be carried as a more continuous flow with pressure variations a hundred times less. However, it is worth noting that the

peak pressure observed (25 mmHg) roughly corresponds to the pressure of the bladder when it is full. If the peak pressure were less, there might be a regurgitation into the ureter when the ureterovesical junction opens.

Because the ureterovesical junction seems to act as a valve, it is not surprising that urine transport is not affected by gravity. The latter simply adds a hydrostatic pressure distribution.

We are not ready to claim that the model we have proposed describes the actual situation, but we think that it does not, so far, contradict any important physiological data. However, further work is necessary to obtain an adequate model of ureteral function. For instance, the mechanics of the muscle and the formation of the bolus at the ureteropelvic junction are not yet understood. This attempt at a hydrodynamic model stresses the need for more conclusive pressure recordings, using possibly smaller catheters, simultaneous measurements of the pressure at the renal pelvis, and the ureterovesical junction and for more information regarding the size and the shape of the lumen.

Appendix

Circular Lumen

The flow induced in the ureter by the passage of a wave is practically inertia-free because of the preponderance of viscous forces. The Poiseuille relation between the volume flow rate Q (positive to the right) and the pressure rise Δp in a cylindrical tube of length L can thus be applied to the flow in each section of the wave (Fig. 3). For a circular lumen of radius R we have

$$Q_i = -(\pi/8\mu)\,(\Delta p_i/L_i)\,R_i^4 \tag{1}$$

where μ is the viscosity, and the subscript i refers successively to sections 0, 1, 2, and 3. For an observer moving to the right at the wave speed c, the ureteral wall shape is the same at all times, and the flow appears steady. Thus, the principle of mass conservation requires that the volume flow rate, q, as seen in the wave frame so defined, must be the same at every cross section. Accordingly, since at any section $q = Q_i - \pi R_i^2 c$, we may write

$$q = Q_1 - \pi c R_1^2 = Q_2 - \pi c R_2^2 = Q_3 - \pi c R_3^2 \tag{2}$$

The flow q is evaluated from the condition assumed earlier that the flow is zero in the inactive part of the ureter, i.e.,

$$Q_0 = q + \pi c R_0^2 = 0 \text{ or } q = -\pi c R_0^2.$$

Hence,

$$Q_i = \pi c (R_i^2 - R_0^2) \tag{3}$$

It follows that Q is positive (i.e., in the wave direction) where R exceeds R_0; but Q is negative in sections where R is less than R_0.

The combination of Eqs. (1) and (3) yields

$$\Delta p_i = -\frac{8\mu c (R_i^2 - R_0^2) L_i}{R_i^4} \tag{4}$$

This agrees with the foregoing result on flow: Where the flow is to the right, the pressure falls in that direction; where the flow is to the left, the pressure falls in that direction (see Fig. 3). The pressure rise across the bolus is, at most, 4×10^{-2} mmHg for $R_3 = 1$ mm and even less if $R_3 > 1$ mm. Thus, it is too small in comparison with the total pressure amplitude to be noticeable on the pressure recording. Then the condition that the amplitude of the pressure pulse is $\Delta p_{\max}$ and that there is no net pressure change across the wave may be written as

$$\Delta p_1 = -(\Delta p_2 + \Delta p_3) = \Delta p_{\max}.$$

Neglecting Δp_3 in comparison with Δp_2 and using Eq. 4, we have

$$\frac{8\mu c(R_0^2 - R_1^2)L_1}{R_1^4} = \frac{8\mu c(R_2^2 - R_0^2)L_2}{R_2^4} = \Delta p_{\max} \qquad (5).$$

It is convenient to rewrite Eq. 5 in terms of the cross-sectional area $S_i = \pi R_i^2$ as

$$A_i S_i^2 - S_i + S_0 = 0 \qquad i = 1,2 \qquad (6)$$

where

$$A_1 = \frac{\Delta p_{\max}}{8\mu c \pi L_1} \qquad A_2 = \frac{\Delta p_{\max}}{8\mu c \pi L_2} \qquad (7).$$

Lobed Lumen

Each lobe is narrow enough to be treated as a two dimensional channel. The Poiseuille relation in this case is

$$Q_i = -\frac{a_i^3 \Delta p_i}{12\mu L_i} \qquad i = 1,2 \qquad (8)$$

where a_i is the thickness of the lobes and Q is the flow per unit span width of the lobe. The flow rate in the wave frame is now

$$q = Q_1 - a_1 c = Q_2 - a_2 c = -a_0 c \qquad (9).$$

When Eqs. 8 and 9 are used, the constraints imposed on the pressure pulse yield the following equation:

$$\frac{12\mu c(a_0 - a_1)L_1}{a_1^3} = \frac{12\mu c(a_2 - a_0)L_2}{a_2^3} = \Delta p_{\max} \qquad (10).$$

References

BARRY, W. F.: Correlation of Cineradiographic Image and Pressure Tracing of Ureteral Activity. Chapt. 11, Urodynamics, Hydrodynamics of the Ureter and Renal Pelvis. New York: Academic Press 1971.

BOYARSKY, S., TANAGHO, E., ZIMSKIND, P., GOTTSCHALK, C. (Ed.). Urodynamics, Hydrodynamics of the Ureter and Renal Pelvis. New York: Academic Press 1971.

BURNS, J. C., PARKES, T.: Peristaltic Motion, J. Fluid Mechanics **29**, 731 (1967).

FUNG, Y. C.: Peristaltic: A Bioengineering Model, Chapt. 15, Urodynamics, Hydrodynamics of the Ureter and Renal Pelvis. New York: Academic Press 1971.

FUNG, Y. C., YIH, C. S.: Peristaltic Transport. J. appl. Mechanics **35**, 669 (1968).

KIIL, F.: The Function of the Ureter and Renal Pelvis. Philadelphia: Saunders 1957.

LYKOUDIS, P. S.: The Ureter As a Peristaltic Pump, Chapt. 16, Urodynamics, Hydrodynamics of the Ureter and Renal Pelvis. New York: Academic Press 1971.

SHAPIRO, A. H., JAFFRIN, M. Y., WEINBERG, S. L.: Peristaltic Pumping with Long Wavelengths at Low Reynolds Number. J. Fluid Mechanics **37**, 799 (1969).

Hydrodynamic Aspects of Urinary Tract Infection

F. Hinman Jr.*

Infection and back pressure are the ultimate agents responsible for urinary tract diseases related to hydrodynamic dysfunction. We have isolated three major hydrodynamic mechanisms involved in the introduction and perpetuation of bacteria in the

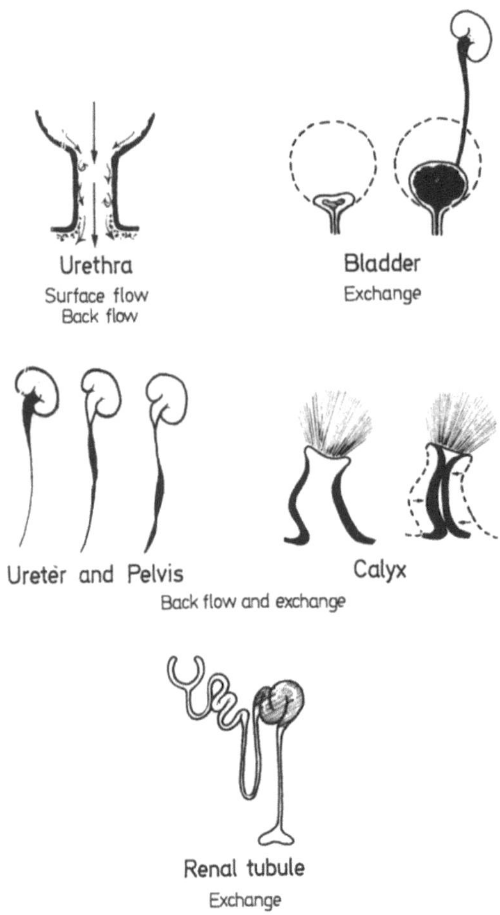

Fig. 1. Hydrodynamic defenses against infection

* Division of Urology, University of California School of Medicine, San Francisco, Calif., USA

urinary tract: 1. *surface flow*, which washes bacteria downward, 2. *backflow*, or regurgitation, which moves them retrogradely, and 3. *exchange* or replacement of the urine residing in the passages with fresh urine from above (Fig. 1). The four levels of the urinary tract utilize one or more of these mechanisms to eliminate bacteria: the *urethra*, surface flow and backflow; the *bladder*, exchange; the *ureter*, *pelvis and calyx*, backflow and exchange; and the *renal tubules*, exchange. (The last will not be discussed here because the renal parenchyma is not a topic of this symposium.)
This paper relates the hydrodynamic factors to the activities of the several levels of the urinary tract and thus to the movement or the retention of bacteria.

Urethra

I have studied the female system more intensively than the male and shall therefore concentrate on the former. Bacteria are normally present in the urethra. Although their number is kept down by the antibacterial activity of the mucosal surfaces (Drach and Cox, 1967; Norden and Kass, 1967) and by interaction between the pathogenic and nonpathogenic forms, the periodic flushing occurring during micturition has a greater effect. However, it has been found that bacteria are not completely washed out by voiding (Helmholz, 1950; Meares and Stamey, 1968; Masih and Hinman). The importance of washout is clear from our studies, which showed increased counts of urethral bacteria after total urinary diversion (Masih *et al.*, 1970).
Surface flow. As urine is forced down the urethra by vesical contraction, those portions of the fluid near the wall ar slowed while those in the center pass freely along, resulting in concentric laminations with centrally progressively greater flow rates. However, the laminations break up and the flow becomes turbulent as it passes beyond the critical Reynold's number of 2500 to between 10,000 and 20,000 (Backman, 1966). Even though turbulent flow is not constantly present throughout the urethra, turbulence always exists in the distal urethra during normal micturition (Ward and Cass, 1969).
The female urethra is neither straight nor of constant caliber (Nordenström, 1952). Changes in its diameter and direction increase the turbulence; they speed up flow in one area at the expense of slowing it down elsewhere. The implication of these observations is that bacteria free on the mucosal surface will be washed out with each micturition, but also that circulation bubbles, laminar sublayers, and other areas of altered flow will retain bacteria or impede their washout.
Since it is known that complete washout does not occur with each voiding, the mucosal surface of the female urethra was studied at cellular level with the scanning electron microscope to see if microscopic organisms could be retained by irregularities not apparent by direct observation (Lloyd-Davies *et al.*, 1970; 1971a; 1971b). The urethral surface was seen to have microfolds of a size capable of holding bacteria, which are smoothed out by hydraulic distention of the urethra, such as would occur during voiding. Induction of urinary infection produced great complexity of the surface (Fig. 2), and bacteria remained on the surface (Fig. 3). Irregularity and adhesiveness of the surface appeared to be factors in the persistence of the bacteria in the urethra. These factors are felt to be important clinically; for example, bacteria have been found to increase in number and pathogenicity in the upper portion of the female urethra prior to episodes of cystitis (Cox, 1966; Cox *et al.*, 1968).

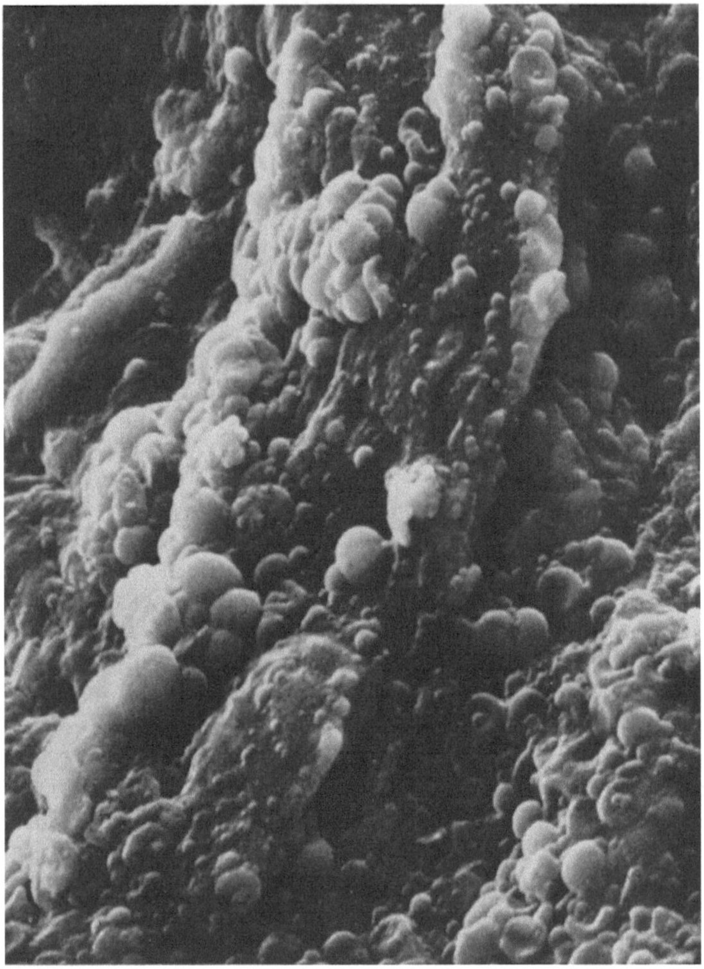

Fig. 2. Increased complexity of canine urethral surface after induced infections. X3,000. (Courtesy of Williams and Wilkins Co)

Backflow. In early studies of female micturition, we observed return of urine from the midurethra to the bladder when urination was voluntarily stopped (HINMAN *et al.*, 1954, 1955) (Fig. 4). Thus bacteria resident in the middle third of the urethra are free to be carried back into the bladder. Hesitant (interrupted) micturition fostered by urethral irritation may be an important factor in recurrent infection in female children (LYON and MARSHALL, 1970; HINMAN, 1966).

Bladder

Exchange. Experiments with a stimulated bladder showed that certain effects of voiding on resident bacterial populations could be duplicated by artificial systems (COX and HINMAN, 1961). Moreover, if growth-inhibiting or growth-stimulating factors in the vesical wall and urine are disregarded (we believe that their role is minor compared to mechanical factors), the effects of vesical emptying on bacterial exchange can be calculated mathematically by a single equation (HINMAN and COX, 1966). The equation states that the volume of urine (Vt) times concentration of bacteria (Ct) in the

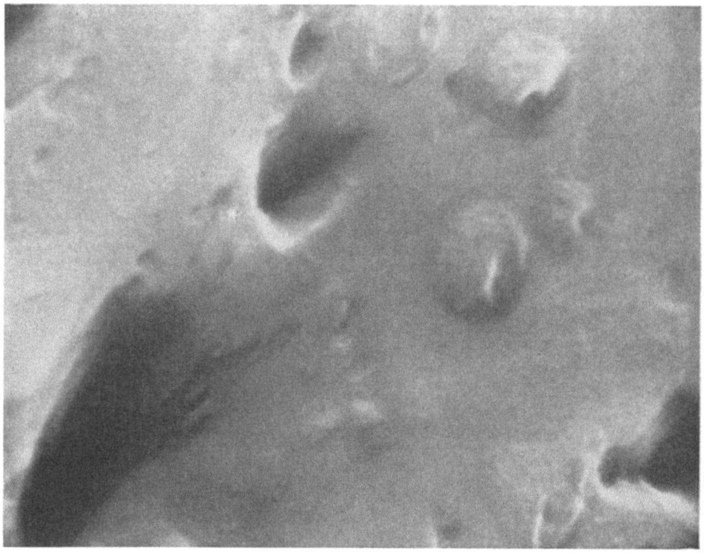

Fig. 3. E. coli on urethral surface. X10,000. (Courtesy of Williams and Wilkins Co.)

bladder at the end of a time interval (t) is equal to the initial volume of urine (Vo) times the initial concentrations (Co) multiplied by a doubling factor for the bacteria that assumes doubling twice every hour (2^{2t}). Thus $Vt\,Ct = Vo\,Co \cdot (2^{2t})$. It is seen that Vo is the expression for residual urine, Vt represents urinary secretion and t the time between voidings. A nomogram can be constructed from which the effects of variations in these three factors can be read.

Further experiments in which bacteria were placed in the bladders of normal human subjects showed that elimination was rapid and complete (COX and HINMAN, 1961). In contrast, in patients with large volumes of residual urine, even specific antibacterial agents which greatly prolong the doubling time of the bacteria are ineffectual since the exchange is inadequate.

Ureter, Pelvis and Calix

Backflow. Vesicoureteral reflux perpetuates infection since it returns urine to the bladder after voiding, resulting in vesical residual urine (HUTCH *et al.*, 1963). The incompetent ureterovesical valve allows flow from bladder to ureter when intravesical pressure rises higher than ureteral peristaltic pressure. Reverse peristalsis also occurs as we have observed during low pressure filling of the bladder (HINMAN *et al.*, 1962).

Recent studies by Weinberg in models simulating peristalsis in two dimensions suggest that, during normal ureteral peristalsis, particulate matter can move against the main stream by a persistent orbiting motion along the undulating ureteral wall (Weinberg, 1971). Whether similar phenomena occur in the normal ureter is not known; it is probable that reflux is the principal mechanism for transport of bacteria to the kidney.

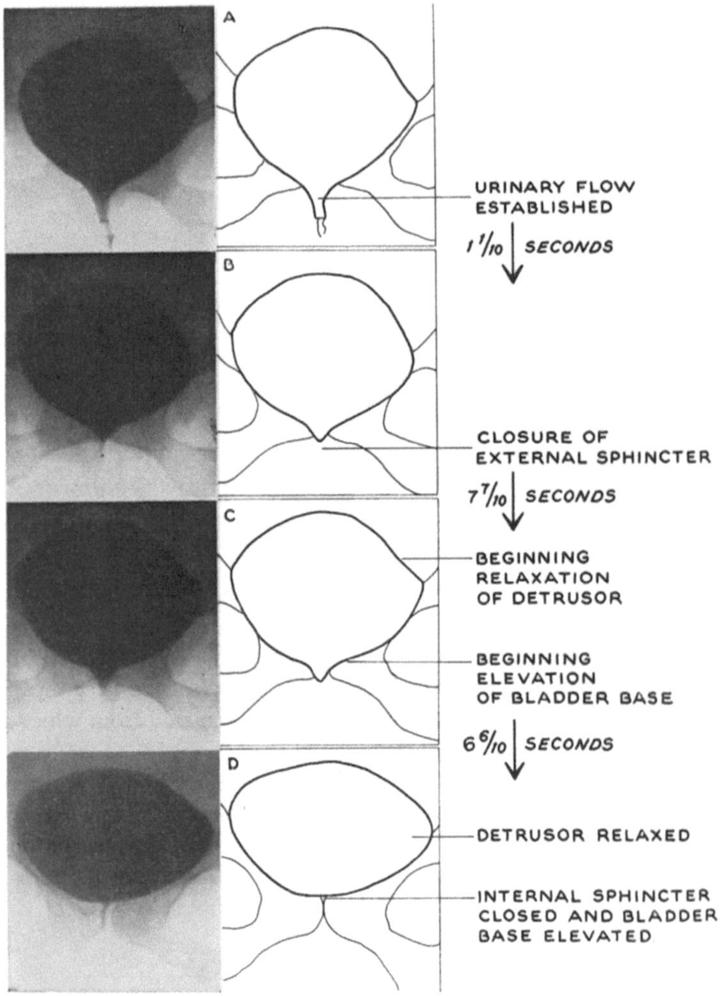

Fig. 4. Rapid sequence radiograph showing interruption of micturition and return of urethral urine to bladder. (Courtesy of Radiological Society of North America, Inc.)

Exchange. Stasis is a classic word that we believe means incomplete exchange of urine in a given space during the doubling time of bacteria (Hinman, 1971). Bacteria double in number approximately every 30 min; thus to maintain a constant number of bacteria in a space, the amount of new urine entering every 30 min must equal (and replace) the amount of old urine contained in the space (Fig. 5). Incomplete exchange can result from one or both of 2 factors: 1. a greater volume in the conduit than the volume passing through half-hourly; 2. incomplete mixing of the new and old urine so that part of the urine, in certain sequestered areas of the conduit, remains for longer than 30 min (O'GRADY and CATTELL, 1966).

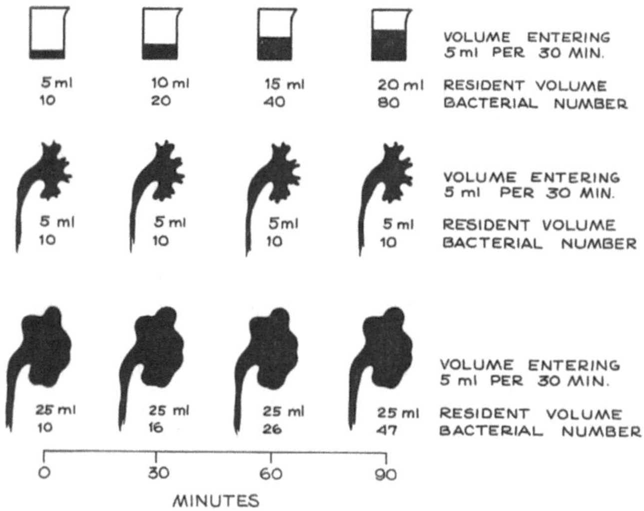

Fig. 5. Exchange of urine in normal renal pelvis and in hydronephrosis, compared to accumulation of urine in bladder between voidings. In normal renal pelvis, volume passing through in half an hour is at least equal to resident volume and thus prevents increased bacterial numbers. In obstructed renal pelvis, resident volume is 5 times that passing through and bacteria accordingly increase in number

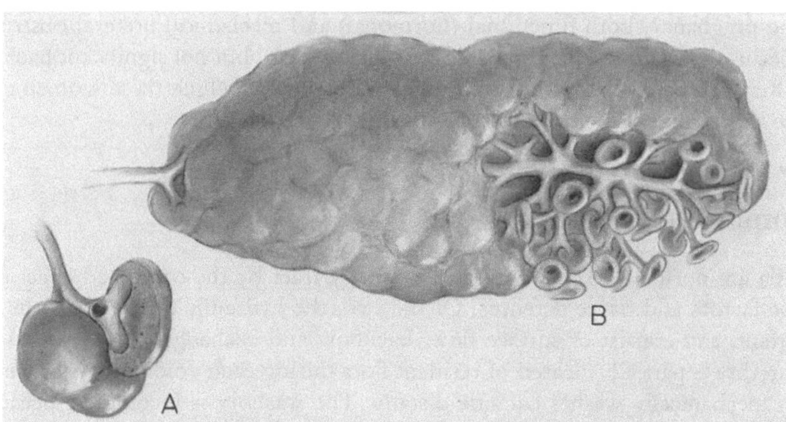

Fig. 6. A Single papillary kidney of rabbit. B Multipapillary kidney of whale, consisting essentially of many unipapillary units draining through infundibulae, which are really "ureters"

1. Disproportionate volume: The urinary collecting systems of man and other mammals have one feature in common: in general, the larger the animal, the more subdivided the renal unit. Small animals have single papillae, and the calyx is the pelvis (Fig. 6). Very large animals, like the whale, have many such kidneys, each attached to a "ureter" that drains into a common ureter (Fig. 6). All the systems maintain a low resident volume.

The normal human pelvio-calyceal complex holds about 6 ml (2 to 12 ml) of urine. With urine produced at 1200 ml per day, an average volume of about 12 ml flow through each 6-ml pelvis in 30 min an exchange ratio of 1:2. The ratio is even better for achieving washout during periods of diuresis. "Stasis" is present when obstruction disproportionately increases the volume in a given space and the ratio falls below 1:1. In the pelvis, this might be when its contained volume exceeds 12 ml.

2. Mixing: For volume exchange to be effective, it must be complete. However, urine passes through the calyx and pelvis at a rate that permits laminar, not turbulent, flow, so that the urine in the center of the stream is exchanged at a normal rate, while the rate decreases progressively to reach zero at the wall itself. As a result, bacteria in the urine next to the wall have ample time to generate even though the exchange volumes are adequate.

Observations of calyceal musculature (Fumagalli and Marchesi, 1944) and of motility (Narath, 1951) lead to the conclusion that calyceal activity is not peristaltic, i.e., it does not propel urine only in an antegrade direction. Rather, it moves the urine both out of and into the calyx, aided by the activity of the renal pelvis. Thus, at least one role of calyceal (and pelvic) activity is to mix the urine to ensure complete washout.

Effect of pathologic states. In acute upper tract infection, endotoxins produced by bacteria can stop smooth muscle activity in the urinary conduits. (Teague and Boyarsky, 1968a; 1968b). Thus bacteria can perpetuate themselves both by paralyzing the pelvio-calyceal mixing mechanism and by reducing ureteral peristalsis. When chronic, the infection not only renders the infundibula of the calyces scarred and relatively inactive (or even stenotic), but makes the calyces themselves immobile and dilated. During pregnancy, both functional (hormonal) and mechanical ureteral obstruction may occur, especially on the right side, producing stasis, but not significant back pressure. Renal function is not impaired unless infection, to which these women are of course more susceptible, supervenes (pyelitis of pregnancy).

Summary and Conclusions

Bacteria are normally expelled from the urinary tract by the combined effect of hydraulic factors and tissue defenses. Of the two, the hydraulic defenses are the more important, and consist of surface flow, backflow and exchange.

The urethra is partially cleared of resident flora during each voiding by surface flow, which mechanically washes bacteria distally. The washout is incomplete because of irregularities of the surface and trapping of bacteria by mucus. During interrupted micturition in the female, resident bacteria can be carried up into the bladder by backflow since the bladder and upper half of the urethra constitute a single chamber upon closure of the external urethral sphincter.

The bladder is cleared of bacteria by the mechanism of exchange, in which the volume of urine remaining behind after each voiding contains too few bacteria to regenerate to the previous concentration because as the bladder refills the relatively larger volume of fresh urine dilutes them faster than they can multiply.
The ureter, pelvis and calyces become infected by backflow (reflux and possibly retrograde movement along the ureteral wall during peristalsis). Infection is perpetuated by inadequate replacement of infected urine with fresh urine from above before the bacteria can double. In addition, complete mixing is required to prevent small portions of the infected urine from remaining in the space for longer intervals even though the major part is washed down.

References

Backman, K. A.: Urodynamics – the hydrodynamics of micturition in normal female subjects. Acta neurol. scand. **42**, suppl. **20**, 79–85 (1966).

Cox, C. E.: The urethra and its relationship to urinary tract infection: The flora of the normal female urethra. Sth. med. J. **59**, 621–626 (1966).

Cox, C. E., Hinman, F., Jr.: Experiments with induced bacteriuria, vesical emptying and bacterial growth on the mechanism of bladder defense to infection. J. Urol. (Baltimore) **86**, 739–748 (1961).

Cox, C. E., Lacy, S. S., Hinman, F., Jr.: The urethra and its relationship to urinary tract infection. II. Urethral flora of the female with recurrent urinary infection. J. Urol. **99**, 632–638 (1968).

Drach, G. W., Cox, C. E.: Immunofluorescence studies of the bladder defense mechanism. I. Localization of Escherichia coli in the human bladder. J. Amer. Geriat. Soc. **15**, 1114–1124 (1967).

Fumagalli, Z., Marchesi, M. G.: Studio anatomico dell'apparato muscolare del calice renale. Arch. ital. Urol. **21**, 24–46 (1944).

Helmholz, H. F., Sr.: Determination of the bacterial content of the urethra: a new method, with results of a study of 82 men. J. Urol. (Baltimore) **64**, 158–166 (1950).

Hinman, F., Jr.: Mechanisms for the entry of bacteria and the establishment of urinary infection in female children. J. Urol. (Baltimore) **96**, 546–550 (1966).

Hinman, F., Jr.: Obstruction: Back pressure or residual volume and laminar flow. J. Urol. (Baltimore) **105**, 702–709 (1971).

Hinman, F., Jr., Cox, C. E.: The voiding vesical defense mechanism: The mathematical effect of residual urine, voiding interval and volume on bacteriuria. J. Urol. (Baltimore) **96**, 491–498 (1966).

Hinman, F., Jr., Miller, E. R., Hutch, J. A., Gainey, M. D., Cox, C. E., Goodfriend, R. B., Marshall, S.: Low pressure reflux: Relation of vesicoureteral reflux to intravesical pressure. J. Urol. (Baltimore) **88**, 758–765 (1962).

Hinman, F., Jr., Miller, G. M., Nickel, E., Miller, E. R.: Vesical physiology demonstrated by cineradiography and serial roentgenography. Radiology **62**, 713–719 (1954).

Hinman, F., Jr., Miller, G. M., Nickel, E., Steinbach, H. L., Miller, E. R.: Normal micturition. Certain details as shown by serial cystograms. Calif. Med. **82**, 6–7 (1955).

Hutch, J. A., Miller, E. R., Hinman, F., Jr.: Perpetuation of infection in unobstructed urinary tracts by vesicoureteral reflux. J. Urol. (Baltimore) **90**, 88–91 (1963).

Lyon, R. P., Marshall, S.: Urinary tract infections and difficult urination in girls: Long-term followup. Trans. Amer. Ass. Genitourin. Surg. **62**, 69–72 (1970).

Lloyd-Davies, R. W., Hayes, T. L., Hinman, F., Jr.: Urothelial micro-contour. I. Scanning electron microscopy of normal resting and stretched urethra and bladder. Trans. Amer. Ass. Genitourin. Surg. **62**, 73–78 (1970).

Lloyd-Davies, R. W., Lee, J. C., Hinman, F., Jr.: Urothelial micro-contour. II. Intracellular crypts as a source for mucosal secretions. J. Urol. (Baltimore) **106**, 75–80 (1971a).

Lloyd-Davies, R. W., Hayes, T. L., Hinman, F., Jr.: Urothelial micro-contour. III. Mucosal alteration by infection. J. Murol. (Baltimore) **106**, 81–83 (1971b).
Masih, B. K., Drouin, G., Hinman, F., Jr.: Voiding and intrinsic defenses of the lower urinary tract of the female dog. I. Effects of episiotomy, colostomy, dilatation and urinary diversion. J. Urol. (Baltimore) **104**, 130–136 (1970).
Masih, B. K., Hinman, F., Jr.: Unpublished data.
Meares, E. M., Stamey, T. A.: Bacteriologic localization patterns in bacterial prostatitis and urethritis. Invest. Urol. **5**, 492–515 (1968).
Narath, P. A.: Renal pelvis and ureter. New York: Grune and Stratton, Inc. 1951.
Norden, C. W., Kass, E. H.: Bladder antibacterial defense mechanisms. In: The proceedings of a workshop on bacterial infections of the male urogenital tract. Washington, D. C.: National Research Council 1967.
Nordenström, B. E. W.: Some observations on the shape and course of the female urethra during miction. Acta radiol. (Stockh.) **38**, 125–132 (1952).
O'Grady, F., Cattell, W. R.: Kinetics of urinary tract infction. 1. Upper urinary tract. Brit. J. Urol. **38**, 149–155 (1966).
Teague, N., Boyarsky, S.: The effect of coliform bacilli upon ureteral peristalsis. Invest. Urol. **5**, 423–426 (1968a).
Teague, N., Boyarsky, S.: Further effects of coliform bacteria on ureteral peristalsis. J. Urol. 720–724 (1968b).
Ward, B. D., Cass, A. S.: Flow development in the urethra. Invest. Urol. **7**, 52–56 (1969).
Weinberg, S.: Experimental evidence of retrograde flow and trapping in two dimensional peristaltic pumping. Read at Annual Meeting, Urodynamic Society, Chicago, May 16, 1971.

Motion Picture: Energy Destroyer in Urology

H. Marberger, H. Madersbacher, and H. P. Swoboda*

When speaking of "energy destruction" in urology one thinks primarily of an obstruction, such as narrowing of the bladder neck or stricture of the urethra. Other factors which can lead to a loss of energy, or to be more precise, to a permanent or temporary transformation of kinetic energy into potential energy, are ignored, despite the fact that they play an essential role in the hydrodynamics of the urinary tract.

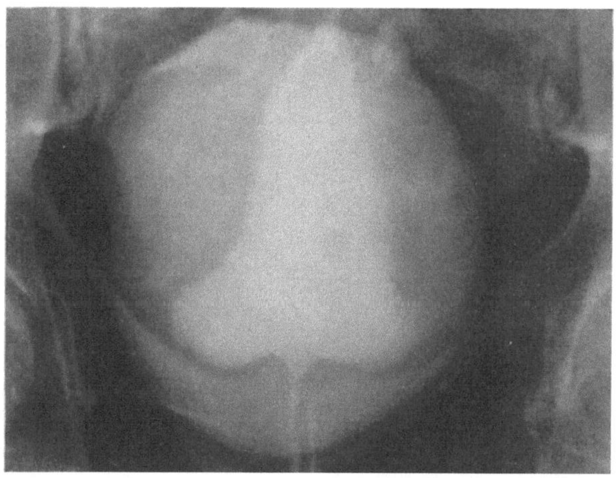

Fig. 1. 35-year-old man with a very poor flowrate: The voiding-cystogram reveals behind the somewhat "pine-tree" shaped bladder a huge, round diverticulum as the "energy-destroyer"

Every day, our own country gives us impressive examples of energy destroyers, of phenomena which have similar effects, though different dimensions, to disorders in the transport of urine. Glaciers represent an immense potential of energy which is utilized and converted into electrical energy in hydro-electric plants. The force of the masses of water coming from glacial rivers and reservoirs is harnessed by technical installations. The spectacle of the water being released from the reservoir magnificently illustrates physical laws which also apply to urology and have a practical significance. As the water is relased into the whirlpool chambers and basins, enormous forces with energies of up to 100,000 HP are harnessed and brought under control.

* Department of Urology (Director: Prof. Dr. H. Marberger) of the Innsbruck University Hospital, Innsbruck, Austria

We have examined and demonstrated in detail some particularly impressive examples of the numerous counterparts to this in urology. Thus we were able to show that one young man needed a quarter of an hour to empty his bladder, despite having a finger-thick muscle wall, because the kinetic energy was lost on a vesical diverticulum the size of a child's head (Fig. 1). With numerous urological patients we were able to confirm our suspicion that changes in the shape and the surface of, for instance the bladder, the bladder neck and the urethra, cause urological illnesses and that in disorders in the conveyance of urine resulting from urethral diverticula, reflux or regurgitation, treatment must be adjusted accordingly.
In a woman, for example, lack of support or anchorage of the bladder when the urethra remains fixed severely impairs normal micturation and leads clinically and flowmetrically to a substantial loss of energy which is practically ignored by urologists. Thus a cystocele is frequently the cause of impeded and insufficient emptying of the bladder.
Flowgraphs and cineccystograms of women with large cystoceles lead to a better understanding of poor micturition, which can be improved to an astonishing degree by the appropriate operative repair (Figs. 2 and 3).

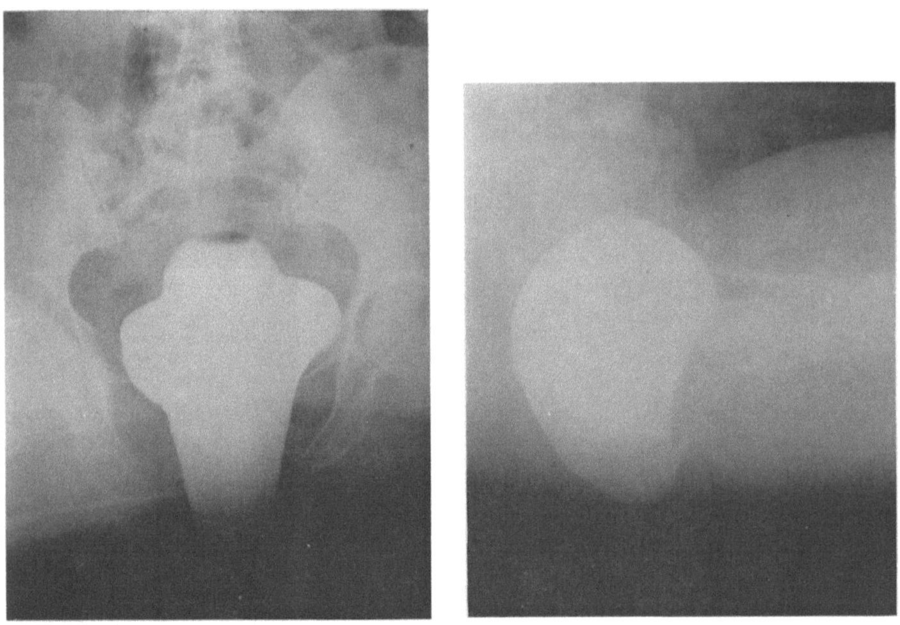

a b

Fig. 2a and b. 48-year-old woman with poor flowrate. Cystogram (Fig. 2a) and voiding cystogram. (Fig. 2b) reveal a marked cystocele

Finally, we attempted to reproduce and examine the information gained on human beings in simple experiments and to explain the underlying principles. Especially impressive were the results of flow tests on tubes of the same calibre but with different shapes and surfaces. Other experiments confirm our assumption that the intubation of a glass tube considerably impedes flow, even when the surface area of flow is not reduced.

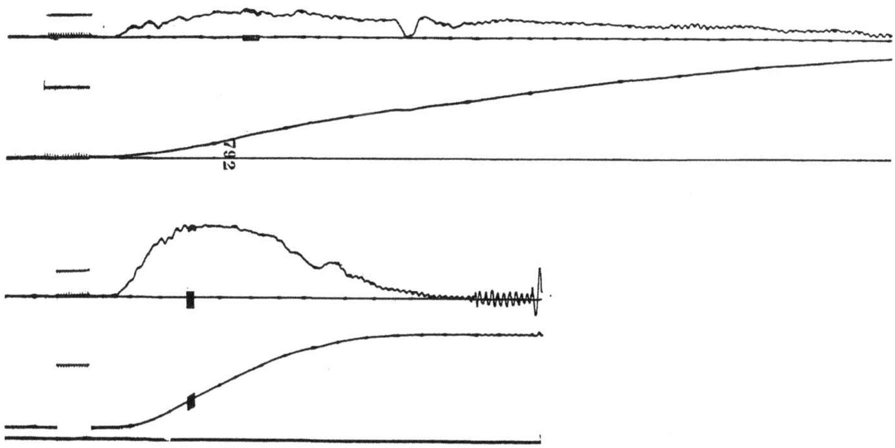

Fig. 3. 56-year-old woman: Preoperative and postoperative diagram of flow and volume. Calibrations (on the left of the diagram) mark a flowrate of 10 cc/sec and a volume of 300 cc. Preoperative flowrate: 6 cc/sec, maximum flow 12 cc/sec. (above). Flowrate after repair of the cystocele: 16 cc/sec, maximum flow 30 cc/sec (below)

These experimental and clinical observations, throw a new light on the etiology and pathogenesis of numerous illnesses which hitherto have been little understood. They show that urodynamic disorders can easily be understood, measured and demonstrated.

The Potential Value of Urodynamic Measurements in Clinical Practice

J. SÖKELAND, P. MAY, K. HAUBENSAK, and K. SCHULTE*

Measurements of pressure and flow in the lower urinary tract have recently been used with increasing frequency in clinical diagnosis, in follow-up checks and in postoperative monitoring.

Urodynamic measurements require elaborate equipment, but this is generally available in the operating room or the intensive care unit.

To detect gradual pressure changes, the older techniques based on the use of a simple manometer can and have been used as a simple recording procedure.

Among more exact methods suitable for hospital use are pressure measurements with pressure sensors, amplifiers and suitable recorders. Apparatus for flow measurements has also been developed.

Urodynamic measurements can be classified as follows:

A. Pressure measurements:
1. Renal pelvic pressure measurements through a renal fistula or a splint;
2. Splint or ureteral pressure measurements;
3. Cystometry;
4. Sphincter pressure and urethral resistance measurements.

B. Flow measurements:
1. Flow measurements in the ureter;
2. Flow measurements in the urethra (uroflowmetry).

A.1. Renal pelvis pressure measurements. One method frequently used to test the patency of a pyeloplasty before removing the tube in the renal pelvis is to inject contrast medium along the tube. This method is unsatisfactory. The appearance of contrast medium in the bladder does demonstrate that a passage exists, but it does not prove that the renal pelvis will be able to empty itself under physiological pressure conditions.

Repeated measurements of renal pelvic pressure during the postoperative period are a valuable aid judging the right time for removal of the drainage tube. If the pressure in the renal pelvis does not exceed 12 mm Hg during the smooth discharge of urine the renal fistula can safely be closed. Higher pressures call for further checks.

---

* Urologische Klinik, Städtische Krankenanstalten Dortmund (Director: Prof. J. SÖKELAND) and the Urologische Universitätsklinik (Director: Prof. Dr. C. E. ALKEN) Homburg/Saar, Germany

Pressure measurements through a drainage tube can be performed as simple pressure checks, as long-term pressure measurements or in association with flushing pressure measurements.

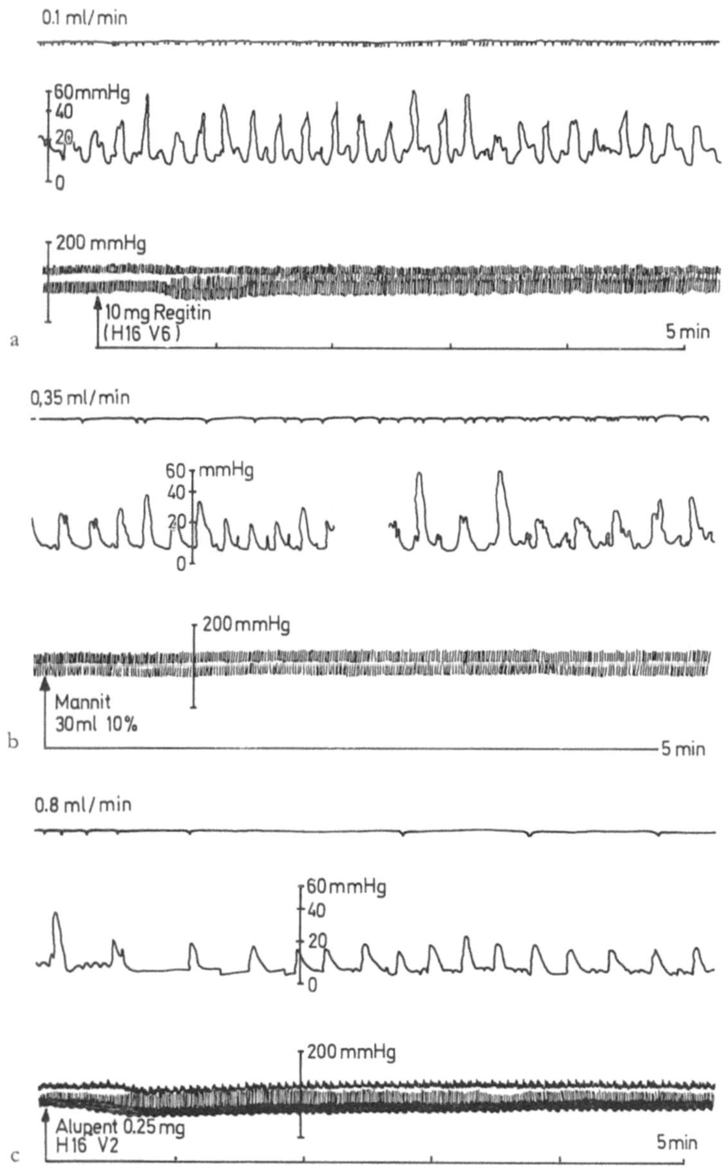

Fig. 1a–c. a Effect of Regitin (10 mg) on urine flow, ureteral pressure and arterial blood pressure. b Effect of Mannitol (30 ml/10 %) on urine flow, ureteric pressure and arterial blood pressure. c Effect of Alupent (0,25 mg) on urine flow, ureteral pressure and arterial blood pressure

Flushing pressure measurement is a new technique in urology. The flushing technique has been made practicable by the introduction of sensitive pressure transducers subject to minimal volume changes. During continous long-term pressure measurements, any blockage of the catheter is prevented by simultaneous perfusion. Any rise above the maximum allowable pressure for a given outflow volume urine production is admittedly disregarded provides clinically reliable information on the transport capacity of the lower urinary passages.
Flushing pressure measurements through a drainage tube are carried out by continuously pumping in fluid with a perfusor pump and measuring the outflow pressure. A pressure transducer (Statham P23) is connected in parallel and the pressure recorded with Siemens monitoring equipment (Sirecust), any pressure changes caused by the system being eliminated by the zero calibration.
Measurements of intrapelvic pressure are recorded at various inflow rates – 0.1 ml per minute, 0.2 ml per minute, 0.5 ml per minute and up to 1 ml per min. If the pressure rises above 15 mmHg measurement must be discontinued. If the pressure level remains within the normal range of 8 to 12 mmHg or below, the transport capacity of the lower urinary passages will be adequate under normal pressure conditions, and an artificial urinary diversion is generally no longer needed.

A.2. Splint pressure measurements or ureteral pressure measurements can be carried out to determine ureteral pressure and urine flow.
In the course of flushing pressure measurements it is possible to determine the ureteral transport volume under given conditions of fluid throughput and to check pressures at the same time.
In animal experiments carried out at the Urological Clinic, University of Homburg/Saar, intraureteric pressure measurements were made during diuresis, e.g. following Mannitol loading, and after administration of Alupent, Regitin etc. (Fig. 1).
Clinical use of the method requires the simultaneous insertion of a splint (during operation) and a ureteral catheter.
In another technique at present still under trial, a flushing catheter inserted into the ureter is used by itself. It seems probable that this will open up new possibilities in urorheography.

A.3. The possibilities of cystometric checks and long-term supervision will be mentioned only in passing. During flushing pressure measurements it is sometimes practicable to record bladder pressure through a catheter of almost capillary dimensions and thus to avert complications after transurethral resection and prostatectomy.

A.4. *Sphincter pressure measurements and urethral resistance measurements.* Especially in neurogenic disorders of bladder emptying, urethral resistance measurements provide useful diagnostic and therapeutic information. Any elevation or loss of resistance can be verified. The indications for sphincterotomy which has been advised very frequently in recent years, can be well evaluated by this method. In obscure cases of abnormal bladder emptying, especially when it is the only abnormality, urethral pressure measurements carried out in addition to uroflowmetry can provide further objective evidence of the disorder.

Technique for measurement of urethral resistance

A penile clamp (Stille) is applied to the penis and 50 ml of saline or contrast medium is continuously introduced by means of a perfusor over a period of 5 min. Simultaneous pressure recording by means of a pressure sensor and the Sirecust equipment (Siemens) make it possible for the intrinsic pressure of the system to be maintained, so that relatively precise urethral pressure measurements can be carried out.

The curves obtained in various clinical conditions demonstrate the normal urethral resistance, and also urethral resistance increased due to prostatic adenoma or sphincter spasm; reduced flushing pressure values are found in patients with atonic or damaged sphincters (Fig. 2).

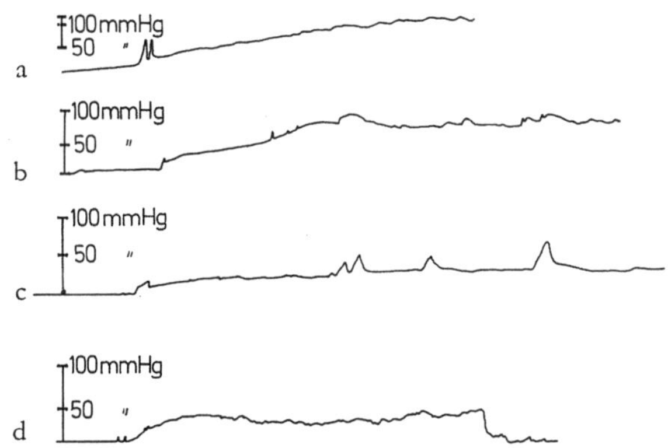

Fig. 2a–d. Flushing pressure measurement in the urethra. a sphincter spasm. b adenoma of prostate. c atonic sphincter. d normal urethral resistance

Flushing pressure measurement is of great clinical value in preoperative diagnosis and postoperative follow up, either short-term or long-term. Its value has been enhanced by the introduction of microcatheters.

Among other advantages of flushing pressure measurement are the simplicity of the procedure, the exactly defined and reproducible results, the absence of serious risk, and the ease of control and supervision.

A New Uro-Rheomanometer

H. Melchior and K. K. Simhan*

Uro-rheomanometry is a method of simultaneous measurement of pressure and flow in the ureter. A method of investigation of this type suitable for clinical use as well as animal experiments, must fulfil the following basic requirements:
1. feasibility of studies in situ under physiological conditions,
2. possibility of long-term observations.

Due to the protected retroperitoneal position of the ureter and to its small size, in the case of uro-rheomanometry these requirements involve:
1. endoluminally applicable recording methods,
2. small dimensions and streamlined shape of the probe-catheter,
3. combination of several measuring systems in one measuring unit,
4. direct analogical conversion of the hydrodynamic parameters into electrical signals.

The main hydrodynamic parameters of peristaltic urinary transport, we believe, are pressure and flow. Although electromanometry (Kiil, 1953) has been very effective for recording endoluminal pressure in the ureter, in 1969 we introduced uro-rheography (Lutzeyer and Melchior) for the measurement of endoluminal flow.
Uro-rheography is based on the heat-conduction method designed for blood perfusion rates in tissues by Hensel (Golenhofen *et al.*, 1963). The method is based on the physical principle according to which a heated body loses heat to its surroundings. The heat conduction is, with some limitations, a function of the flow velocity of the surrounding medium. Such a heat conduction element has the following schematic structure:
Inside the probe tip, which is filled with araldite, there is a heating coil (HW), energized by a constant heating current. The rise in temperature of the probe tip in relation to the surroundings is measured by a battery of three heat-conduction elements of copper and constantan wires. Its three soldered measuring points are located in the probe tip within the heating coil, and its three soldered reference points are located near the probe shaft. From the temperature rise and the heating intensity of current the heat conduction coefficient can be calculated, which is then a function of the flow velocity. The heat-conducting element has an outer diamter of 1.2 mm and, mounted on a normal No. 4 French ureteral catheter, can be introduced into the ureter with a cystos-

* Abteilung für Urologie (Director: Prof. Dr. W. Lutzeyer), Medizinische Fakultät, and the Institut für Ärodynamik (Director: Prof. Dr. A. Naumann), Rheinisch-Westfälische Technische Hochschule, Aachen, Germany

cope transurethrally. Combined with electromanometry, uro-rheography forms the basis of our programme for the study of urodynamics. Both catheters are introduced into the same ureter and therefore simultaneously allow pressure- and flow-measurements at each point of the ureter.

This method, in our opinion, fulfils the main methodical requirements:

1. It permits direct studies in situ.
2. It permits studies under normal conditions without anesthesia and without premedication.
3. It causes minimal interference with the physiological dynamics of the ureter, owing to the 8-shape diameter of the two No. 4 French catheters, which cause less disturbance than only one slightly bigger catheter, as Kiil (1957) showed.
4. It ensures readily reproducible results which allow even interindividual comparisons.

Example: Pressure and flow in the ureter are measured simultaneously before, during and after hypertonic bladder contraction with voiding. During the hypertonic bladder contraction the intravesical pressure rises, thus forming a functional outlet obstruction. The ureter tries to overcome this obstruction by an increase in the force and frequency of contractions; thereby the flow velocity is accelerated, a sign of compensation (Fig. 1).

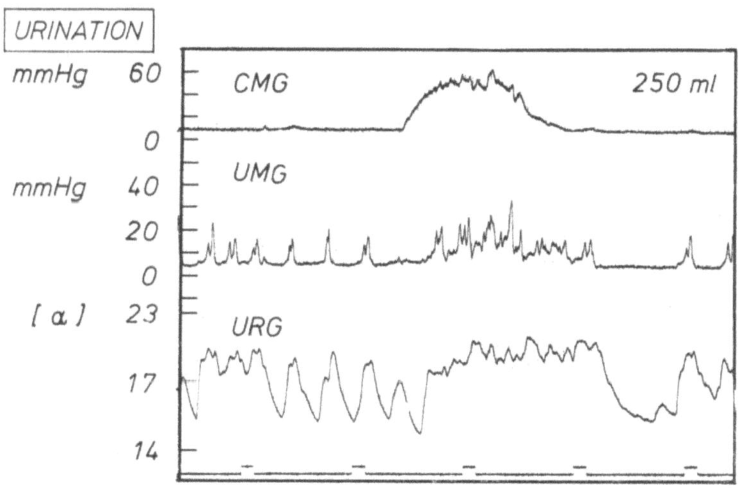

Fig. 1. The dependence of ureteral dynamics on the bladder dynamics. (*CMG* = cystomanometrogram, *UMG* = uretero-manometrogram, *URG* = uro-rheogram)

In spite of all its advantages, the combination of uro-rheography and electromanometry for simultaneous monitoring of pressure and flow was not completely satisfactory. Besides the fact that this system does not facilitate direct pressure recording because of the imposed hydrostatic column of fluid, it was desirable to combine both measuring systems in a single probe. In collaboration with the Institute of Aerodynamics of the RWTH Aachen we have developed a probe which combines these measuring system for endoluminal pressure and flow (Fig. 2). The flow-measuring system is located in the probe tip and the pressure system in the stem. Flow-measurement is made by heat conduction in this probe too. The temperature rise, however, is not recorded by thermocouples but by *N*egative-*T*emperature-*C*oefficient-Elements (NTC). NTC-elements are semi-conducter elements, their electrical resistance being very dependent on temperature (Fig. 3). These NTC-elements form a Wheatstone bridge. For endoluminal pressure measurement, four metal strain gauges are glued onto a thin gold membrane in such a way that two strain gauges are exposed to the peristaltic pressure changes by elliptical openings on the left and right side of the stem. For temperature compensation, two more strain gauges form the other arms of the Wheatstone bridge. Pressure compensation in the stem is accomplished by the catheter.

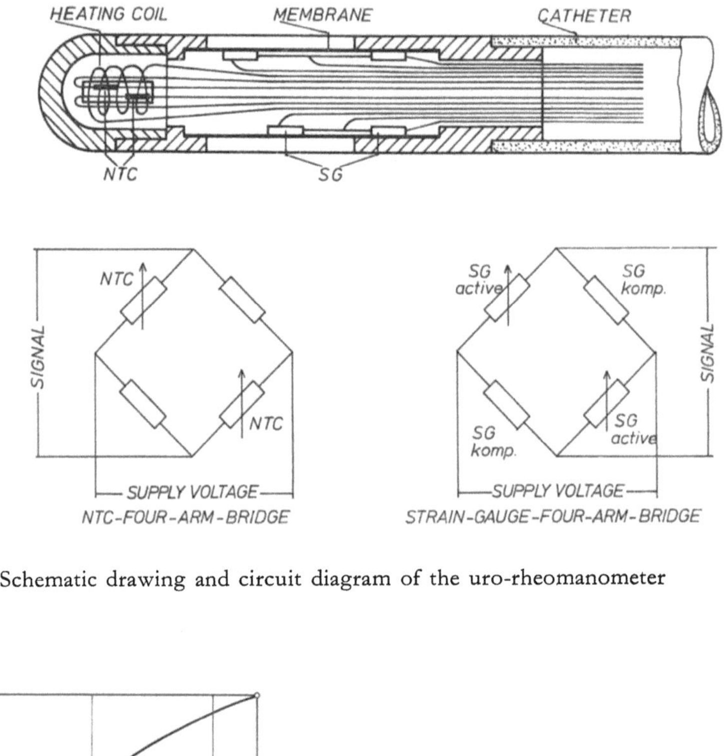

Fig. 2. Schematic drawing and circuit diagram of the uro-rheomanometer

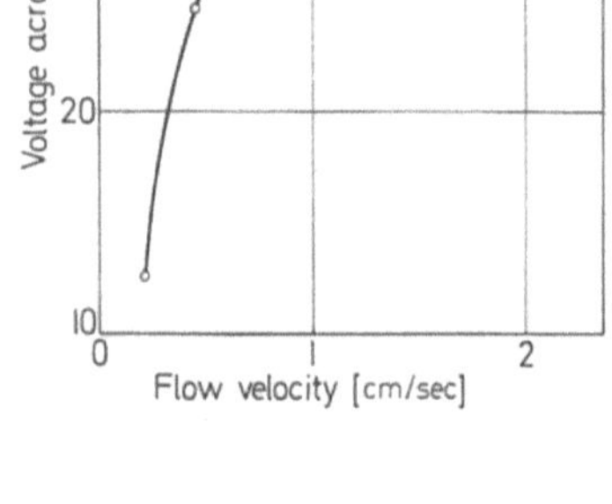

Fig. 3. Calibration curve of the new uro-rheomanometer

This uro-rheomanometer has an outer diameter of 1.5 to 1.6 mm, corresponding to a No. 5 French catheter (Fig. 4). It can also be introduced into the ureter through the urethra by a cystoscope and it makes simultaneous continuous monitoring of pressure and flow possible. Whereas the pressure system allows linear measurements in the range of $\pm$ 50 mmHg, the limitations for lower flow measurements are given by the relative increase of convection at heat transfer and for higher flow measurements by deviation from the laminary flow characteristic. Such high flow velocities, however, are practically never obtained in the ureter.

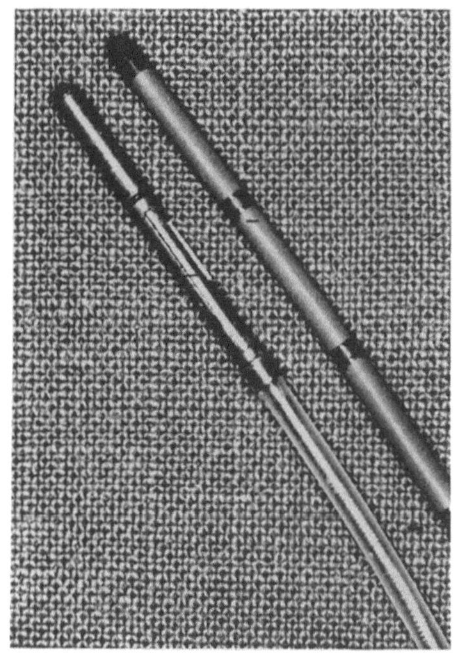

Fig. 4. Prototype of a new uro-rheomanometer corresponding to a No. 5 French catheter

Summary

Uro-rheomanometry is a method of simultaneous measurement of pressure and flow in the ureter.

The combination of electromanometry and uro-rheography has formed the basis of our programme for the study of urodynamics for the last three years. Information is given on the main urodynamic parameters of peristaltic urinary transport: tonus or basal pressure, amplitude and frequency, flow velocity and efficiency of peristaltic contractions.

A new-rheomanometer, which combines the pressure and flow sensors in a single probe, has been developed in collaboration with the Institute of Aerodynamics of the RWTH Aachen. This uro-rheomanometer also allows direct pressure measurements in situ.

References

Golenhofen, K., Hensel, H., Hildebrandt, G.: Durchblutungsmessungen mit Wärmeleitelementen. Stuttgart: Georg Thieme 1963.

Kiil, F.: Pressure recordings in the upper urinary tract. Scand. J. clin. Lab. Invest. **5**, 383 (1953).

Kiil, F.: The function of the ureter and renal pelvis. Philadelphia-London: W. B. Saunders 1957.

Lutzeyer, W., Melchior, H.: Urorheographie. Urologe **8**, 208 (1969).

Melchior, H.: Uro-Rheomanometrie (Simultane Urorheographie und Elektromanometrie). In: Lutzeyer, W., Melchior, H.: Ureterdynamik, S. 125. Stuttgart: Georg Thieme 1971.

Technique of Routine Cine/Pressure/Flow Cystography

C. G. WHITESIDE*

The increasing awareness that there may be little relation between the symptoms of disordered micturition and underlying dysfunction of the bladder is reflected in the increasing number of reports in the literature of visual (TURNER-WARWICK) and dynamic (TANAGHO *et al.*) methods of assessment. In this paper a technique of combining these methods on video tape is described (BATES and CORNEY).

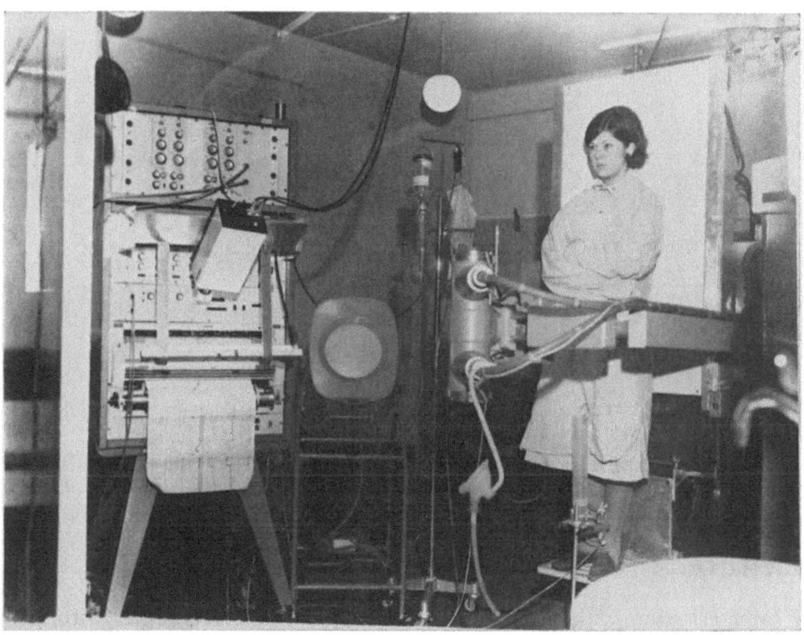

Fig. 1. Apparatus, showing 6 channel recorder, television camera, monitor, mobile transducer stand, tilting table, Stanmore urinal and voiding cylinder

Apparatus (Fig. 1)
A standard tilting x-ray table (LIEBEL FLARSCHEIM) is fitted with an overcouch tube and image intensifier. The Devices 6 Channel Pen Recorder is fed by signals from four

* The Middlesex Hospital, London W. 1, Great Britain

transducers, two mounted on a mobile platform which are connected by fine polytetrafluorethylene catheters to the rectum and bladder respectively. These are filled with a 1/500 Detergicide solution from a bottle hanging on the same stand; also shown is a plastic bag which contains 35% Diodone, the contrast medium for the bladder. The latter is supported by a strain gauge which records the filling volume, while the two transducers on the platform record the total bladder pressure and the abdominal (rectal) pressure. The fourth transducer records the volume voided from the base of the measuring cylinder (Von Garrelts 1956), into which males void by a simple funnel and females by a Stanmore urinal and/or on a commode with similar facilities for measuring flow and pressures.

Two further channels record 1. the flow rate, which is differentiated electrically from the voided volume on a time base and 2. the intrinsic (detrusor) pressure by electrically subtracting the rectal (abdominal) pressure from the total bladder pressure. The television camera picks up these three recordings (total pressure, intrinsic pressure and flow rate) and mixes them with the x-ray image of the bladder on a common monitor and also on a videotape recorder (Fig. 2).

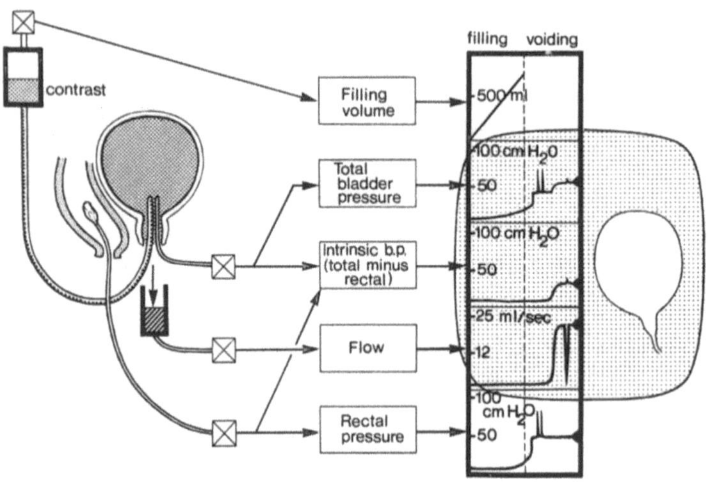

Fig. 2. Diagramatic summary of technique. Note television coverage

Procedure

After catheterisation with a standard No. 12 FG catheter (together with the small pressure catheter), the bladder is filled at about 150 cc/min and a running cystometrogram recorded on the paper. After filling, the 12 FG catheter is removed, the table tilted erect and the x-rays and recorder switched on. The patient coughs to test 1. the pressure apparatus 2. for stress incontinence 3. the stability of the intrinsic pressure which should remain unaltered through all these proceedings, until the patient passes urine. When the peak flow has been reached the patient is asked to stop; this tests the action of the external sphincter and also the distal urethral "milkback" mechanism. Voiding is then completed and the apparatus switched off.

The Filling Cystometrogram

This reflects the hydrostatic function of the bladder as a storage organ. In a series of normal cases the pressure rose only 5.6 cm of water for a mean volume of 490 cc, in spite of quite rapid filling. This final intrinsic pressure remains unchanged on standing while, of course, the total pressure rises due to the increased abdominal tension. Apart from overt neurological disease, abnormal cystometrograms may be seen in about half the males with outlet obstruction and in a number of females complaining of stress and/or urge incontinence, especially on standing and coughing (Figs. 3 and 4).

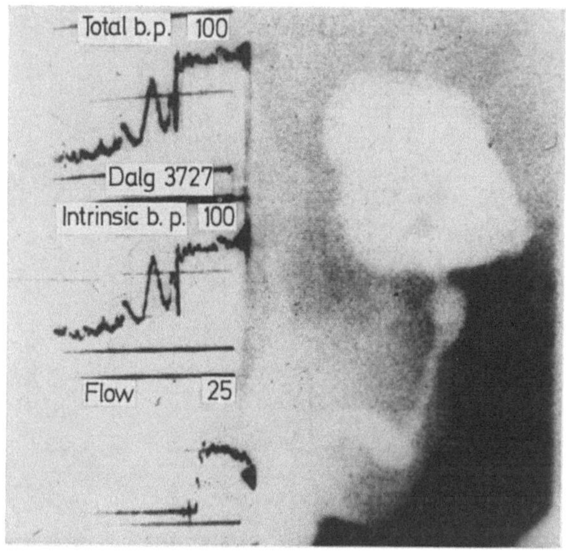

Fig. 3. Combined recording in a case of bladder neck obstruction, showing a climbing systolic cystometrogram, a high intrinsic voiding pressure and poor flow. Relief of the obstruction returned all these parameters to normal

The Flow Rate

This parameter is extremely sensitive both to the pressure/resistance conditions of micturition and, unfortunately, to the artificial conditions under which the examination takes place, such as the passage of catheters and, in the female, the standing position. Consequently, it is wise to retest it with the patient passing urine into the flow meter in private on a separate occasion if it is suspected of being artificially low. In the female this is quite a common problem, affecting about 20% of cases. It is our practice to transfer such cases to the commode once the outlet has been visualised in order to obtain a true physilogical representation of the act of micturition; in such cases the rate may double or even be increased three times (Fig. 5).

Apart from the height, the shape of the flow rate curve is important. It should rise and fall vertically at the beginning and end of micturition; in outlet obstruction compensated by a high intrinsic pressure, the peak rate may approach normal values, but the second half of the curve will slope gently away as a manifestation of the outlet problem.

Bladder Pressure

The intrinsic pressure is a direct manifestation of detrusor activity and is consequently the parameter of importance – the total pressure is of no value except to demonstrate straining by the patient and to confirm that the intrinsic trace is valid. In males the normal pressure seldom exceeds 50 cm of water and in females 40 cm. Outflow obstruction in both sexes usually exceeds these levels (Fig. 3). Early cases in males present either as a low flow with normal or slightly raised pressures only or, less frequently, as higher pressures which may compensate and return the flow to near normal; in the latter, however, the characteristic slope of the flow trace in the later part of micturition will be evident. In all cases of outlet obstruction the visual image will indicate the nature and position of the obstruction.

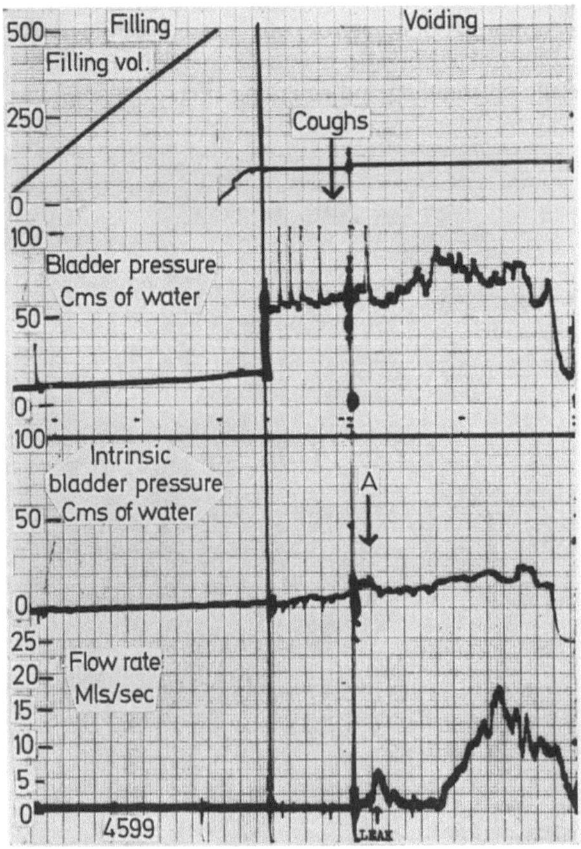

Fig. 4. Female complaining of incontinence. The trace shows normal cystometrogram-coughing, however, causes the detrusor to contract with leakage A↓. The voiding pressure is lower than average

Residue

Only in decompensated obstruction will a residue be present, associated with low flow and high pressure which falls prematurely and also with trabeculation. False residues may result on a standard excretory pyelogram due to reflux or to an unduly long interval between micturition and the final bladder film; conversely, the absence of a residue does not preclude the presence of outlet obstruction of quite severe degree, but only absence of decompensation.

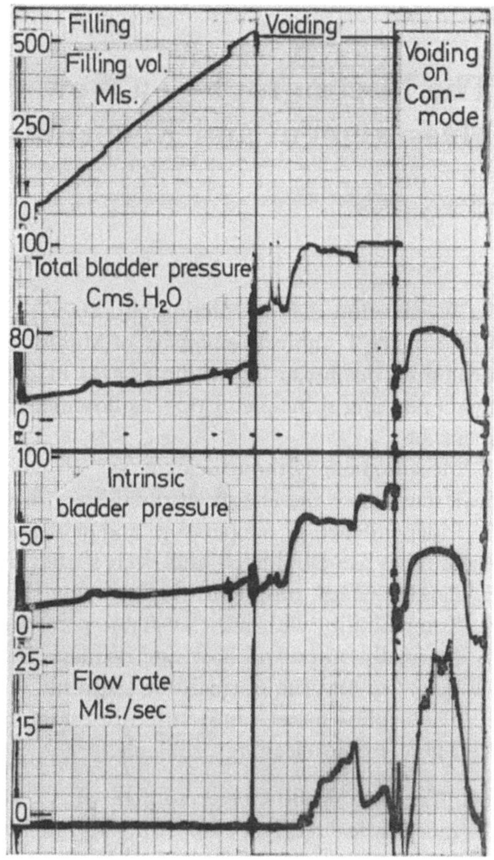

Fig. 5. Normal female, with poor flow and high pressure while standing, followed by good flow at normal pressure on the commode

Conclusion

At the Middlesex Hospital this test is routinely used in selected cases and run as a clinic by one Radiologist and a Radiographer. The results presented to the clinicians supply irrefutable facts which serve to underline the direction which treatment should follow and also objective data of the results of such treatment.

References

Bates, C. P., Corney, C. E.: Brit. J. Radiol. **44**, 44 (1971).
Garrelts, B. von: Acta chir. scand. **112**, 326 (1956).
Tanagho, E. A., Miller, E. R., Lyon, R. P., Fisher, R.: Brit. J. Urol. **43**, 69 (1971).
Turner-Warwick, R. T., Whiteside, C. G.: Modern Trends in Urology – 3, p. 295. London: Butterworth 1969.

The Mictiograph — a New Principle in the Measurement and Recording of Urinary Flow

H. TAMMEN*

A. We developed the "Mictiograph 7005", a new uroflowmeter, in the Urological Clinic of the Technical University of Munich on the suggestion of the first meeting of urologists on measuring uroflow, held in December 1969 in Aix-la-Chapelle. The new operating principle of this uroflowmeter fulfils all conditions for exact measurement, i.e.:
1. Measurement without delay;
2. Exclusion of the energy of the urinary flow bounding upon the uroflowmeter;
3. Independence of temperature and acidity of the urine.
4. Bubbles are eliminated and even if they are not, it is unimportant.
5. Psychic effects on micturition are only slight.
6. Compatibility with most available recorders.
7. Normally output on flow rate is linear, but if required the pathologic range may be sprawled and the normal one reduced.
8. The uroflowmeter can be used in examintation of men, women and children.
9. There are no cleaning problems.

B.I. The new uroflowmeter consists of
1. a mechanical receiver comprising a funnel with adjustable height, a disk rotated by a special motor, with a vessel for collecting urine and a cylinder for measuring quantity and density if required;
2. an electronic component with the switch for selection of one of three possible expected micturition rates, 25, 50 and 100 g/sec, the recorder, the indicator, the key for starting the recorder and the integrator of micturition volume.
This electronic component is sited in the urologist's room and the receiver is installated in the toilet. The two are connected by a long cable, so that there are no psychic influences on the patient.

II. The function of the flowmeter can be seen from the illustration.
1. Mechanical component. When a lamp on the tripod is lit the patient voids into the funnel, which has no neck. This avoids any delay or integration of flow. The urine falls directly on the rotating disk and the energy of fall is destroyed; the urine attains the

* Urologische Klinik des Klinikums rechts der Isar der Technischen Universität (Director: Prof. Dr. W. MAUERMAYER), München, Germany

velocity of the rotating disk and is thrown against the wall of the vessel and gathered in the cylinder where the quantity is measured.

2. Electronic component. The velocity of the disk is kept constant by the electronic equipment, so that the electric energy used by the motor gives an exact measure of the mass of urine flowing through. The area below the recorded curve is integrated

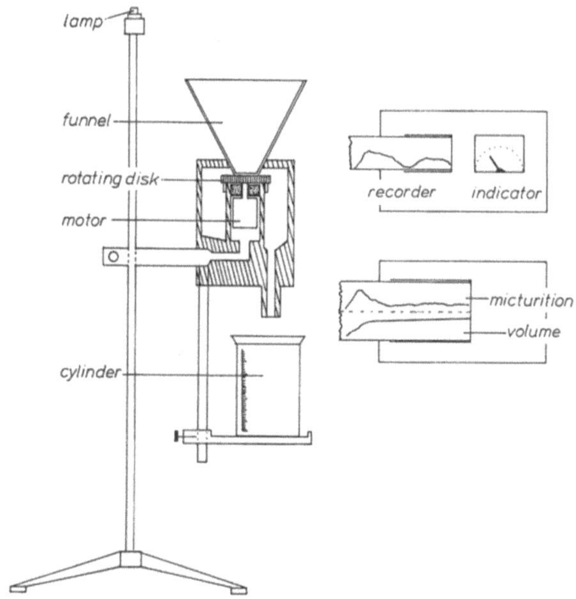

Fig. 1. System of the uroflow-meter "Mictiograph 7005"

by an integrator in the electronic component so that the total amount of the urine voided is measured. The velocity of the paper is increased by pushing a special button and the total amount of urine in the cylinder is written by the recorder. The normal velocity of recording paper is 2,5 mm/sec, which yields a manageable size of uroflow recording.

III. Results. The "Mictiograph 7005" yields more information about patients with urinary tract pathology. It shows the "hesitating time" from the order to start voiding to the actual beginnung of micturition. Then we get the curve without delay and without partial integration, we see dribbles, 'bulbocavernosus jet' and subsequent flow. The integrator records the mass of urine voided. If a special exit output attachment is used, it is possible to write this mass of urine by means of a recorder with two or more channels. By integrating the area under the recorded curve we can relate the contents of the bladder to the flow rate. The residual urine is not registered. Thus the "Mictiograph 7005" advantageously supplements the methods of examining patients with urinary obstructions.

Descending Simultaneous Measurement of the Urinary Bladder and Sphincter Pressures during Continuous Perfusion of the Urethra after Operations for Incontinence

F. Schreiter*

Descendirg simultaneous measurement of bladder and sphincter pressures with continuous urethral perfusion (Heidenreich and Beck) is a method we employ to determine, and plot graphically, urethral resistance before and after surgery for incontinence in males and females in order to obtain an objective standard against which to assess postoperative results.

The graph of the sphincter pressure in a totally incontinent woman remains constantly at zero, parallel to the curve of bladder pressure. Following successful fascioplastic sling surgery, the curve indicates a distinct increase in sphincter pressure which, with continence, is always higher than 20 mmHg.

At our clinic, fascial sling operations are carried out quite successfully on femal patients suffering from incontinence. Surgical treatment in males, however, still provides unsatisfactory results.

Recently we have used Puigvert's method for surgical treatment of male incontinence. Initial results have been promising.

Surgical Procedure

The corpora cavernosa are prepared from the perineum. After lateral mobilisation, taking good care to protect the medial vessels, the corpora cavernosa are drawn tight to form a pad underneath the urethra. Because of the extreme danger of ischemic necrosis of the cavernous tissue, great acre must be taken to avoid complete mobilisation of the corpora cavernosa and crossing of the individual crura.

Surgery (Puigvert's method) was performed in a 60-year-old male suffering from severe incontinence due to stress following prostatectomy. A lesion on the sphincter externus was apparent on urethroscopy. The constriction of the urethra at the site of the corpora cavernosa "padding" is clearly visible on the urethrograph. After the passage of the catheter through the pars membranacea (sphincter externus), the curve

* Urologische Klinik of the Johannes-Gutenberg-Universität (Director: Prof. Dr. R. Hohenfellner), Mainz, Germany

of sphincter pressure (Fig. 1), recorded concurrently with a serial cystourethrograph, showed an increase in pressure up to 15 mmHg, which corresponds to that of the insufficient sphincter externus in stress incontinence. When the catheter passed the urethra at the "padding" site, another increase of 15 mmHg in pressure is noted. The patient's continence was fully restored.

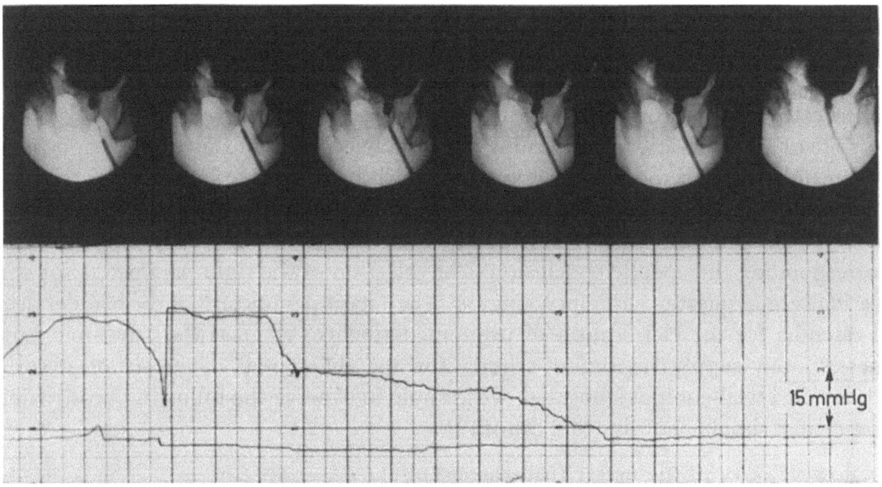

Fig. 1. Serial cystourethrograph and curve of sphincter pressure of male incontinence following prostatectomy after surgery (Puigvert's Method). Increase of pressure up to 30 mgHg. Continence is fully restored

References

BECK, L.: Gynäkologe **4**, 59 (1971).
HEIDENREICH, J., MELCHERT, F., WENDT, B., BECK, L.: Arch. Gynäk. **211**, 325 (1971).
MORGAN, J. E.: Amer. J. Obstet. Gynec. **106**, 396 (1970).
PUIGVERT, A.: Mem. Acad. Chir. **96** 675 (1970).

Discussion

Problems of the Urodynamic Examination

The spectrum of urodynamic methods of examination is relatively broad as concerns the functional exploration of the urethra and bladder, but rather restricted as regards the renal calyces and the ureter (Lutzeyer). The dimensions of the lower urinary tract are bigger and can therefore be measured more easily (Marberger). Twenty years ago the function of the urinary tract was thought to be that of a "gutter" (Marberger). Nobody even suspected hydrodynamic laws. Only since the introduction of modern methods of examination which permit visualization of the urinary tract we have learned to regard peristaltic activity functionally. A technical procedure, however, can never solve diagnostic problems; it will always be important to interpret the results of the measurements in the right way. Mainly, we have to consider what factors are causing the disturbance.

Today there are fairly exact methods of measuring and a high standard has been reached, especially in manometry. The difficulty is that the physiology of an organ system is changed by the insertion of measuring instruments (Boyarsky). The environment and the system are artefactual, as the reflexes observed are influenced by psychological effects. It is not only the use of technical apparatus which produces misleading results. The very atmosphere of the laboratory affects the patient, and psychic influences are essentially disturbing factors (Boyarsky). Hinman even thinks that a sensitively taken case history is often better than all the measurements.

In spite of all reservations concerning the instrumentation of functional diagnostics, it is mainly these methods of examination which have been responsible for the high standard of the urodynamic examination today. If you know the limits of a method of investigation, you will be able to interpret the results correctly and draw the corresponding therapeutic conclusion. The selection of the right parameters for answering certain questions is important, too. The "static parameter" of the bladder neck is essentiai for the clarification of urine incontinence, whereas the „dynamic parameters" can supply important criteria for the differential diagnosis of dysuria. Today, in a well-equipped hospital, we are able to observe the following urodynamic parameters simultaneously (Melchior):

1. flow velocity in the ureter (uro-rheography)
2. variations in the peristaltic pressure in the ureter (uretero-manometry)
3. bladder contents
4. intravesical pressure (cysto-manometry)
5. pressure in the urethra (sphinctero-manometry, urethra-profile)
6. activity of the muscles of the pelvic floor (electro-myography)

Radiological examination methods have a much wider range of indications than functional examination techniques. Both Turner-Warwick and Whiteside feel that these two techniques can be usefully combined to provide more evidence. On all patients referred for voiding cystography we measure the total and the substracted-detrusor pressures of the bladder synchronously. The method takes only a little more time than routine micturition cystography and it does not worry the patient, though it does involve more work for the examining doctor.

Electromyographic studies of the smooth muscles had first to be done in animal experiments, but they now represent valuable methods of examination, mainly in combination with pressure and flow measurements (Malatinskky).

As regards functional diagnosis in the area of the upper urinary tract, the perfused-catheter method has the advantage of measuring intravesical pressure under differing fluid load. This technique measures the transport capacity of the urinary system by perfusion (Sökeland).

In hospital the indication for a functional exploration is generally that no clear morphological substrate nor mechanical injury is found to account for functional disturbances such as functional ureteral stenosis, functional or neurogenic disturbed micturition, and incontinence of urine (Melchior). Patients with vegetative symptoms as well as patients suffering from prostatitis need a functional clarification

(LUTZEYER). Furthermore, the development of a suitable operating technique is promoted by urodynamic examinations: for example the importance of adjusting the voiding mechanism after cystoplasty (TURNER-WARWICK). The risk associated with instrumental examinations is relatively small if due care is taken. In more than 200 instrumental examinations of the upper urinary tract, only 5 per cent of out patients had colic or infections in spite of prophylactic administration of antibiotics and supporting diuresis (MELCHIOR).

Urodynamic Investigations without Instrumental Treatment

Every instrumental intervention interferes with the physiology. A reliable method of examination, establishing urodynamic facts without instrumental examination of the patient, should be developed (HINMAN). For the present, such a method of examination is possible for the urethra only.

The external urinary stream gives a great deal of information about the bladder and urethra. RITTER shows this in a film made with a high-speed camera (3000 frames/sec). The stream is shown to break into drops shortly after exit from the urethral meatus. If such stream is passed through a thin light plane, the light is interrupted by the urine drops. These interruptions can be converted into electrical signals which can be stored electronically and analysed. The number, volume and velocity of the urine drops are measured to give a distribution pattern which produces characteristic pictures for normal individuals and typical deviations for patients with meatal stenosis or prostatic hypertrophy for other conditions of disturbed micturition (RITTER).

Another related method of examination is based on the turbulent pressure fluctuations in the fluid which are transduced into far-field sound by the wall of the urethra. This sound can be recorded by a small microphone placed near the urethra. Theoretical considerations indicate that the spectrum of this sound can be used to determine both the diameter of the urethra and the average velocity of the fluid. The most appealing aspect of this technique is the high sensitivity of the sound intensity to the diameter: the intensity is proportional to the 8th power of the diameter of the tube (STERLING).

Spontaneous Generation of Excitation in the Pyeloureteral System and the Effect of Adrenergic Substances*

K. GOLENHOFEN and J. HANNAPPEL**

Summary

1. In unanesthetised and undisturbed guinea pigs, the frequency distribution of peristaltic intervals is multimodal. The peak value of the first mode is 10 sec, which can be taken as the basic period duration. From this value, the normal peristaltic frequency of the guinea pig can be determined as 6/min.
2. The normal pacemaker is localized in the utmost renal ends of the renal pelvis. A theory of „multiple coupled oscillators" is derived for the pacemaker process.
3. The main difference between pyeloureter of dog and rat compared to that of guinea pig is that the variability of the frequency is much greater in the former group; peristaltic frequency can easily be increased through frequency multiplications. A theory is presented that there is an „ideal norm" for ureteral peristalsis, which is equal to the frequency of stomach peristalsis in all mammals.
4. Species differences as well as topographic differences for α- and β-adrenergic effects on pyeloureter are described.

"There is sufficient evidence that the cause of the periodic movements, which start in the renal pelvis, is an automatic excitation of the muscular substance of renal pelvis. – The cause of the peristaltic progress of the movement is that the excitation is conducted, without mediation of ganglion cells and nerve fibres, from muscle cell to muscle cells. In other words, under normal conditions, the ureter is a single hollow organic muscle fibre. – The pressure which acts on the walls has, at least within certain limits, no noticeable influence on the frequency." (ENGELMANN, 1869).
ENGELMANN's paper is the foundation of the myogenic theory of ureteral peristalsis. Some of his correct statements were doubted in later years. For example, recent literature is dominated by the contrary opinion that the pyeloureteral junction is the origin of peristalsis (BARR, 1971; WEISS, 1971). Other fundamental questions have remained unsolved over the last 100 years. The normal frequency of ureteral peristalsis, for example, is usually given only as a broad range. KIIL (1957), in his excellent

* This work was supported by a grant from the Deutsche Forschungsgemeinschaft (Go 130/13)
** Physiologisches Institut der Universität Marburg/Lahn, Germany

monograph, gives the frequency of 1–5/min as normal for the human ureter, adding that frequencies as low as 0.25/min might be observed but rates over 10/min only rarely. An analogous statement in the field of cardiology would be that human heart rate is between 20 and 100/min, sometimes even as low as 5/min, but seldom over 200/min. Either the activity of the upper urinary tract is very much more irregular than that of the heart or the determinations on the pyeloureter are much less precise. We began with comparative studies in guinea pigs, rats and dogs, measuring ureteral activity on unanesthetised animals as well as on isolated preparations, the main object being the collection of more precise information in the normal situation. In-situ studies under normal conditions were found to be the basis necessary for all further analytical measurements in isolated preparations. In connection with a short report of the results, some theories will be offered which partly go beyond the solid foundations of experimental data.

1. Normal Ureteral Activity of the Guinea Pig

In 13 guinea pigs, fine wire electrodes were implanted at the ureter and the electrical activity was monitored in the unanesthetised animal for several days (GOLENHOFEN and HANNAPPEL, 1973). A frequency distribution of peristaltic intervals is given in Fig. 1, a summation of 8 individual histograms of 7 guinea pigs in which it was possible to record a long enough undisturbed time period under resting conditions on the first post-operative day. The multimodal character of the histogram is evident. All individual histograms had the several modes in similar positions. The peak value

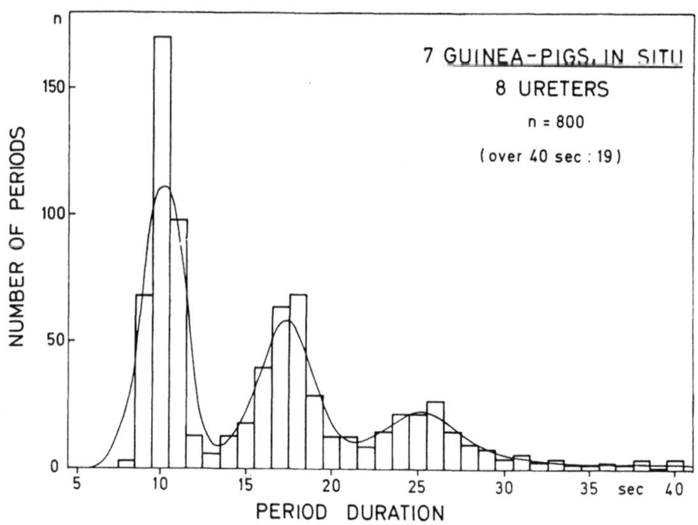

Fig. 1. Frequency distribution of period durations of ureteral peristalis; summation of 8 individual histograms obtained from unanesthetised guinea pigs on day after implantation of electrodes. For each individual histogram 100 successive peristaltic intervals were measured in the recording of electrical activity taken when the animal was quiet and undisturbed. Continuous curve: smoothed histogram obtained by averaging three adjacent classes. (From GOLENHOFEN and HANNAPPEL, 1973)

of the first mode was always around 10 sec. The position of the second and third modes can be determined at 17.5 and 25 sec. The deviation from the precise double or triple of the peak value of the first mode is significant; it was present in all individual histograms included in Fig. 1 and was seen regularly in many other histograms in this series. The simplest explanation for such a multimodal histogram is that the pacemaker process is oscillatory in nature and that not all oscillations reach the threshold for a conducted persistaltic wave. The shortest period duration – the peak value of the first mode at 10 sec – can be taken as the basic period duration of the

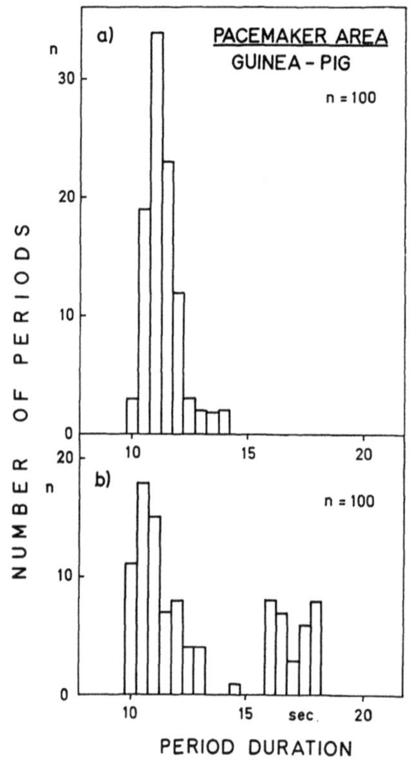

Fig. 2. Two histograms of peristaltic period durations from two successive time periods. Experiment with tension recording from the proximal end of a complete pyeloureter preparation of a guinea pig. Part a: period immediately after the proximal 3 mm of the preparation were clamped off; the frequency of the mechanical activity was equal to the activity before clamping. Part b: subsequent period with appearance of the first multiplied period durations

pacemaker process. There are two possible reasons for the deviation of the other modes from the precise multiplications of the basic period duration. The first mode could be a selection of the longer period durations, the shorter ones being smaller and having a lesser chance of triggering conducted excitation. A superthreshold excitation could, on the other hand, have an inhibitory reinfluence on the pacemaker, thereby producing a phase shift for the next wave.

Fig. 2 shows that the second possibility is the main reason for the deviation. The mechanical activity was recorded from a preparation of proximal renal pelvis only. Fig. 2a shows a frequency distribution of the period durations from a period with regular activity, with a manifestation ratio of the basic period duration of 1:1. The successive time period, in which the first longer period durations appeared, was worked out separately and the result is given in Fig. 2b. As a result of the selection

theory, the period durations of the second mode in Fig. 2b should be at least 20 sec. Altogether, the normal frequency of ureteral peristalsis in the guinea pig, as calculated from the basic period duration, is 6/min.
Multimodal frequency distributions with precise multiplication also occur; they seem to be typical conduction blocks in more distal regions and were more often observed on other species, e.g. by BARRY et al. (1971) and CONSTANTINOU (1971) on the dog.
The multimodal frequency distribution of peristaltic intervals means that the average frequency of ureteral peristalsis can by influenced in two different ways, either by continuous changes of the basic period duration or by whole numbered multiplications.

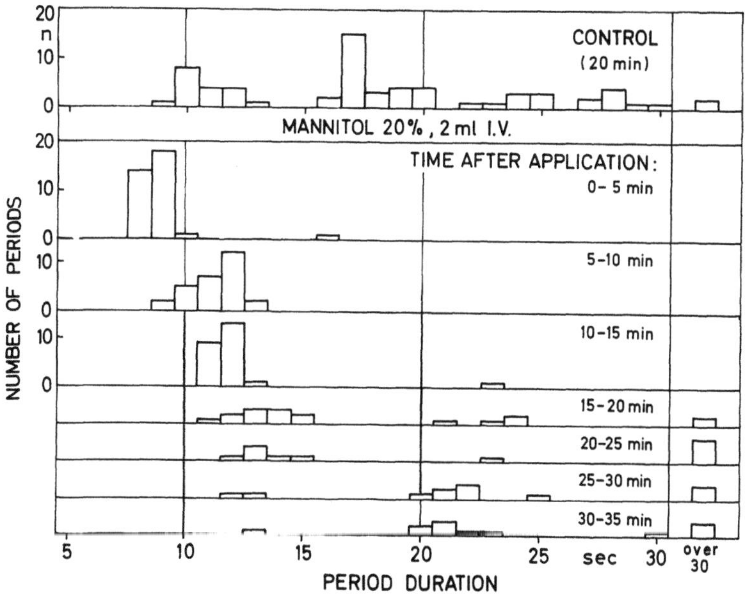

Fig. 3. Frequency distributions of peristaltic period durations from an unanesthetised guinea pig; for a control period, above, and for a series of successive 5-min periods after intravenous application of mannitol. (From GOLENHOFEN and HANNAPPEL, 1973)

There is no doubt that the latter way is of greater importance in the guinea pig. Discrimination is only possible if a series of histograms is made for successive time periods as in Fig. 3, which shows a reaction to administration of mannitol. Immediately after administration, a full 1:1 manifestation of the basic oscillations occurs, all intervals falling in the first mode compared with the multimodal histogram under control conditions. A slight reduction of the basic period duration also seems to be involved, but this is no longer present in the second 5 min period, and a slight reactive prolongation of the basic period duration seems to occur in the later phases. The example in Fig. 3 is one of the strongest reactions seen in a series of infusions of mannitol and hypotonic and hypertonic NaCl solutions in unanesthetized guinea pigs. During adrenaline infusion, similar reactions to those in Fig. 3. could be observed, but in most cases the changes were so slight that it was impossible to be sure they were not spontaneous fluctuations. The most pronounced shortening of the basic period duration (down to 6 sec) was found to be connected with the stress of the operation. On the whole, it can be concluded that the basic period duration of ureteral peristalsis is remarkably stable in the guinea pig; the variations under physiological conditions are probably not greater than $\pm$ 10%. The wide variations which do occur in the manifest frequency are usually caused by multiplications, i.e. by changes in the conduction ratio from the pacemaker to the ureter.

2. Isolated Pyeloureter of the Guinea Pig

Spontaneous activity in an isolated preparation comparable to the activity in situ is achieved by excision of the ureter together with parts of the renal pelvis up to the utmost renal ends which are visible with the naked eye (Golenhofen and Hannappel, 1973). About 1/4 to 1/3 of the whole proximal circumference of the renal pelvis is necessary to obtain a full activity over several hours, and no differentiation was detected between different parts of the circumference. The frequency distribution of peristaltic intervals for such a complete pyeloureter preparation is multimodal, as in situ, and the position of the modes is the same if it is corrected for the temperature difference (35 to 38 °C). The normal pacemaker can therefore be localized in the most proximal pointed ends of the renal pelvis, which is in keeping with results obtained in dogs by Sleator and Butcher (1955). This localization was confirmed by a series of clamping and ligature experiments using isolated pyeloureters (cf. Golenhofen, 1971). If the first proximal millimeters are clamped off, the proximal part continues at the normal rate and the distal part acts at a slower rate or stops completely, depending on the position of the clamp. The effects of stretch or catecholamines on the normal frequency of a complete pyeloureter are negligible (see Fig. 6).

The ureter alone produces no spontaneous activity under normal conditions. But ureter smooth muscle is not very far from the threshold of producing spontaneous activity. If the generation of excitation is stimulated in an isolated ureter alone by some alteration such as storage in an unoxygenated solution for several hours, the frequency of this activity is increased by stretch or by the application of adrenaline. Thus, reactions of the natural pacemaker, secondary pacemakers and stimulated latent pacemaker tissue can be completely different. Before conclusions applicable to the physiological regulation of ureteral motility can be drawn from the results in isolated preparations, it must be proved that the preparation is driven by the natural pacemaker.

3. Pacemaker Process for Ureteral Peristalsis

When Bozler (1942) published his fundamental studies on pacemaker processes in smooth muscle, including ureter, he could still hope "that the processes underlying automaticity are essentially the same in all muscles". This hope is no longer justified since the differences in the triggering processes for spontaneous activity between heart, stomach, intestine and ureter are too great. The fact that in the pyeoloureteral system of the guinea pig even a part of the proximal renal pelvis containing as few as 3 or 4 of the pointed ends is enough to produce normal ureteral peristalsis, and that no significant difference has been detected between different parts on the whole circumference of the proximal renal pelvis, indicates cooperation of several pacemaker subunits with similar potencies. A diagrammatic illustration of this theory of multiple coupled oscillators is given in Fig. 4. It is assumed that there are coupling forces (mainly electrical) between the subunits, which tend to synchronize the oscillations, thereby enlarging the amplitudes by summation and increasing the chance of triggering a conducted peristaltic wave. Each subunit may contribute a similar amount to the total pacemaker process but alone it is insufficient to trigger peristalsis.

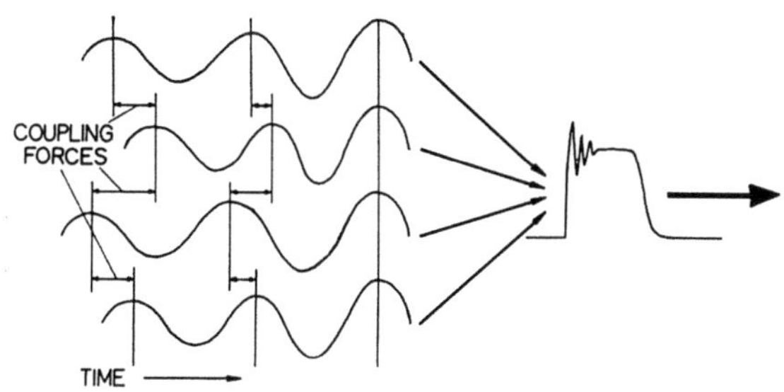

Fig. 4. Diagram illustrating a pacemaker process, consisting of multiple coupled oscillators

In the guinea pig, with its relatively simple monopapillary kidney, there may be about ten such subunits. In other species with more advanced kidneys and a complex calyx system, the number of subunits may be larger, or greater subunits may again be subdivided, thus making the pacemaker process more complicated; but we assume that the principle of the theory described is valid for all mammals. The observation that in patients with resections of the upper, middle or lower part of the kidney and adjacent sections of the renal pelvis, no significant changes occur in rate and rhythm of ureteral contractions (Kiil, 1957; Kiil and Kjekshus, 1967), is consistent with this view. That in humans the contractions of the calyces and the renal pelvis usually appear as independent (Kiil, 1957), does not exclude the possibility that the natural pacemaker is sited in the calyces. One or two synchronized calyces may be sufficient for triggering pelvic contraction, the others contracting independently and perhaps becoming the pacemaker at another time.

4. Three Components of Pyeloureter Concentration

The pyeloureter is a functional unit, but not every excitation in the pacemaker is conducted over the whole system. The excitation may be arrested at any point, but there are preferred areas for a conduction block. In the guinea pig there are two critical points for conduction, so that a three-component character of the pyeloureteral contraction is often evident in the tension record. An example is given in Fig. 5. At high amplification, a very small contraction (2.3% of the full contraction) can be seen before each main contraction (lower part of Fig. 5). Bozler (1942) found that the subthreshold pacemaker depolarization is combined with a small contraction, about 1/50 of the full peristaltic contraction. Thus this very small first component of the contraction in Fig. 5 is probably due to the pacemaker process. Fig. 5 clearly shows two further components in the subsequent main contraction. The second component (first part of the main contraction) was always quite similar in size, sometimes occurring alone (middle of upper record) and sometimes followed by a further component, which

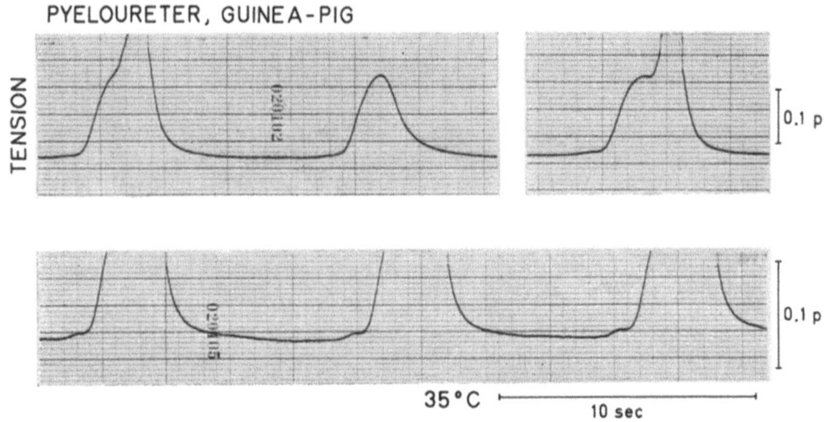

Fig. 5. Three-component character of pyeloureter contraction. Recordings of tension development of an isolated preparation at two different sensitivities. The whole tension amplitude was 0.3 pond

again was an all-or-nothing reaction. It is most likely that components 2 and 3 are the contractions of the renal pelvis and the ureter, and that the pyeloureteral junction is an area where the conduction is critical. Therefore the pyeloureteral contraction consists of three components: 1. The pacemaker process; 2. the renal pelvis contraction and 3. the ureteral contraction.

5. Ureteral Peristalsis in Rats and Dogs

Engelmann noted as early as 1869 that ureteral peristalsis in the rat is particularly frequent, the intervals being 1–2 sec. Tindall (1972) found that the frequency is usually about 25/min in the anesthetised rat, with maximum values up to 62/min. In isolated pyeloureters of rat, we also found very fast activity during the initial period (10–30 min), with period durations of 2.5–4 sec. But the first multiplications of period duration appeared very early and a steady state with a frequency of about 5/min was reached and maintained for several hours. We assume that this value is the normal one, but the pyeloureter of the rat differs from that of guinea pig in that it is easily irritable, and therefore high frequency activity occurs under various influences (see Fig. 6). In-situ studies in unanesthetised and undisturbed animals are necessary for a definite decision.

It is also more difficult to determine the normal rate of ureteral peristalsis in dogs than it is in guinea pigs. In several unanesthetised dogs, in which the electrical activity of the ureter was followed over several days, multimodal histograms of peristaltic intervals were regularly found. One mode usually appeared at 5–7 sec and a second between 10 and 15 sec. Experiments with isolated pyeloureter preparations of dogs are difficult to evaluate since the wall is quite thick and the oxygen supply by diffusion may not be sufficient for a normal function. We assume that the normal value for the dog is approximately 5/min.

The main difference between guinea pig (and rabbit and some other species), in contrast to dog and rat (and many other species including man) seems to be that in the first group the normal frequency is more or less the upper frequency limit, and in the latter group, large increases are possible due to frequency multiplications. Fig. 6 can be taken as an example; this compares the reactions of complete pyeloureters of guinea pig and rat to adrenaline. There is no change in frequency in the guinea pig preparation, but the rat preparation, which has the same initial rate, increases its frequency roughly fourfold.

6. Normal Frequency of Ureteral Peristalsis

In spite of the great variability of peristaltic frequency in the same species and the great differences between different species, we feel justified in deriving the theory that the normal frequency of ureteral peristalsis is equal to the frequency of stomach peristalsis in all mammals. This means that both systems contain a process of equal frequency, in their functional basis, which manifests itself very precisely and clearly in the stomach function but is more masked in the pyeloureteral function due to interference with other processes leading together to great variability, this again differing considerably from species to species. If this concept could be proved by further measurements, a very precise "ideal norm" for man could be given at 3/min.

7. Effect of Catecholamines

It is now well established that catecholamines have a dual effect on the pyeloureteral system: an excitatory effect which is inhibited by α-blocking agents, and an inhibitory effect which is suppressed by β-blocking agents (Boyarsky et al., 1971; Boyarsky and Labay, 1969; Deane, 1967; Deane and Garry, 1966; Malin et al., 1970; Melchior et al., 1971; Melchior and Rathert, 1971). But there are still many inconsistencies in the literature. As we found in comparative studies in rats, guinea pigs and dogs (details published in a subsequent paper), these inconsistencies are based on the following reasons: 1. Differences in species; an example is given in Fig. 6. 2. Normal pacemaker and secondary pacemaker areas differ in their reactions, as described in part 2. 3. The reaction depends on the initial value of ureteral peristalsis, and precise information on the normal resting value is necessary for a description of normal catecholamine reactions. 4. α- und β-adrenergic components are not equally distributed over the whole length of pyeloureter, systematic changes existing from proximal to distal parts. 5. α- and β-sensitivities also vary with time, in correlation with other functional parameters. 6. α- and β-adrenergic receptors have varying degrees of differentiation in different species, with differences in the affinity to adrenaline and noradrenaline as well as to blocking substances.

An example for point 5 is given in Fig. 7. In one dog, in which the electrical activity of the ureter was followed over 17 days, all types of reactions were observed during noradrenaline infusion. Part a) shows an excitatory reaction on day 8, when the dog was unanesthetised. The inhibitory reaction in part b) was measured during Nembutal-induced anesthesia on day 7. The difference between the two reactions is not dose-dependent; during an infusion of 5 μg/kg/min shortly before the section of Fig. 7b, the inhibitory effect had also been predominant, and excitatory reactions were also re-

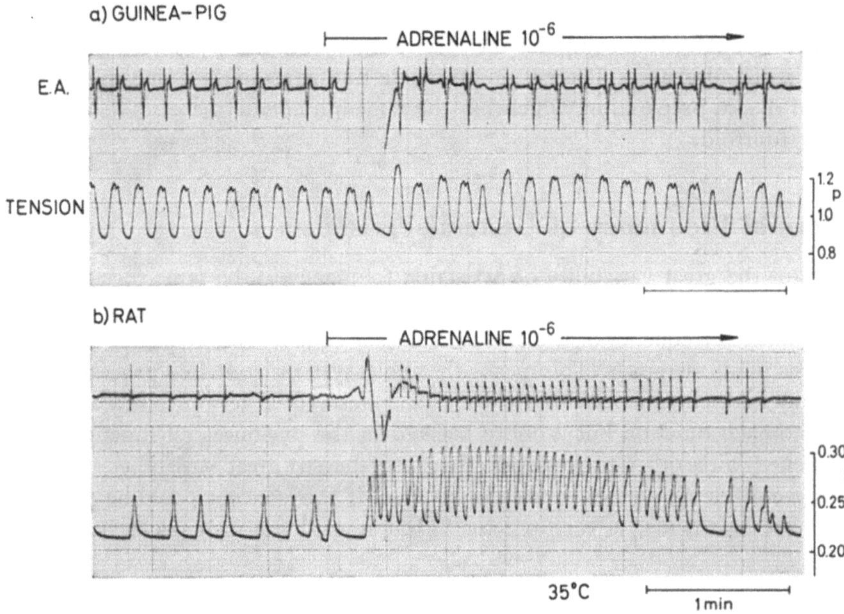

Fig. 6a and b. Reactions of isolated pyeloureter preparations on application of adrenaline 10^{-6} g/ml, a preparation from a guinea pig, b from a rat. Electrical activity (E.A.) of the ureter above, tension development below, calibrated in pond

corded during infusion of 10 μg/kg/min noradrenaline. Biphasic reactions were observed as well as records showing no significant effect on peristaltic frequency. Uniform reactions during noradrenaline infusion were only observed in the dog when one component was selectively blocked: excitatory reactions after β-blockade (propranolol) and inhibitory reactions after α-blockade (phentolamine or phenoxybenzamine).

α-adrenergic activity is strongest in proximal regions (but in some species, such as the guinea pig, the natural pacemaker itself is not affected) and declines towards distal parts; conversely, β-adrenergic activity increases towards distal regions (point 4). These topographical differences were found repeatedly and are consistent with observations published by other authors (e.g. FINBERG and PEART, 1970; TINDALL, 1972). In isolated pyeloureter preparations of the rat, for example, we observed during blockade of the α-component (with phenoxybenzamine) that the activity in the renal pelvis continued at the normal rate during administration of adrenaline, while the activity in the distal ureter was completely suppressed due to a conduction block. The variability of the noradrenaline reactions in the dog is also a result of the uneven distribution of α- and β-receptors, combined with the fact that, in dogs, both types of receptors have a similar affinity to noradrenaline. Using a proximal electrode it was thus possible to measure an increase in frequency during noradrenaline infusion, due to the dominance of α-receptors in this region, while the frequency declined in the distal ureter due to the predominance of β-receptors which produced conduction blocks in distal areas.

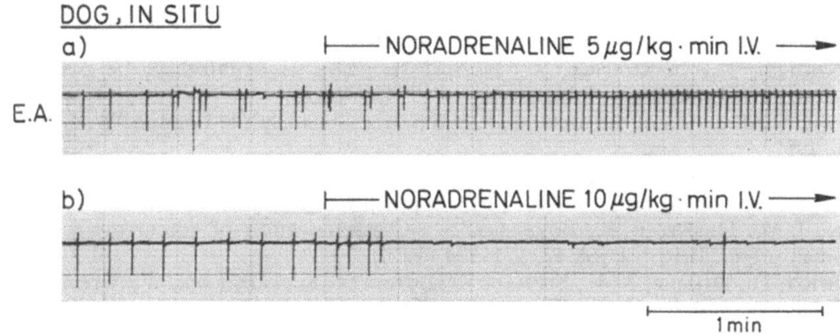

Fig. 7. Peristaltic activity of dog ureter in situ during intravenous infusion of noradrenaline. Electrical activity (E.A.) recorded with electrodes implanted at the ureter. a Dog, unanesthetized, day 8 after implantation; b the same dog, during Nembutal-induced anesthesia, day 7 after implantation

Finally, an example for point 6 should be given. In dogs, α-receptors can be blocked selectively with phentolamine and phenoxybenzamine, β-receptors with propranolol; β-receptors have a stronger affinity to noradrenaline than to adrenaline, while α-receptors have a similar affinity to both substances. In rat pyeoloureter, however, effects of adrenaline and noradrenaline are usaully indistinguishable. Phentolamine blocks both α- and β-receptors in this preparation, and only phenoxybenzamine, which is generally the more specific α-blocking agent, produces a selective α-blockade. This together indicates a lesser degree of α-receptor differentiation in the rat.

References

Barr, L.: Some biophysical aspects of ureter smooth muscle. In: S. Boyarsky *et al.* (Ed.): Urodynamics, pp. 49–53. New York, London: Academic Press 1971.

Barry, W. F., Absher, R., Boyarsky, S.: Modal activity of ureteral peristalsis. In: S. Boyarsky *et al.* (Ed.): Urodynamics, pp. 399–410. New York, London: Academic Press 1971.

Boyarsky, S., Gottschalk, C. W., Tanagho, E. A., Zimskind, P. D. (Ed.): Urodynamics. Hydrodynamics of the ureter and renal pelvis. New York, London: Academic Press 1971.

Boyarsky, S., Labay, P.: Ureteral motility. Ann. Rev. Med. **20**, 383–394 (1969).

Bozler, E.: The activity of the pacemaker previous to the discharge of a muscular impulse. Amer. J. Physiol. **136**, 543–552 (1942).

Constantinou, C. E., Dale, R. L., Briggs, E. M., Govan, D. E.: Electrophysiological methods in the study of ureteral dynamics. In: S. Boyarsky *et al.* (Ed.): Urodynamics, pp. 143–150. New York, London: Academic Press 1971.

DEANE, R. F.: Functional studies of the ureter: Its behaviour in the domestic pig (Sus scrofa domestica) as recorded by the technique of Trendelenburg. Brit. J. Urol. **39**, 31–37 (1967).
DEANE, R. F., GARRY, R. C.: The behaviour of the ureter of the pig recorded by the Trendelenburg technique. J. Physiol. **183**, 36P–37P (1966).
ENGELMANN, T. W.: Zur Physiologie des Ureter. Pflügers Arch. ges. Physiol. **2**, 243–293 (1869).
FINBERG, J. P. M., PEART, W. S.: Function of smooth muscle of the rat renal pelvis-response of the isolated pelvis muscle to angiotensin and some other substances. Brit. J. Pharmacol. **39**, 373–381 (1970).
GOLENHOFEN, K.: Physiologie der Uretermuskulatur. In: W. LUTZEYER and H. MELCHIOR (Ed.): Ureterdynamik, pp. 46–59. Stuttgart: Georg Thieme Verlag 1971.
GOLENHOFEN, K., HANNAPPEL, J.: Normal spontaneous activity of the pyeloureteral system in the guinea-pig. Pflügers Arch. (1973, in press).
KIIL, F.: The function of the ureter and renal pelvis. Philadelphia, London: W. B. Saunders Co. 1957.
KIIL, F., KJEKSHUS, J.: The physiology of the ureter and renal pelvis. Proc. 3rd. int. Congr. Nephrol. Washington 1966, Vol. 2, pp. 321–335. Basel, New York: Karger 1967.
MALIN, J. M., DEANE, R. F., BOYARSKY, S.: Characterization of adrenergic receptors in human ureter. Brit. J. Urol. **42**, 171–174 (1970).
MELCHIOR, H., DIEMER, K. F., SIMHAN, K., LUTZEYER, K., LUTZEYER, W.: Ureterdynamik. II. Der Einfluß des autonomen Nervensystems auf die Ureterdynamik. Z. Urol. **64**, 93–101 (1971).
MELCHIOR, H., RATHERT, P.: Die Steuerung der Ureterperistaltik. In: W. Lutzeyer and H. MELCHIOR (Ed.): Ureterdynamik, pp. 32–43. Stuttgart: Georg Thieme Verlag 1971.
SLEATOR, W., BUTCHER, H. R.: Action potentials and pressure changes in ureteral peristaltic waves. Amer. J. Physiol. **180**, 261–276 (1955).
TINDALL, A. R.: Preliminary observations on the mechanical and electrical activity of the rat ureter. J. Physiol. (Lond.) **223**, 633–647 (1972).
WEISS, R. M.: Physiological organization of function with reference to a pacemaker. In: S. BOYARSKY *et al.* (Ed.): Urodynamics, pp. 261–282. New York, London: Academic Press 1971.

Discussion

Has the Ureter a Pacemaker?

NOTLEY holds the opinion that mechanical stretching of the ureteric smooth muscle is the stimulus for contraction, this being an important characteristic of all smooth muscle. He finds it difficult to accept the pacemaker theory, which has been developed by WEISS, WAGNER and HOFFMANN. It seems more probable that the pelvi-ureteric segment of the ureter is the first part of the ureter which is actually able to propel the urine onwards, the renal pelvis being merely a compliant reservoir with no propulsive power even though it shows contractions.

GOLENHOFEN and HANNAPPEL: The pacemaker theory described here does not contain special statements about the importance of renal pelvis contraction for urine transport. The existence of such a pacemaker is independent of whether or not a morphologcial substrate can be demonstrated by an electron microscope (BOYARSKY). There are sufficient physiological proofs for the existence of a pacemaker in the area of the renal calyces.

Urinary Flow and Ureteral Peristalsis

F. Kiil*

Two Modes of Urinary Transport

The only known function of the renal pelvis and ureter is to convey urine from the kidney to the urinary bladder. This transport is achieved in two ways: 1. By peristaltic activity. A urine portion is pushed ahead of a contractile wave, which almost completely obliterates the tubular lumen. 2. By a pressure gradient in an open system. These two modes of transport are shown in Fig. 1. During the first type of transport renal pelvic pressures are low ($<$ 10 mmHg); ureteral pressures during diastole are even lower, but show a considerable local increase during contraction. The second type of transport, through a wide-open ureter, may be present during large urine flow; the pressure head in the renal pelvis is not caused by contraction of the smooth muscles

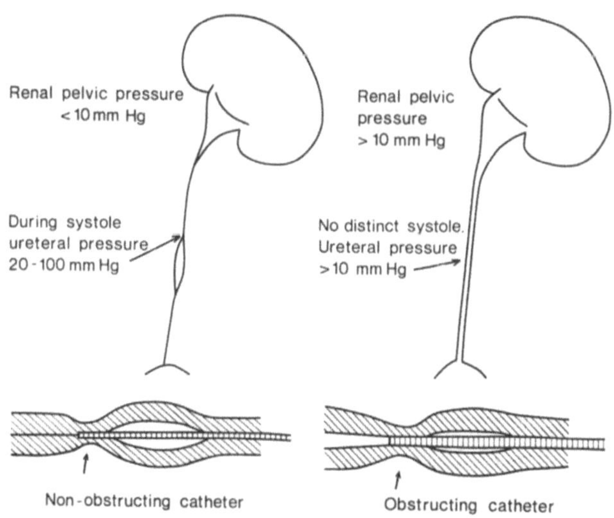

Fig. 1. Peristaltic transport (left) and urine transport by a pressure gradient in an open system (right). Lower part of diagram illustrates peristaltic activity along a non-obstructing catheter and along an obstructing catheter producing retention of urine

* Institute for Experimental Medical Research, University of Oslo, Ullevaal Hospital, Oslo, Norway

of the upper urinary tract but generated by the flow of urine, and therefore reflects arterial blood pressure. When the test subject assumes an upright position, the pressure gradient along the ureter may be increased to some extent by gravity.

A change from the first to the second type of urine transport occurs when the ureter is partly obstructed, as it may be in both pathological and physiological conditions. The flow into the bladder is obstructed either when intravesical pressure is raised or when ureteral resistance is high during bladder distention without high bladder pressure.

Ureteral activity may transiently obstruct urinary transport. Retrograde transport of urine may be achieved either by reflux into an open ureter or by peristaltic contractions starting in the lower end of the ureter and progressing towards the renal pelvis. This simple description of the types of ureteral activity is comparatively new. At urological meetings in Germany and France between the wars, the control of ureteral peristalsis was a popular subject. Several hundred papers published at that time deal with a postulated nervous control system, which was thought not only to govern the frequency of ureteral contractions, but also to guard against reflux into the renal parenchyma by co-ordinated activity of a series of sphincters localized at the level of the calyces, the pelvio-ureteral junction, and the ureter. The ureter was believed to be separated into a series of small „bladders", or cystoids, each of which pushed the urine to the next lower cystoid by co-ordinated activity of sphincters. These and many other hypotheses were based on the interpretation of radiograms, and particularly on pyeloscopic observations. Much of the early literature has been reviewed by NARATH (1951) and by FEY and QUENU (1965). With the advent of methods permitting simultaneous recordings of ureteral pressures and urographic examinations, most of the old philosophies were abandoned (KIIL, 1957). It was found that the renal pelvis was a low-pressure system, and although urine may be retained at various levels of the upper urinary tract, no evidence was produced indicating that these sites of retention had any functional specificity. Most of the earlier ideas were based on the erroneous assumption that the smooth muscles of the upper urinary tract were capable of rapid, strong contractions, and that high pressures might be generated briefly in the renal pelvis. This misconception still exists, and accounts for the popularity of cineradiographic methods of high picture frequency.

How the Renal Pelvis is Emptied

Figure 2 shows the radiographic outlines of the renal pelvis and upper part of the ureter during the resting phase of the renal pelvis (diastole) and subsequent systole. During contraction the renal pelvis changes in shape, and the renal pelvic pressure is slightly raised. In an extrarenal pelvis of normal size and with normal urinary flow the pressure increase during systole may be less than the pressure fluctuations mediated by respiratory movements. The gradient for flow of urine from the renal pelvis and into the upper part of the ureter is therefore often no steeper during renal pelvic systole than during diastole. At high urine flow, and in the case of an intrarenal pelvis, pressure may increase by 5–10 mmHg during systole. When the contractile wave of the renal pelvis has reached the pelvio-ureteral junction (defined as the highest level of the ureter where the tract is obliterated during contraction) urine distal to the ureteropelvic junction is moved along the ureter by peristalsis. The contraction of the smooth muscle increases the wall tension. To a first approximation, wall tension is proportional to both the radius of the tube and the transmural pressure (Laplace's law). A similar rise in wall tension during contraction would thus create higher pressure in a tube with a small radius, such as the ureter, than in a wide tube, such as the renal pelvis.

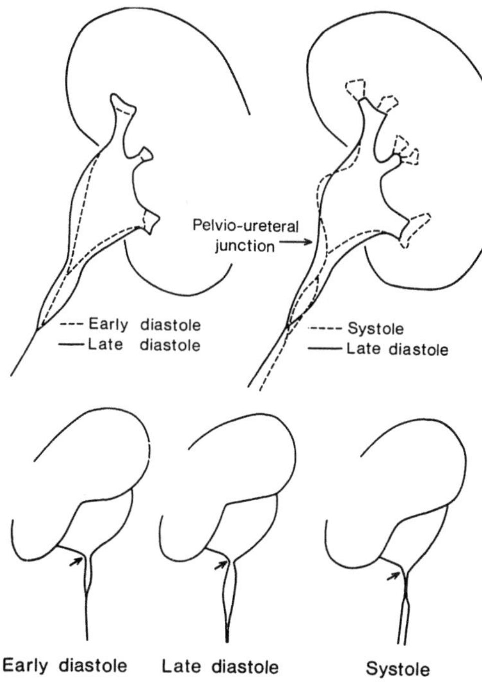

Fig. 2. Above: Radiographic outlines of a normal renal pelvis during systole and diastole. Below: Schematic drawing of a renal pelvis with stricture at the pelvio-ureteral junction (*arrow*). Under both conditions, urine accumulates in the upper part of the ureter during diastole and is transported distally by the subsequent contractile wave

Ureteral Pressure and Frequency of Contractions

Local pressure rises rapidly during occlusion around the catheter in the ureter, and pressures exceeding 100 mmHg have been recorded. It is apparent from Laplace's law that a thin catheter may record higher pressures than a catheter of larger diameter. The absolute pressures recorded are thus clearly a function of the instrument introduced, and should be interpreted with caution.

In an analysis of the undisturbed function of the upper urinary tract under physiological and pathological conditions, it is essential to have a clear idea of the disturbances caused by the instruments introduced. Cystoscopic examinations show that the ureters sometimes contract with high precision, and send an urine spurt into the urinary bladder at regular intervals. The frequency of contractions varies considerably between the two ureters, and it is only on rare occasions that both contract simultaneously. The amount of urine entering the urinary bladder with each contraction varies widely, as every cystoscopist is aware. From these observations it is clear that there is no close correlation between urine flow and frequency of ureteral contractions. By counting the frequency of contractions it is possible to show that the introduction of a closed catheter for pressure measurements does not usually change the frequency of contractions. The wave of peristaltic contraction obliterates the ureteral lumen, pushing the urine portion along the outside of the catheter, and no change in peristaltic frequency occurs.

Ureteral Activity during Distention

Retention of urine in the ureter occurs in physiological, pathological and artificial conditions, such as too large a catheter is inserted (Fig. 1). Whatever the cause of distention, the transition from normal peristalsis to the distended ureter follows a highly reproducible pattern (Fig. 3). The diastolic or resting pressure increases, frequency of contractions is raised, and the pressure amplitude is reduced to zero when the ureter is sufficiently distended. Pressure fluctuations disappear at a diastolic pressure of 10–15 mmHg. At this pressure the peristaltic mode of contraction becomes inefficient.

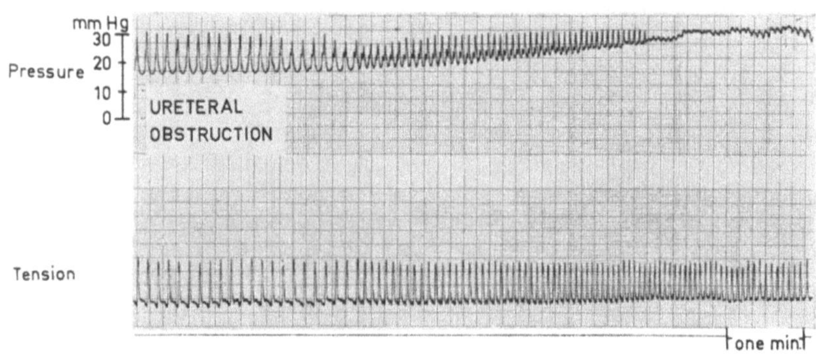

Fig. 3. Ureteral distention changes. Upper tracings show pressure recordings indicating increasing ureteral pressure and decreasing pressure fluctuations as urine is retained. Lower recording shows that local contractions proceed at unchanged force but with increasing frequency as the ureter distends

Absence of pressure fluctuations does not imply that contractions are no longer propagated. Action potentials proceed in the usual way and even at unchanged speed. Actual measurements have not been made to determine whether shortening of the ureteral muscles is less when inside pressure builds up, but it is reasonable to assume that the muscle contraction becomes more and more isometric. A local contraction which does not obliterate the lumen, however, does not raise the ureteral pressure. In the lower part of Figure 3, the force against an open cylinder placed in the ureter has been measured. By means of silicon strain gauges this instrument measures the local contractile force at a constant ureteral diameter (= tension). It is shown that the tension developed at this constant diameter remains almost unchanged during ureteral distention. Distention changes are more likely to occur with high urinary flow; they can easily be induced by rapidly injecting saline into the ureter.

Anterograde and Retrograde Contractions

Several investigators have suggested that stretch plays an important role in the propagation of the contractile wave along the ureter. Normally, the ureter is slightly distended by the urine portion pushed ahead of the contractile wave. It is therefore feasible that distention is the stimulus to contraction, a difficult hypothesis to rule out. One argument against the distention hypothesis is that peristaltic waves may be

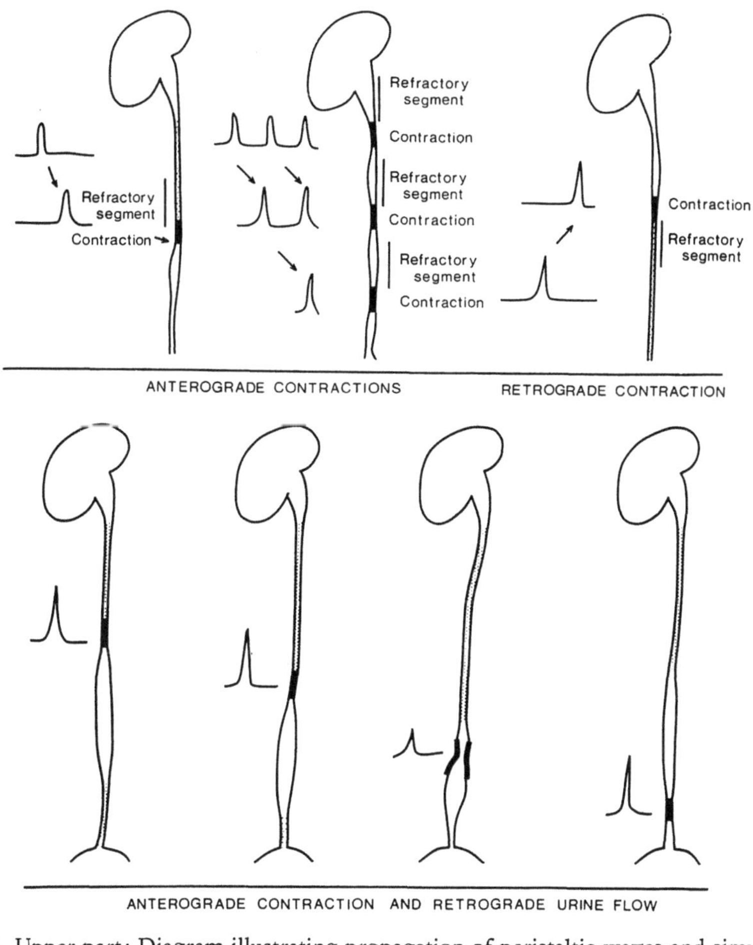

Fig. 4. Upper part: Diagram illustrating propagation of peristaltic waves and simultaneous pressure recordings from different levels of the ureter. Arrows connect pressure complexes caused by the same anterograde or retrograde peristaltic wave. Lower part: Diagram showing anterograde contractions and retrograde urine flow from a urine spindle in the lower part of the ureter

propagated through the empty ureter, although such contractile waves are likely to fade out. However, even in the empty ureter, stretching may be caused by contraction of adjacent cells.

It is a widespread clinical belief that the completely or partly obstructed ureter is capable of hyperperistalsis. When obstructed, a large part of the ureter is distended, and simultaneous contractions might be elicited from all levels of the distended segment. Rapid injections of saline or radiographic contrast medium result in uncoordinated ureteral activity, and not coordinated contraction of the distended segment. What probably happens is that such stimuli, as well as sudden obstructions, elicit ectopic ureteral contractions which proceed in both anterograde and retrograde directions. The absolute refractory period of contraction ranges from 0.5 to 5 sec. In intact dog ureters, the highest contractile frequencies observed after distention, or after injection of drugs (histamine, noradrenaline), are less than 20 per min. The excitability has been found to recover gradually, and the relative refractory period may last as long as one minute. Because of this long refractory period, the ureteral muscles do not contract tetanically. The long refractory period of ureteral muscles may therefore account for the absence of simultaneous contraction of the distended segment of the ureter. This response is different from that of the urinary bladder, in which rapid distention elicits simultaneous contraction of all muscles; they have the same refractory period, but also recover from it simultaneously. In the normal, unobstructed ureter a contractile wave usually starts in the proximal end of the ureter, while the lower part is in the refractory period of the preceding wave. At a high frequency of contractions more of the ureter will be in a refractory period, and the likelihood of eliciting a simultaneous contraction of the whole distended part will be even further reduced (Fig. 4). Parts of the ureter are frequently filled with urine even during the resting phase. Since the contractile wave passes along at constant speed, the urine column is thus pushed in the retrograde direction when the contractile wave has reached the lower part of the column (Fig. 4). Retrograde transport of urine may thus be caused by an anterograde contractile wave.

Even when all possible precautions have been taken during insertion of the catheters, retrograde contractions may still be recorded. Whether they should be considered as an instrumental artefact is a debatable point. Whatever the explanation, the fact that retrograde contractions are so easily elicited might be of urological significance.

Ureteral Response to Bladder Filling and Micturition

On account of lack of circular muscles in the intramural and submucosal segments of the ureter, the transport of urine into the bladder is mainly a function of the extravesical part of the ureter. No transport of urine can therefore occur if the intravesical pressure is higher than the ureteral pressure. The intramural part of the ureter is narrow, and thus represents appreciable hydrodynamic resistance. This becomes more pronounced when the ureter is stretched and the lumen narrowed during filling of the urinary bladder.

Retention of urine in the extravesical part of the ureter is therefore more likely to occur at large bladder volume, even when the intravesical pressure is not raised (Fig. 5).

The pressure head needed to drive urine along the intramural and submucosal segments of the ureter is proportional to urinary flow. A copious flow of urine therefore predisposes to distention of the extravesical part of the ureter. The driving force determining flow is the pressure during systole of the juxtavesical part of the ureter. Pressures exceeding 60–70 mmHg are frequently recorded, and continued transport of urine into the bladder has been observed in adult man at bladder pressures of 40 mmHg and at bladder volumes exceeding 1000 ml (PIERCE and BRAUN, 1960). There is no evidence of any compensatory mechanism increasing contractile force in the lower ureter during bladder distention. When intra-ureteral resting pressure exceeds 10–15 mmHg, peristaltic transport of urine is no longer possible (Fig. 1). This transition to an open ureter may occur in the two ureters at different degrees of bladder distention.

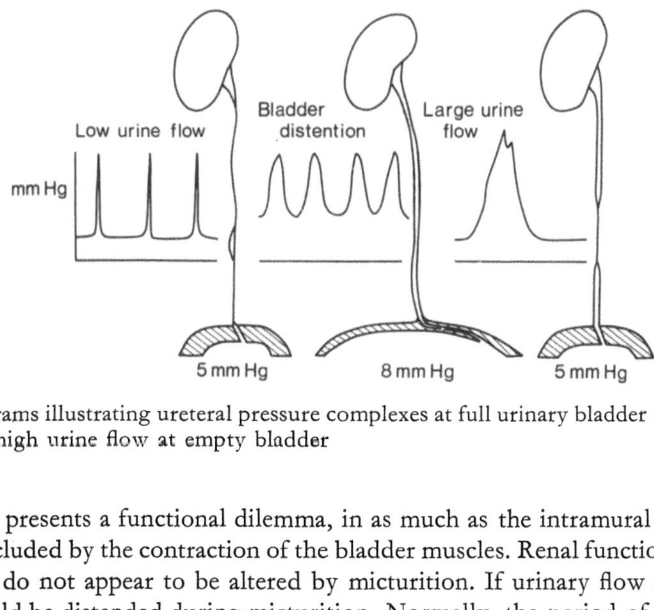

Fig. 5. Diagrams illustrating ureteral pressure complexes at full urinary bladder (middle) and at low and high urine flow at empty bladder

Micturition presents a functional dilemma, in as much as the intramural part of the ureter is occluded by the contraction of the bladder muscles. Renal function and urine production do not appear to be altered by micturition. If urinary flow is high, the ureters would be distended during micturition. Normally, the period of micturition is too brief to result in ureteral distention changes. In experiments on anesthetised dogs, high ureteral pressures may be rapidly built up by stimulating bladder detrusor activity when urinary flow is sufficiently copious, as during ethacrynic acid diuresis (KIIL and KJEKSHUS, 1967).

Renal Function at High Renal Pelvic Pressure and High Urinary Flow

High intrapelvic pressures may be expected to influence the function of the kidneys, a problem of great concern to many urologists. There is no doubt that chronic high pressure in the renal pelvis may result in irreversible changes, although kidney function is usually completely restored, even after several weeks of obstruction. Acute increments in renal pelvic pressure are counteracted by important regulatory mechanisms. Several investigators have noted that glomerular filtration rate is singularly unaffected when pelvic pressure is raised by 50–60 cm water, and that renal blood flow either remains constant or increases. As long as there is a flow of tubular fluid along the nephron, the pressure in the proximal tubules and in the Bowman's capsule must be higher than in the renal pelvis. If the flow of urine is copious, a considerable pressure head is needed to drive tubular fluid along the nephron, even if renal pelvic pressure is normal. Micropuncture studies during Mannitol diuresis have shown that proximal tubular pressure may rise to 40–50 mmHg (Gottschalk and Mylle, 1956). These observations lead to some interesting reflections. Normally,

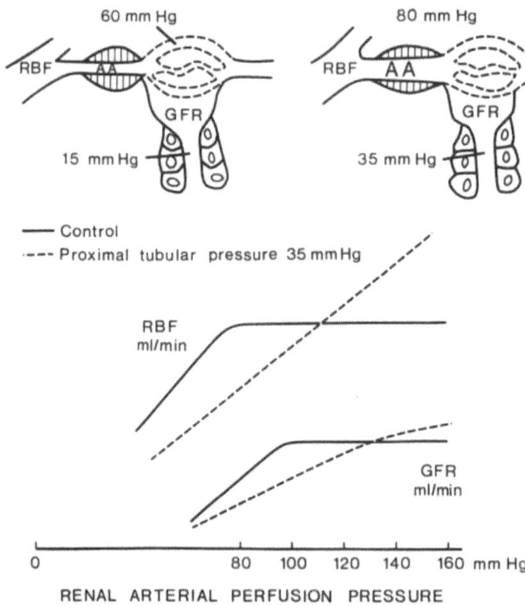

Fig. 6. Upper diagram illustrates how a rise in proximal tubular pressure from 15 to 35 mmHg is compensated by dilatation of afferent arterioles. As shown in the lower part of diagram, the effect is an increase in GFR and RBF, if renal arterial perfusion pressure is high, and a decrease if renal arterial perfusion pressure is low. Data from Kiil, Kjekshus and Løyning, 1969, and unpublished observations

glomerular capillary hydrostatic pressure is about 60 mmHg, colloid osmotic pressure 25 mmHg, and proximal tubular pressure 15 mmHg, leaving a pressure head of 20 mmHg to drive the filtrate across the glomerular membrane. A rise of 20 mmHg in tubular pressure would accordingly stop glomerular filtration if there were no compensatory mechanisms. As the glomerular filtration rate is unchanged, even with a rise of 30–40 mmHg in pressure, glomerular capillary pressure must necessarily have been correspondingly increased. These reflections therefore lead to the hypothesis that an increase in tubular pressure by some autoregulating mechanism culminates in dilatation of the afferent arterioles. This is supported by the observation that at high ureteral pressure renal autoregulation has been found to be abolished (Kiil, Kjekshus and Løyning, 1969) (Fig. 6).

A rise in ureteral pressure leads to dilatation of the nephrons and a slight increase in sodium reabsorption. Urine sodium concentration falls, but since the glomerular filtration rate remains essentially unchanged, urea and creatinine concentrations increase with the reduction in urinary flow. Osmolal concentration will fall or rise, depending on whether the main osmotic constituent of the urine before obstruction was sodium or urea. Under conditions of antidiuresis, osmolal concentration of urine may rise from 600 to 1200 mOsm/kg H_2O.

To summarize: When the ureteral pressure exceeds 10–15 mmHg the ureter becomes an open tube and the driving force for urinary flow is henceforth related to the glomerular capillary pressure, which can be increased by a compensatory autoregulation mechanism. This mode of emptying applies when copious flow of urine is combined with bladder distention or micturition.

Dilatation of the Renal Pelvis and Ureter

Renal pelvis and ureter may increase in size when exposed to high intraluminal pressure, but dilatation often occurs in both pathological and physiological conditions with no measurable rise in pressure. In cases of hydronephrosis, renal pelvic pressure may not be noticeably raised. The simplest explanation of this phenomenon is that muscle fibres and other tissue elements can be slowly stretched without a rise in tension. A stretching of the structures involved is probably of considerable importance, not only in the regulation of ureteral activity, but also in regulating renal function during increasing urinary flow.

The delivery force for urinary flow from Bowman's capsule to the renal pelvis is the hydrostatic pressure gradient. The hydrodynamic resistance to the flow is determined by the tubular dimensions. Tubular dimensions are dependent on the transmural pressure across the tubular wall. Any inhibition of tubular sodium reabsorption induced, for instance, by a diuretic, will increase the pressure inside the tubules due to the concomitant reduction in water reabsorption. The tubules are distended, and the pressure head for driving urine along the nephron therefore both increases the driving force and reduces the hydrodynamic resistance of the nephron (Fig. 7). When this latter effect is sufficiently marked the proximal tubular pressure may remain essentially constant, despite a large increase in urinary flow. The same principles also apply to the ureter and renal pelvis. Even when the bladder is empty, a large and sudden increase in urinary flow after administration of a potent diuretic, ethacrynic acid, may result in a rise in renal pelvic pressure to 20–30 mmHg in some dogs, whereas in other dogs the increase is no more than 5–10 mmHg. The large increase in pressure is transient. In the course of a few minutes, renal pelvic pressure in all dogs returns to a level slightly above controls, despite continued high urinary flow. This reduction in ureteral resistance signifies a stretching of the ureteral wall and probably also of the ureteral smooth muscles.

When dilatation has occurred, peristaltic activity is resumed. In contrast to distension, the resting ureteral pressure is low in the dilated ureter. The contractile wave obliterates the ureteral lumen and empties the fluid-filled ureter (Fig. 5). During the resting period, urine continues to issue from the ureteral orifice, since between contractions the whole ureter is an open system. In such a dilated upper urinary tract, stronger stimuli are evidently needed to elicit contractions. Frequency of ureteral contractions is therefore usually less than before dilatation.

There is a smooth transition between the transport of a small urine portion and the completely dilated ureter (Fig. 8). Provided the volume of the urinary bladder is low, the response in both dog and man seems to be dilatation, even at the highest obtainable urinary flows (KIIL and KJEKSHUS, 1967). There is accordingly no relationship between urinary flow and frequency of contractions, as long as the bladder is empty. On the other hand, if distention changes are elicited, a rise in urinary flow may be associated with increased frequency of ureteral contractions.

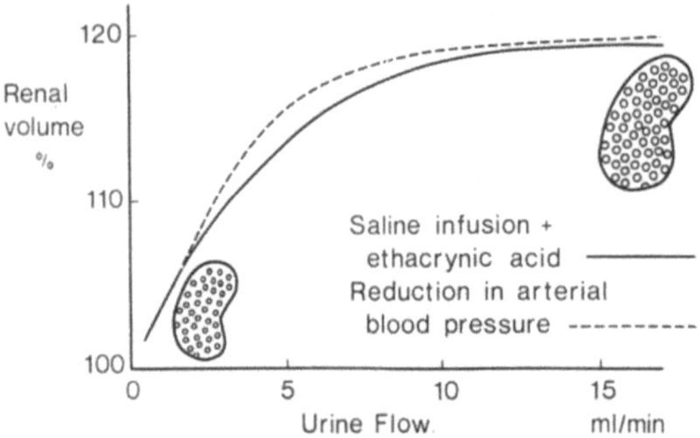

Fig. 7. Relationship between renal volume and urine flow. Inserts indicate that the increase in renal volume is mainly caused by dilatation of tubules. Data from OMVIK, RAEDER and KIIL, Amer. J. Physiol. **221**, 1560–1567 (1971)

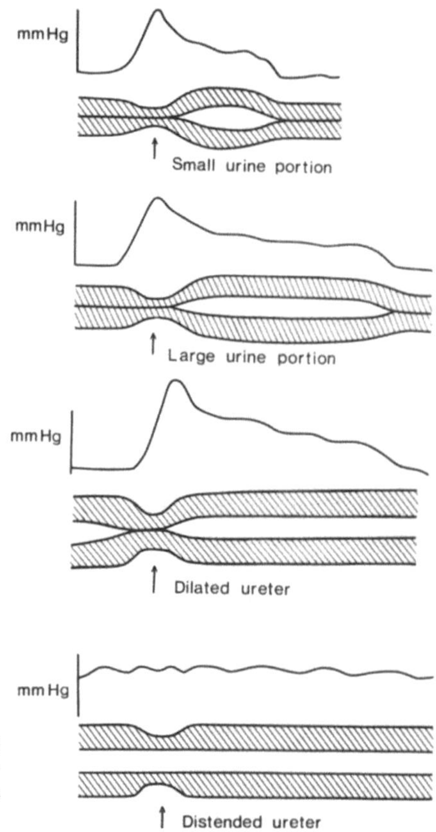

Fig. 8. Pressure complexes during increasing dilatation (upper three panels) and during distention with high diastolic ureteral pressure (bottom panel). Arrow at contractile wave

Pacemaker Activity

In man, most if not all of the contractile waves are propagated throughout the upper urinary tract, whether it is dilated or empty between contractions. Differences in frequency of contraction therefore reflect changes in pacemaker activity. It has been well established that all parts of the ureter, pelvio-ureteral junction, renal pelvis proper and calyces may act as pacemakers. Calyces contract largely independently of the renal pelvis, but some of the contractions originating in a calyx may be propagated further. Even a slight increase in renal pelvic pressure during contraction may result in regurgitation of contrast material (and probably urine) into the calyces (Fig. 2). These structures therefore seem to be the most easily distensible, and accordingly the ones where muscle fibres are subjected to the highest degree of stretch. In some ureters the activity is regular. In other ureters, contractile activity occurs in spasms interrupted by long periods of rest; in yet others, the activity is irregular and unpredictable. Regular activity suggests that one pacemaker is dominant. All contractile waves probably propagate from the same pacemaker, undergoing constant metabolic changes during the refractory period, and becoming accessible to new stimuli after constant intervals. Irregular activity suggests competition between several pacemakers. Since the dilatation response is associated with a reduction in frequency of contractions, slow stretching raises, rather than lowers the threshold for a stimulus leading to the propagation of a contractile wave.

Numerous studies suggest that nervous stimulation affects pacemaker activity; histamine, acetylcholine, serotonin and α-stimulants increase peristaltic frequency, whereas it is depressed by papaverine and β-stimulants. None of these drugs has been shown to influence the force of ureteral contractions (KIIL and KJEKSHUS, 1967), and it has not been established whether nervous and humoral stimuli within the physiological range affect frequency of contractions.

Concluding Remarks

The general patterns described here are mainly based on my own experience in experiments on man and dog. Numerous urologists have measured ureteral pressures, and confirmed the distention changes, but so far the dilatation pattern has not been acknowledged. The dilatation response will obviously not be obtained if a catheter inserted from the urinary bladder seriously obstructs the ureter. Although it has been well established that ureteral activity may be accurately measured by careful insertion of catheters, it should not be assumed that *any* recording of pressure fluctuations reflects normal activity. In the absence of simultaneous recordings from two levels of the ureter it is not possible to judge whether the pressure fluctuations recorded represent anterograde or retrograde contractions. More than 90% of the published papers on ureteral function report only one pressure recording; there are practically no reports of serial urograms or other methods of simultaneous and independent measurement of ureteral activity. Considerable work is therefore needed before the relationship between urinary flow and ureteral peristalsis is definitely established.

Summary

A rise in urinary flow leads either to distension or to dilatation of the ureters. Distension is characterized by a rise in basal pressure and increased frequency of peristalsis, which becomes inefficient at basal ureteral pressures of 10–15 mmHg. This response requires a ureteral obstruction which is present, physiologically, during micturition and at large bladder volume.

Ureteral dilatation is characterized by low *basal* pressure and maintained ureteral peristalsis, usually at a lower frequency than at control urinary flow. When the whole ureter is open, urine also flows into the bladder between ureteral contractions.

References

Fey, B., Quenu, L.: Physiologie normale et pathologique des voies urinaires supérieures. In: Handbuch der Urologie, Bd. II. Berlin-Heidelberg-New York: Springer 1965.

Gottschalk, C. W., Mylle, M.: Micropuncture study of pressures in proximal tubules and peritubular capillaries of the rat kidney and the relation to ureteral and renal venous pressures. Amer. J. Physiol. **185**, 430–439 (1956).

Kiil, F.: The function of the ureter and renal pelvis. Philadelphia: W. B. Saunders Co. 1957.

Kiil, F., Kjekshus, J.: The physiology of the ureter and renal pelvis. In: Proceedings of the Third International Congress of Nephrology, Vol. 2. Basel-New York: S. Karger 1967.

Kiil, F., Kjekshus, J., Løyning, E.: Role of autoregulation in maintaining glomerular filtration rate at large urine flow. Acta physiol. scand. **76**, 24–39 (1969).

Narath, P. A.: Renal pelvis and ureter. New York: Grune & Stratton 1951.

Omvik, P., Jr., Raeder, M., Kiil, F.: Determinants of renal cortical volume. Amer. J. Physiol. **221**, 1560–1567 (1971).

Pierce, J. M., Braun, E.: Ureteral response to elevated intravesical pressures in humans. Surg. Forum **11**, 482–484 (1960).

Discussion

Regulation of Peristaltic Urinary Transport

The generation and conduction of excitation are the characteristics of peristalsis. Various theories to regulate the peristalsis are discussed concerning the ureter as well as other transport organs.

1. Mechanical distension of the smooth muscle fibres is the stimulus for generation and conduction of peristaltic waves.
2. The excitation is generated in a pacemaker area, on a myogenic basis, and is transmitted from cell to cell by electrical mechanisms.

Theory of Mechanical Distension

Tanagho is of the opinion that mechanical distension of smooth muscle fibres is the basis for the induction and conduction of a peristaltic stimulus. By manometric, cinefluoroscopic and myographic investigations, he found that the peristaltic wave dies out in case of ureter obstruction. It also dies out in the area of a functional obstruction. However, if a peristaltic wave were transmitted by an electrical muscular potential and were linked to the distension stimulus of the volume, it might not die out at the point of the obstruction. Melchior admits that peristaltic activity can be induced by mechanical distension. After the ureter has been cut and an alloplastic junction created, peristalsis is transmitted. But this transmission is clearly delayed. Furthermore, a secondary peristalsis induced by distension shows different characteristics from a normal one.

Theory of Electrical Transmission

Investigations on the isolated ureter have shown that distension by a bolus of urine is not necessary for the development and transmission of a peristaltic wave and the muscular potentials (GOLENHOFEN). Furthermore, simultaneous pressure and potential measurements have shown that in the case of a peristaltic wave the electrical reactions are always faster than the mechanical ones (MALATINSKY). To induce an effective peristaltic wave a sufficiently strong stimulation is necessary, because otherwise the wave would die out after a few centimeters (KIIL).

The distension of the smooth muscle fibres causes a contraction: this theory is rather convincing and well-known. Investigations of isolated and complete renal pelvio-ureter preparations by GOLENHOFEN and co-workers provided evidence that a pacemaker in the renal calyces is the center where the excitation starts. In the guinea pig its frequency is not influenced by a distension stimulus. Conduction can, however, be improved by distension. In man, distension of the pyelo-ureteral junction seems to be rather important in this connection. The filling pressure, therefore, can regulate the contraction frequency of the ureter at constant pacemaker frequency by controlling conduction in the pyelo-ureter. This would also explain the tendency to multiplications of peristaltic intervals.

Recent investigations with tetrodotoxin indicated that the generation and conduction of peristaltic waves is myogeneic. After blockade of all nervous elements by tetrodotoxin the peristalsis remains unchanged in the isolated pyelo-ureter preparation (GOLENHOFEN), although catecholamines can be shown to have some effect on isolated preparations as well (BOYARSKY).

Electromyographic investigations of the isolated renal pelvio-ureter preparation by GOLENHOFEN and co-workers showed that in the region of the renal pelvis potentials of plateau type dominate and are superimposed by small spike components, whereas in the distal ureteral segments the spike components are more important.

Renal Function and Ureteral Peristalsis

(Is the ureter under nervous control?)

The question whether the ureteral peristalsis is under nervous control has been much discussed. There is general agreement on optimal contraction frequency for the ureter as a muscular system, thus permitting the ureter to evolve its maximum effectiveness (BOYARSKY). If diuresis is increased, the contraction frequency may become smaller (KIIL).

Overall, ureteral peristalsis is exposed to multiple influences. For methodical reasons it is difficult to analyse every component and to evaluate its importance. The interdependence of bolus volume, contraction frequency and force of contractions has already been demonstrated by MORALES (BOYARSKY). BOYARSKY and MELCHIOR are of the opinion that the ureter is under nervous control; that would mean it is controlled by the sympathetic nerve. α-receptors have stimulating and β-receptors inhibitory influences. The difficulty is that renal function, too, is influenced analogously by nervous influences. In addition to direct impulses, indirect impulses, too, have an effect on ureteral peristalsis by nervous irritation, both glomerular filtration and sodium output being regulated by the sympathetic nerve.

Kiil demonstrated that a reduction in ureteral contraction frequency at high urine flow implied a dilatation of the ureter and presented evidence indicating that dilatation of the more proximal part of the urinary tract, the tubules and collecting ducts, played a fundamental role in renal adaptation to high urine flow.
Kiil further observed that high doses of noradrenaline are required to increase ureteral contraction frequency. But even high doses of catecholamines had no influence on ureteral contraction force as measured by strain gauges inserted in the intact ureter. The presence of chronotropic and absence of inotropic effects suggested to Kiil that catecholamines in pharmacological doses might act as pacemakers but would not regulate ureteral contractility. He therefore considers there is little evidence that either nervous stimulation or circulating catecholamines in physiological doses influence ureteral contractility.
Boyarsky and co-workers, however, were able to demonstrate that there is a connection between contraction frequency and muscular tension at the ureter. If the contraction frequency is increased, there will be an increase in the tonus a little later. They showed, as did Melchior and co-workers as well, that the tonus of ureter musculature is directly influenced by noradrenaline and isoproterenol.

The Pressure Gradient in the Ureter

For methodical reasons it is difficult to carry out systematic investigations in connection with the pressure gradient. We do not have sufficiently small measuring catheters (Boyarsky). Based on a comprehensive series of tests on monkeys, Roberts was able to demonstrate that the pressure amplitude in the distal ureter is higher than in the proximal ureter. Boyarsky and Kiil believe that this fact is not very important. But Melchior showed that in general there is a pressure gradient between renal pelvis and bladder which has to be overcome by the pumping mechanism of ureteric peristalsis.

Infections and Ureteral Peristalsis

Before you can answer the question: do infections influence the ureteral peristalsis, you have to ask some more questions. Is it an infection, an inflammation, or a contamination; a suspension of bacteria, bacterial substances of bacterial toxins? For example: if you inject *Escherichia coli* in high concentration into the ureter, it loses its motility and dilates. There is a minor effect from *Aerobacter* and *Proteus*. In general, intravenous or intrarterial injection of bacterial cultures reverses these effects. However, these are pharmacological questions (Boyarsky).

Functional Parameters of Ureteral Peristalsis

H. MELCHIOR and W. LUTZEYER*

The interpretation of observations recorded during functional examinations requires parameters that facilitate a qualitative and quantitative determination of the functional state of an organ and, at the same time, interindividual comparisons. Thus, a large number of biotechnical studies have been performed with a view to analysing the relationships between the peristaltic contractions of the ureteral muscles on the one hand and the flow velocity and the endoluminal pattern of pressure distribution on the other (SHAPIRO *et al.*, 1969, SIMHAN 1971, FUNG 1971). The physiology of the ureter, however, is characterized by a number of parameters, some of which are connected with one another like dependent and independent variables of a mathematical equation.

We feel that the classical parameters of pressure alone, i.e. basic pressure, pressure amplitude and contraction frequency (KIIL 1957), do not allow an adequate assessment of the ureteral dynamics (LUTZEYER and MELCHIOR 1969a). For a rising frequency of contraction, an increase of the basic pressure with diastolic loss and variations in the contraction complexes with longer contraction periods and a change from a regular to an irregular contraction process have all been described both with increasing diuresis and with stasis (LAPIDES, 1948; NARATH, 1951; MORALES *et al.*, 1952; KIIL, 1957; CHAILLET, 1957; BOYARSKY and MARTINEZ, 1962; WEINBERG and PENG, 1962; RUTISHAUSER and GRABER, 1963; LUTZEYER, 1963; BOYARSKY, 1964).

The wall stress in a cylindrical tube is calculated by the Laplace laws in a longitudinal direction by the equation:

$$t_1 = \frac{p\,D}{4\,S}$$

and in a transverse direction by the equation:

$$t_2 = \frac{p\,D}{2\,S}$$

where t_1 = stress in the longitudinal direction, t_2 = stress in the transverse direction, p = internal pressure, D = diameter, S = wall thickness.

In situ, however, calculation of the tonus, or wall tension, has so far been impossible, as under physiological conditions neither the internal pressure p nor the diameter D are constant parameters, but vary more or less periodically depending on the supply

* Abteilung Urologie der Medizinischen Fakultät der Rheinisch-Westfälischen Technischen Hochschule (Director: Prof. Dr. W. LUTZEYER) Aachen, Germany

of liquids and flow resistance. For this reason, the internal pressure alone cannot be regarded as a quantitative measurement of the wall tension while simultaneous determination of the diameter is impossible.

Since the introduction of urorheography (LUTZEYER and MELCHIOR 1969b) we have been able to measure changes in the flow velocity as well as endoluminal pressure variations.

The results of simultaneous measurements of pressure and flow in the ureter suggest the probability of a close functional connection among the various hydrodynamic pressure and flow parameters. Using our own clinical investigations and experiments with animals as a basis we have therefore attempted to establish empirically a mode of interconnection which facilitates the calculation of information on the current functional condition of the ureter.

The ureter may be substantially regarded as a pump. The capacity of a pump may be physically defined as follows:

$$L = \frac{p \cdot V}{t}$$

where p = pressure, V = volume, t = time.

By means of the measuring methods currently available to us, we are, however, unable to determine the rate of flow V/t in situ under physiological conditions. However, the pressure velocity product

$$P = p \cdot v$$

where v = the flow velocity,

seems to us to be a comparable parameter quite suitable for use as a measurement of the functional condition. This pressure-velocity product is given, by definition, the dimension of a *work per time and area unit:*

$$\left[P\right] = \frac{kp}{\text{cm}^2} \cdot \frac{\text{cm}}{\text{sec}} = \frac{kp\,\text{cm}}{\text{sec}} \cdot \frac{1}{\text{cm}^2}$$

The pressure-velocity product is calculated by multiplying the mean pressure of ureter by the mean flow velocity:

$$P = \frac{\int_0^T p(t)\,dt \int_0^T v(t)\,dt}{T^2}$$

While the pressure-velocity product covers both the peristaltic activity by way of the mean pressure of ureter and the effectivity of the ureteral peristalsis by way of the flow velocity, this alone yields no information on the relationships of the individual factors to one another. For this reason, in addition to the pressure-velocity product, we have defined, a "urodynamic quotient":

$$Q = \frac{\int_0^T v(t)\,dt}{\int_0^T p(t)\,dt}$$

This urodynamic quotient rises as the flow velocity increases, and falls as the flow resistance increases. The mean flow velocity and mean pressure of the ureter are determined conventionally by planimetry. As for technical reasons the erection of a suitable calibrating model required to calibrate the urorheograph has not yet been feasible , we enter the heat transfer coefficient α as measurement of the flow velocity (MELCHIOR *et al.* 1971)[1]:

$$[v] = 10^{-4}\,\text{cal}^\circ\,\text{C}^{-1} s^{-1}\,\text{cm}^{-2} = E$$

As a result of 173 clinical examinations conducted in situ, the standard values of the (pessure-velocity product P and the urodynamic quotient Q were established, together with other urodynamic standard values (Table 1).

Table 1. Urodynamic Standard Values

Urodynamic Parameters			
Mean pressure of ureter	$\bar{p}$	=	12.6 ± 0.6 mmHg
Diastolic pressure	p_{diast}	=	9.5 mmHg
Systolic pressure	p_{syst}	=	23.5 mmHg
Contraction frequency	f	=	3.7 min^{-1}
Mean flow velocity	$\bar{v}$	=	$17.8 \pm 0.6\ E$
Diastolic flow velocity	v_{diast}	=	16.2 E
Systolic flow velocity	v_{syst}	=	20.4 E
Pressure-velocity product	$\overline{P}$	=	$87 \pm 7\ E$ mmHg
Urodynamic quotient	$\overline{Q}$	=	$0.76 \pm 0.23\ E\ \text{mmHg}^{-1}$
E		=	10^{-4} cal° $\text{C}^{-1}\ s^{-1}\ \text{cm}^{-2}$
			(n = 173)

Examples

1. The Influence of Diuresis and Stasis on the Ureteral Dynamics

The ureter reacts to an increased supply of liquid by a rising contraction frequency), increase in the contraction pressure, decrease in the pressure amplitude due to diastolic loss associated with a rising basic pressure, and acceleration of the flow velocity (Fig. 1). Accordingly, the pressure-velocity product will increase as an expression of rising transport capacity, while the urodynamic quotient – after an initial rise – diminishes again, which is indicative of an increase of the flow resistance (Fig. 2).
In physiological conditions the ureter reacts to an increased intravesical pressure, with a rising contraction frequency and contraction force as well as acceleration of the flow velocity (Fig. 3). Accordingly, the pressure-velocity product increases while the urodynamic quotient as an expression of compensation remains practically unchanged (Fig. 2). In pathologic conditions, however, the ureteral dynamics decompensate with rising intravesical pressure, so that despite growing ureteral activity the flow velocity in the ureter diminishes: the pressure-velocity product changes only to a minimal extent, the urodynamic quotient becomes small (Fig. 2, 4).

[1] The conditions of an isotherm of a flowing liquid and a surrounding medium at 37° C can be fulfilled in a model only at considerable instrumental expense. An appropriate calibrating model is, however, being designed at the Institute of Aerodynamics of the Rheinisch-Westfälische Technische Hochschule Aachen (Director: Prof. Dr. phil. A. NAUMANN).

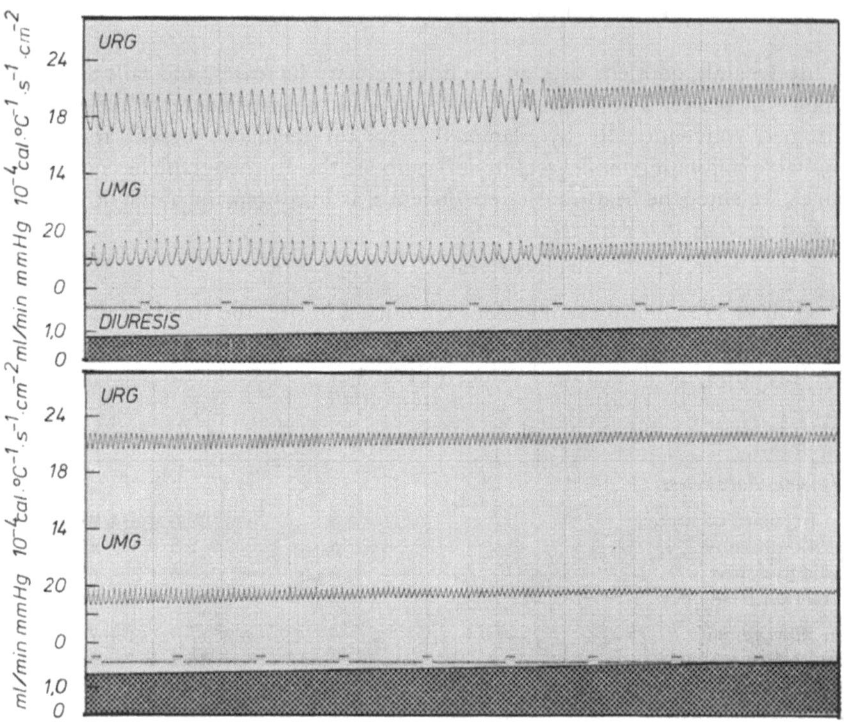

Fig. 1. The dependence of the ureteral dynamics on the diuresis

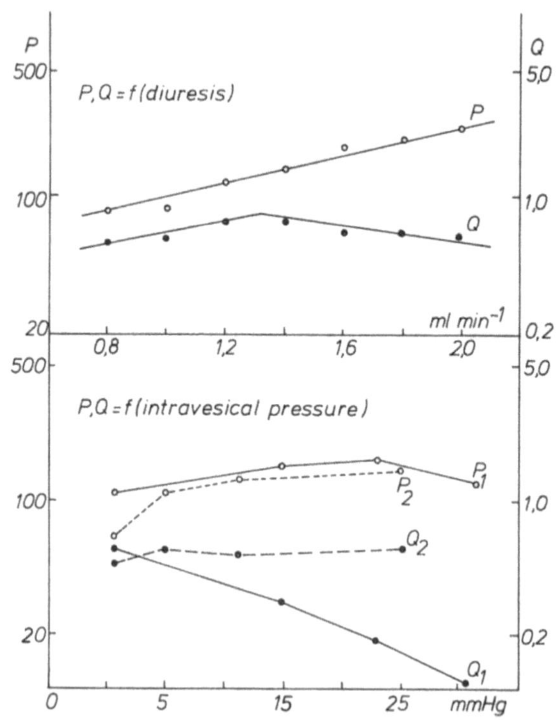

Fig. 2. The dependence of the pressure-velocity product (*P*) and the urodynamic quotient (*Q*) on diuresis (see Fig. 3) and intravesical pressure (see Fig. 5 and 6)

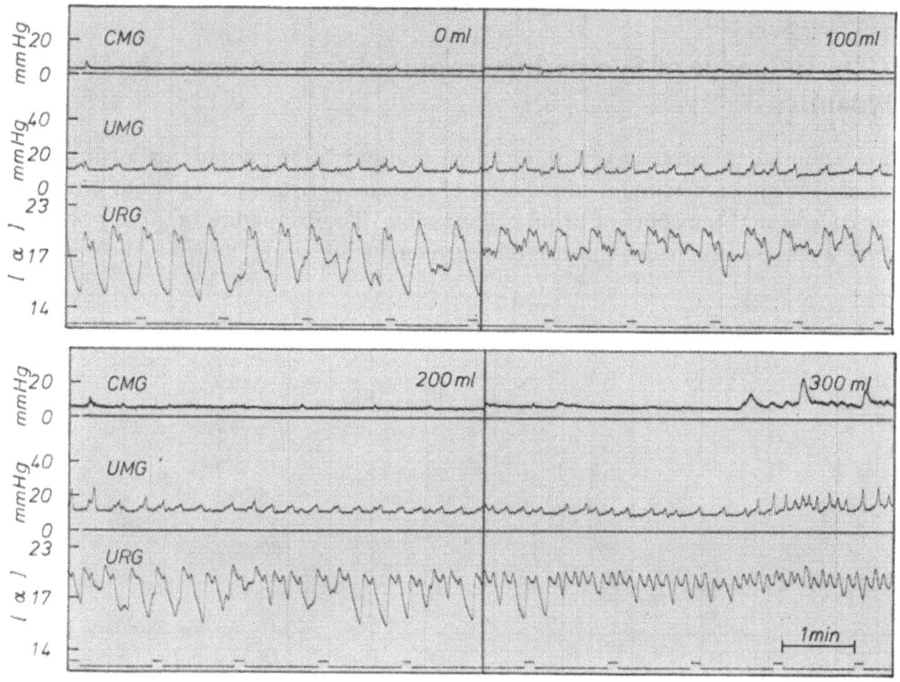

Fig. 3. The influence of the bladder dynamics on the ureteral dynamics (I: physiological conditions). (*CMG* = cysto-manometrogram, *UMG* = uretero-manometrogram, *URG* = uro-rheogram)

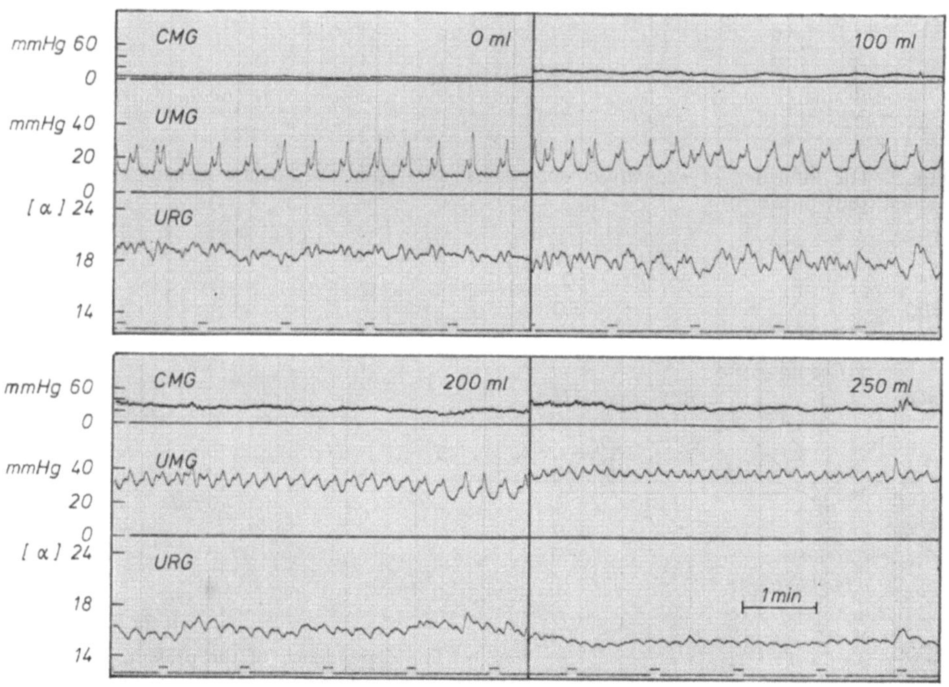

Fig. 4. The influence of the bladder dynamics on the ureteral dynamics (II: pathologic conditions). (*CMG* = cysto-manometrogram, *UMG* = uretero-manometrogram, *URG* = uro-rheogram)

2. The Influence of Sympathicotropic Substances upon the Ureteral Dynamics

Intravenous administration of 1.0 mg adrenaline and 1.0 mg noradrenaline in animal experiment causes maximum stimulation of ureteral peristalsis over about 7 min, so that coordinated transport of urine is impossible. The admixture of 10 mg phentolamine 2 min p.i. adrenaline interrupts the ureteral stimulation, which is succeeded by

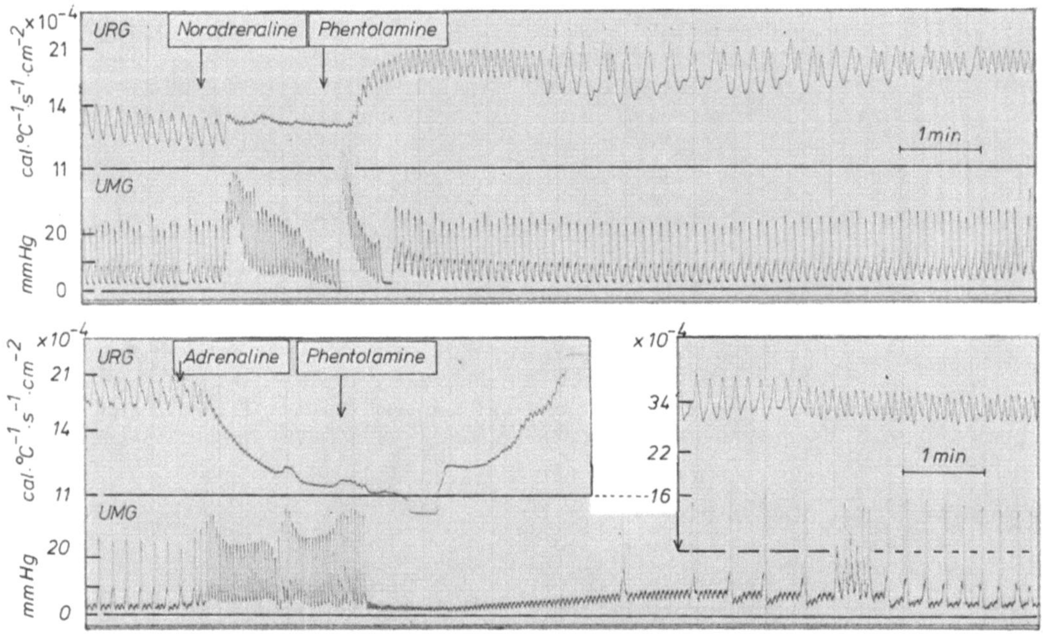

Fig. 5. The influence of sympathicotropic substances upon the ureteral dynamics (Experimental study on dogs)

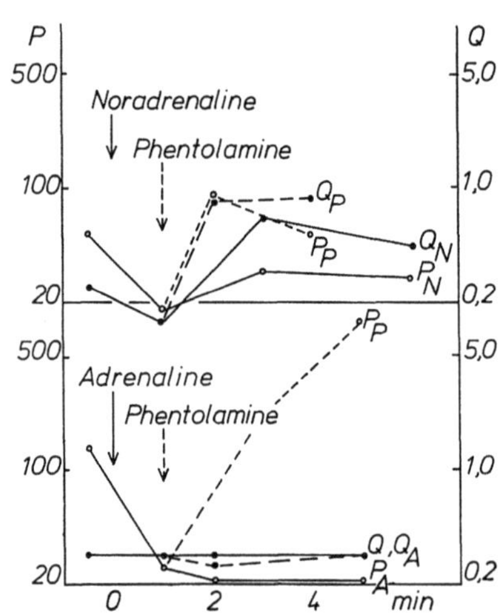

Fig. 6. The dependence of the pressure-velocity prodcut and urodynamic quotient on sympathicotropic substances

an aperistaltic phase with maximum acceleration of flow velocity. Following the administration of 10 mg phentolamine 2 min p.i. noradrenaline, the initial peristalsis is practically resumed within one minute (Fig. 5). This difference is reaction in the behaviour of the pressure-velocity product and the urodynamic quotient: While after noradrenaline plus phentolamine the pressure-velocity product 3 min p.i. phentolamine has recovered its initial value, the urodynamic quotient is still increased, indicating the restraining influence of the liberated β-adrenergenic components of noradrenaline. In contrast, the pressure-velocity product rises steeply after phentolamine plus adrenaline, indicating an increased supply of liquid. Nevertheless, the urodynamic quotient changed only negligibly, so that the presence of strong restraining influences (β-adrenergenic valencies) may be assumed (Fig. 6).

3. The Influence of Psychic Exitement and Autogenous Training on Ureteral Peristalsis

Psychic excitement, such as is caused by the announcement of a disagreeable injection or questions about family and occupation, stimulates ureteral peristalsis. Moreover, a close functional dependence of the ureteral and vesical dynamics on mental concentration was observed in a female patient who had acquired the technique of autogenous training to help her to cope with a nervous bladder (Fig. 7).

Evaluation of the pressure-velocity products and of the urodynamic quotients show that stimulation caused by mental excitement is due primarily to an increased muscular activity accompanied by increased tension. Autogenous training, however, will

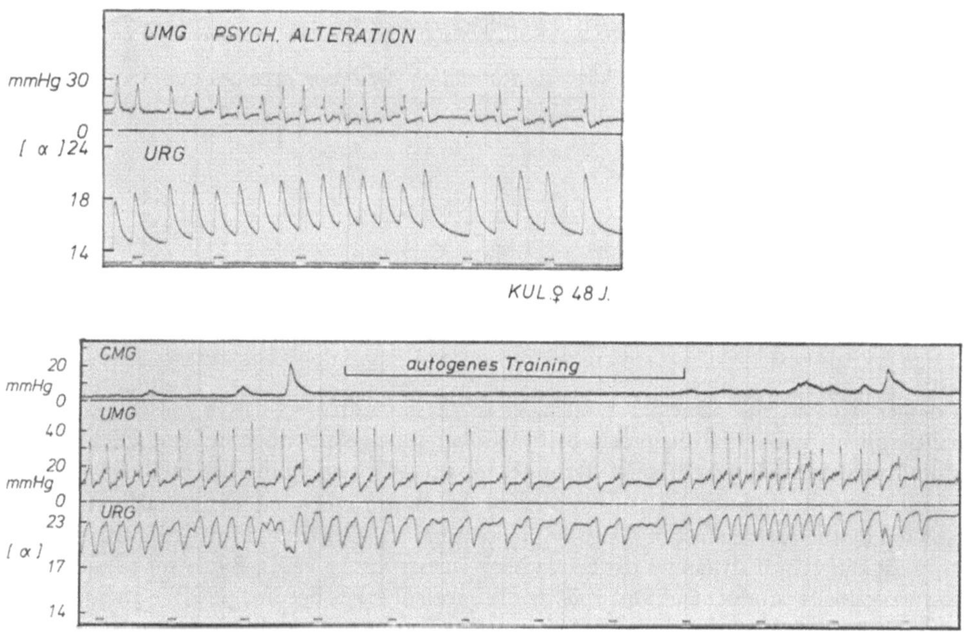

Fig. 7. The influence of psychic excitement and autogenous training on the ureteral peristalsis

increase the pressure-velocity product and the urodynamic quotient, which suggests an increased ureteral output at a diminished flow resistance (Fig. 8).

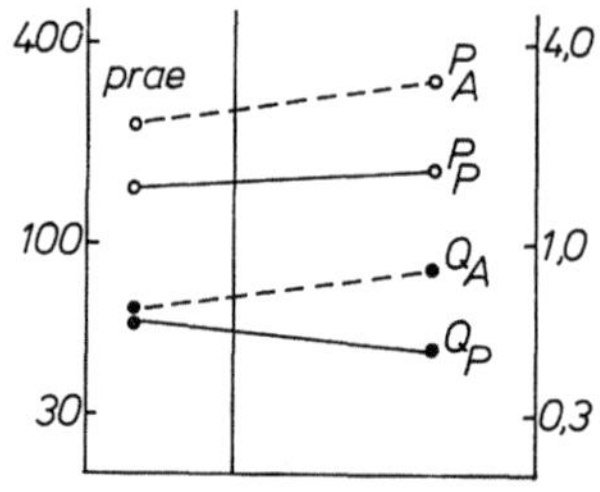

Fig. 8. The dependence of the pressure-velocity products and the urodynamic quotients on psychic excitement and autogenous training.

Discussion

The parameters we have introduced, pressure-velocity product and urodynamic quotient, enable us to supply information on the functional state of the ureter, which is not yielded by the common urodynamic parameters. For example, a high endoureteral pressure or strong peristaltic activity may be due as much to an obstruction of passage as to an increased supply of liquid (polyuria) or a primary muscular stimulation, while a low flow velocity may be ascribed to an extensive paralysis of the ureteral muscles with a distended cavity system (low ureteral tension, diminished peristaltic activity) or to obstruction of the passage by a stricture or stone. Pressure-velocity product and urodynamic quotient, however, establish a quantitative relation between the two values endoureteral pressure and flow velocity.

PARAMETERS OF URODYNAMICS

	diuresis	obstruction	hypertony	hypotony
frequency	++	+++	++	-
pressure	+++	+++	+	--
amplitude	+/-	-	++	--
flow velocity	++	---	+	-
pressure-velocity-product	+++	+/-	++	-
urodyn. quotient	-	--	0/-	0/+

Fig. 9. The value of the functional parameters of ureteral peristalsis

In our experience, the principal importance of these parameters lies in the clinical application in differential diagnosis of functional passage obstructions and in the field of basic research into ureteral dynamics in situ. Particularly in pharmacological investigations the question arises as to how far the reactions of ureteral peristalsis are triggered by the direct effect of drugs on the ureteral muscles or, indirectly, by changes in the diuresis or dynamics of the bladder (Rutishauser 1970). By means of the pressure-velocity product and the urodynamic quotient is has become possible to answer questions to which the classic parameters of basic pressure, amplitude, contraction frequency and flow velocity do not yield adequate explanations (Fig. 9).

Summary

The pressure-velocity product and the urodynamic quotient are introduced as new supplementary parameters of ureteral dynamics and discussed with examples.

References

BOYARSKY, S.: Surgical physiology of the renal pelvis and ureter. Monogr. surg. Sci. **1/2**, 173 (1964).

BOYARSKY, S., MARTINEZ, J.: Ureteral peristaltic pressures in dogs with changing urine flows. J. Urol. (Baltimore) **87**, 25 (1962).

CHAILLET, B.: L'étromanométrie urinaire. Thèse faculté de médicine, Paris 1957.

FUNG, Y. C.: Peristaltic pumping: a bioengineering model. In: BOYARSKY *et al.*: Urodynamics of the renal pelvis and ureter. New York: Academic Press 1971.

KIIL, F.: The function of the ureter and renal pelvis. Philadelphia-London: W. B. Saunders 1957.

LAPIDES, J.: The physiology of the intact human ureter. J. Urol. (Baltimore) **59**, 501 (1948).

LUTZEYER, W.: Harnleiterdruckmessungen. Urol. int. (Basel) **16**, 1 (1963).

LUTZEYER, W., MELCHIOR, H.: Die Wertigkeit urodynamischer Untersuchungsmethoden. Urol. int. (Basel) **24**, 289 (1969a).

LUTZEYER, W., MELCHIOR, H.: Uro-Rheographie. Urologe **8**, 208 (1969b).

MELCHIOR, H.: Uro-Rheomanometrie. Urodynamisches Symposion, Aachen 1969. In: LUTZEYER, W., MELCHIOR, H. (Ed.): Ureterdynamik. Stuttgart: Georg Thieme 1971.

MELCHIOR, H., SIMHAN, K. K., RATHERT, P., LUTZEYER, W.: Uro-Rheomanometry: Simultaneous pressure and flow monitoring in the ureter. Symposium on Flow, Pittsburgh, Pennsylvania/USA (1971).

MORALES, P. A., CROWDER, C. H., FISHMAN, A. P., MAXWELL, M. H.: The responses of the ureter to changing urine flow. J. Urol. (Baltimore) **67**, 484 (1952).

NARATH, P. A.: Renal pelvis and ureter. New York: Rune & Statton 1951.

RUTISHAUSER, G., GRABER, P.: Zur Harnwegsdynamik bei wechselndem Urinfluß. Urol. int. (Basel) **16**, 16 (1963).

RUTISHAUSER, G.: Das elektromanometrische Bild der Harnwegsdynamik unter physiologischen und pathologischen Bedingungen. Fortschr. Med. **83**, 919 (1965).

RUTISHAUSER, G.: Druck und Dynamik in den oberen Harnwegen. Darmstadt: Dietrich Steinkopf 1970.

SHAPIRO, A. H., JAFFRIN, M. Y., WEINBERG, S. L.: Peristaltic pumping with long wavelengths at low Reynold's number. J. Fluid Mech. **37**, 799 (1969).

SIMHAN, K. K.: Über ein theoretisches Modell zur Erfassung des peristaltischen Transportvorganges. In: LUTZEYER, W. und H. MELCHIOR (Ed.): Ureterdynamik. Stuttgart: Georg Thieme 1971.

WEINBERG, S. R., PENG, B. B. K.: Experimental production of obstructed ureters and evaluation of their activity. Invest. Urol. **2**, 181 (1964).

Complex Analyses of the Dynamics of the Renal Cavity System

F. Balogh and L. Szomor*

Two kinds of methods are known for the observation of the active and passive draining function of the upper urinary tract urodynamics: the intra- and the extra-cavitary methods. The methods requiring endoscopic manipulation, uro- or pyelo- and ureteromanometry etc., are seemingly more objective and demonstrable, but their physiological character can be questioned and their indications and application demand due consideration because of the indubitable possibility of infection.
Considering the several fundamental discoveries to which they have led, the value of these direct methods is not challenged here. Preference of the indirect methods is, however, justifiable; though probably more difficult to evaluate, more expensive and complicated, they have the advantage of involving only minor risks. Most of these methods are based on roentgenological or radiological procedures.
The various types of intravenous urography can provide valuable date on both morphological and functional dynamic points of view. Approximate information on the initiation and advance of the peristaltic spindle and on regular periodical filling of the ureters can be inferred from conventional intravenous urography, and even more from serial intravenous urography; the method of direct visual observation with image intensifier and moving pictures would be almost perfect.
But exposure of the gonads demands particular care. This is one or two orders of magnitude higher in the latter types of examination than in regular intravenous urography, and is inevitable even when the most modern equipment and the most sensitive film are used. The general and particulary the medical radiation-load of the population is increasing, but this problem deserves especial attention because the diagnosis and treatment of developmental anomalies, which are the main indication for indirect methods of examination, are shifting towards the years of childhood.
These methods are applied mainly when functional and morphological information is required simultaneously, and when the value of the information is greater than the risk of radiation load.
One of the conditions for taking moving pictures is a good intensity of the promptly excreted radioopaque substance. The estimation of function is worth the risk of a radiation-load of moving pictures, which is the most plastic and informative method, if the age and renal function of the patient permit it. But other methods should be attempted in children and in women of child-bearing age, e.g. isotopic methods.

* Department of Urology (Prof. Dr. F. Balogh) School of Medicine, University of Pécs, Hungaria

Isotope-renography and reno-cystography seem to be useful methods because the radiation-load involved is lower than that in intravenuos urography by 2–3 orders of magnitude and 4–5 orders of magnitude lower than is involved in X-ray moving pictures. The advantages of these methods are that they do not influence physiological conditions and the third phase of the renographic curve, the excretory phase, shows increased sensitivity to the condtitions of the cavity system. At least three nuclear channels are required for simultaneous recording of the activities of the two kidneys and the bladder. The choice of certain technical parameters and strict standardization of the conditions of examination are also indispensable.

In our experience it seems desirable to perform the examinations in both sitting and supine postures.

Most investigators perform radiorenography with the patient in the sitting posture for two reasons, as localization of the kidneys and the examination itself can be performed more easily and quickly, and it is assumed that stagnation due to ureteral kinking caused by ptosis or any other cause becomes more conspicuous. But in one group of our patients stagnation or excretory defects occurred more often in the supine posture, and it was in this group that the functional results obtained by other types of examination were better.

Recently we have been using radio-renocystography routinely and X-ray cinematography optionally for urodynamic examination. Two of the several indications are examination of the function of the cavity system after plastic surgery and examination of renal complaints for which conventional routine methods yield inadequate explanations.

The methods of postoperative examination following plastic surgery are described below.

The figures show the results of examination in patients with advanced pyelo-ureteral stricture and renal pelvic dilatation, who were subjected to the Hynes-Anderson operation. These subjects were selected because of the possibility of resultant damage or destruction of Pieper's ganglion-system; it is often supposed that this would probably exert a decisive influence on the motility of the renal pelvis or the upper urinary tract.

A female patient 38 years of age, underwent Hynes-Anderson's operation for serious pyelo-ureteral stricture and renal pelvic dilatation on the left side 4 years ago (Fig. 1). The urogram demonstrates a renal cavity of normal size and normal excretion. The contour of a resected renal pelvis and the outline of pyelo-ureteral anastomosis are also visible. The pyelo-ureteral junction shows no filling defect.

In our clinic radiorenography is performed with a 4-channel apparatus. The 4th curve, 3 rd from the top in the figures, is the blood clearence curve. In the case illustrated in Fig. 1 the investigation lasted 45 min because of pronounced stagnation. The left side of the figure shows the results recorded with the patient supine and the right those recorded with the patient sitting. The radio-renocystography of the side operated upon performed while the patient was supine demonstrates stagnation with normal excretion on the left side of the figure above; the activity of the left kidney does not begin to decline until after 20 minutes, and after 45 minutes it is only half the maximum. Discontinuous irregular excretion appears on the contralateral intact side, which is also reflected in the cystogram and can be seen in the lowest curve on the left of the figure. Better excretion is seen when the patient is sitting; the curve

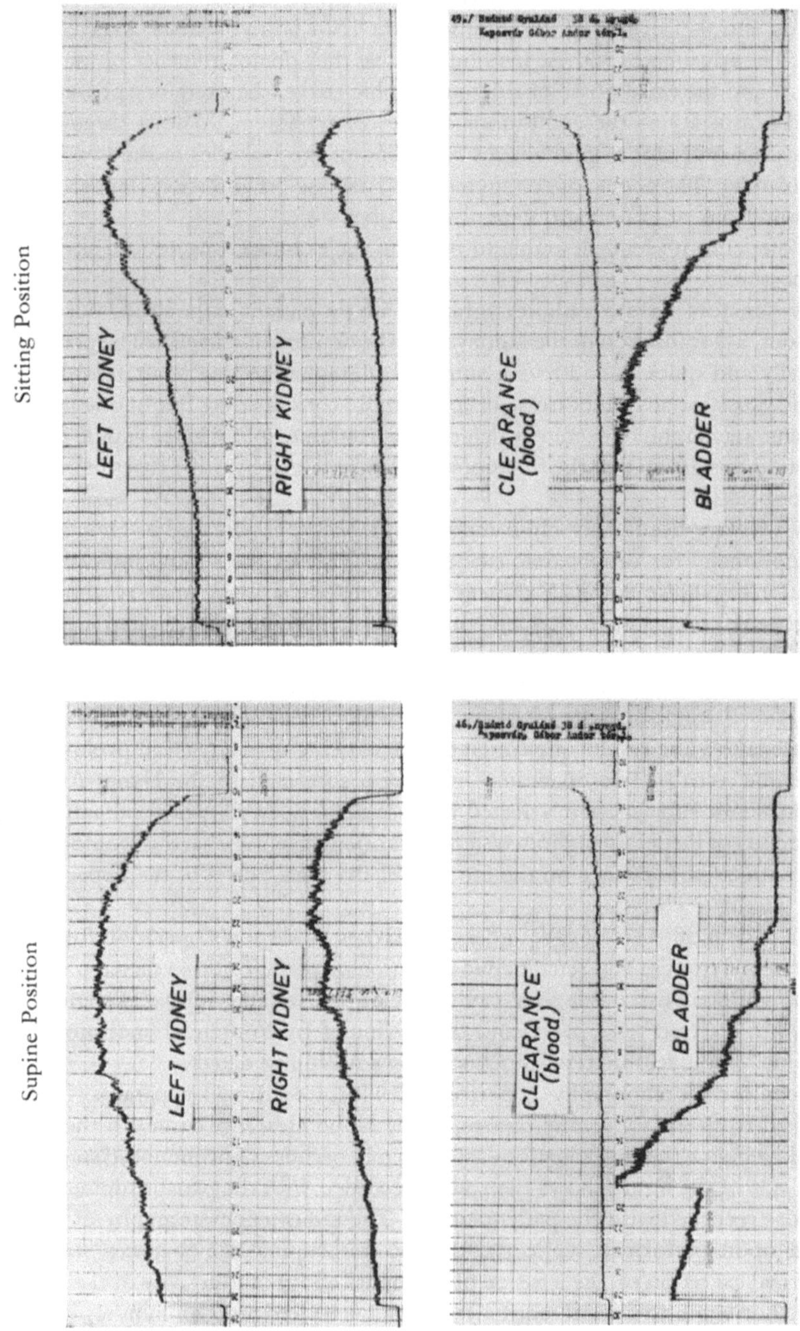

Fig. 1. Four-channel radiorenograms of the same female patient. The examination lasted 45 min

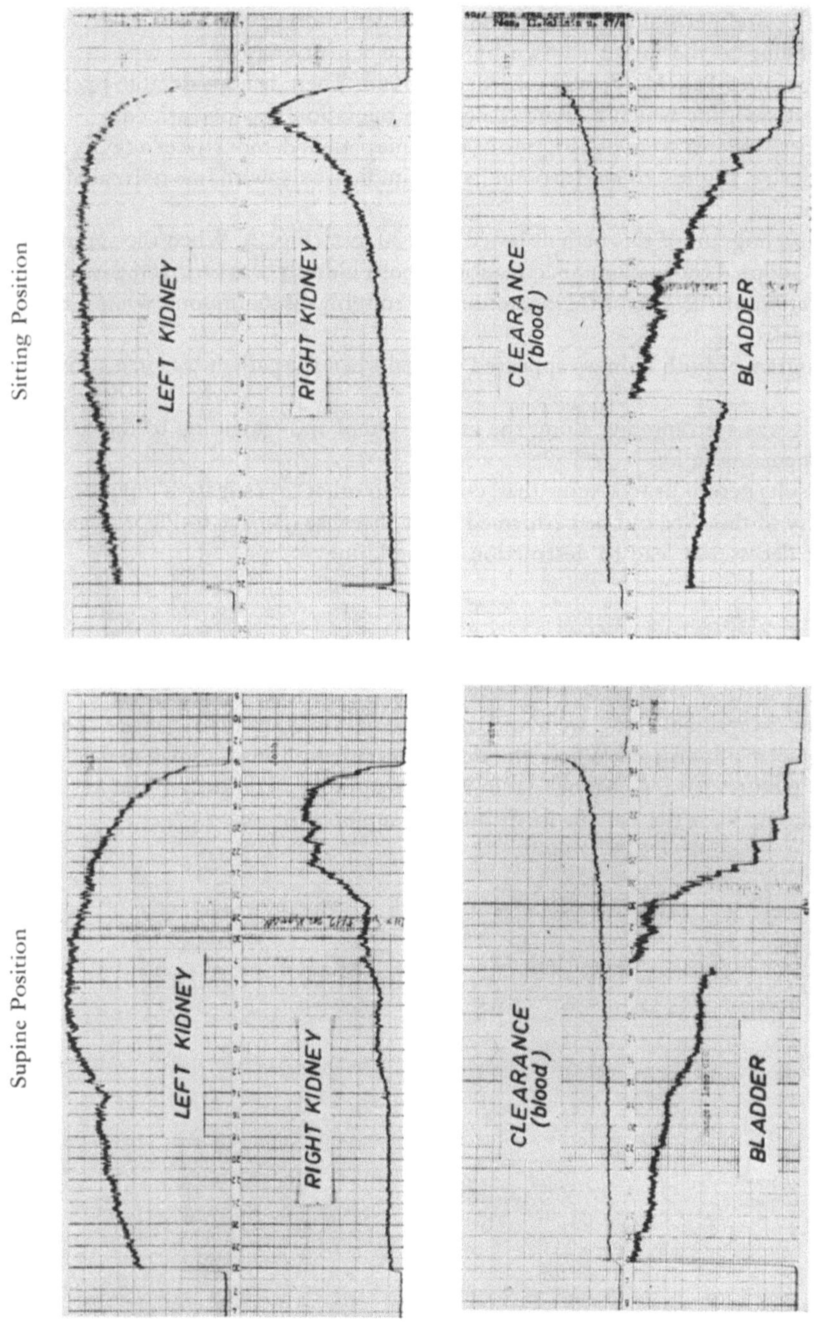

Fig. 2. Four-channel radiorenogram of Pat. M. V. This examination lasted only 30 min

recorded from the intact area is approximately normal, while excretion on the operated side can also be considered good.

There is a good peristaltic action, passing repeatedly along the renal cavity system and extending to the ureter (shown by X-ray cinematography).

The urogram of Pat. M. V. was recorded in 1968. There is considerable pyelectasia on the left side, for which a Hynes-Anderson operation was performed.

A postoperative intravenous urogram of the same patient demonstrates a cavity system of normal size, slightly rotated around its longitudinal axis, with free normal filling of the ureter.

The radiorenogram of this patient lasted only 30 min (Fig. 2). When the supine posture was adopted the maximum activation of both kidneys was protracted or delayed and excretion or the side of the operation was roughly discontinuous, while secretion was normal.

The renogram of both kidneys appeared normal when the patient was in a sitting position.

Peristalsis was seen moving along the cavity system and extending to the ureter by X-ray cinematography.

These results permit us to assume that, contrary to other investigators' opinions, plastic surgery of this type can be performed safely, because damage to Pieper's ganglion does not necessarily lead to destruction of the kidney.

Summary

Except in exceptional cases, we avoid manometrial or other examinations requiring endoscopic manipulation. Where the degree of radiation load allows we use X-ray moving pictures and 4-channel radio-renocystography, comparing the results of examintations with the patient in sitting and supine postures.

Electron Microscopic Observations on Human Ureteric Structure

R. G. NOTLEY*

The function of the ureter is to transport urine from the kidney to the bladder and the coordinated muscular contraction propagated along the ureter is the mechanism by which this function is discharged. The nature of ureteric peristalsis has been the subject of much speculation and study and by using observations from a wide variety of sources it is possible to construct a simple hypothesis describing the spread of peristalsis along the ureter. No satisfactory anatomical basis has hitherto been provided to support any such hypothesis and with this problem in mind I have undertaken a study of the detailed anatomy of the human ureter using the electron microscope. The material studied comprised 25 normal human ureters.

There is no doubt that the initial contraction of ureteric peristalsis arises in the renal pelvis, but even careful study of the microscopic innervation of the ureter does not reveal how it is propagated. There are many unmyelinated nerve fibres in the wall of the ureter (NOTLEY, 1968, 1969, 1970) and nerve terminals can be found in association with the muscular walls of blood vessels and the ureteric smooth muscle. These are presumably efferent in function. There are in addition many terminal axons in the submucosal layer which are probably afferent. This innervation of the muscle layer of the ureter is unlikely to be responsible for the propagation of peristalsis for 3 reasons: the nature of the action potentials associated with ureteric peristalsis and their rate of propagation is not compatible with nerve fibre conduction (PROSSER, SMITH and MELTON, 1955); the number and distribution of terminal axons in the muscular layer is inadequate; normal peristalsis can be demonstrated in the isolated ureter. The ureter, unlike the gut, has no intrinsic nerve cells in autonomic ganglia within its wall. Such ganglia described in the ureteric wall by optical microscopists are not nerve cells in my opinion. No ganglia have been found in or on the ureter in this investigation although their electron microscopic appearance is well known and striking.

If the nerve fibres of the ureter are not responsible for the propagation of peristalsis, it is probable that the excitation passes directly from muscle cell to muscle cell in some way, as the ureter behaves as a "single unit" visceral smooth muscle (BOZLER, 1938, 1942) and the rate at which the action potentials associated with contraction pass along the ureter support a muscular propagation (PROSSER, SMITH and MELTON, 1955).

Examination of smooth muscle from other organs with the electron microscope has shown points of fusion between the walls of adjacent muscle fibres (DEWEY and BARR,

---

* Royal Surrey County Hospital, Guildford, Great Britain

1964). Such a point of fusion has been termed a nexus (DEWEY and BARR, 1962) and it has been shown that a nexus offers an area of low resistance which will permit the spread of electrical excitation from one cell to another (BARR, BERGER and DEWEY, 1968). In this investigation many intimate contacts between ureteric muscle cells have been demonstrated (Fig. 1). At these points the cell walls lie in close apposition with no intervening basement membrane. In my own observations the cell walls are separated by a gap about 200 Å wide, but this is probably an artefact produced by fixa-

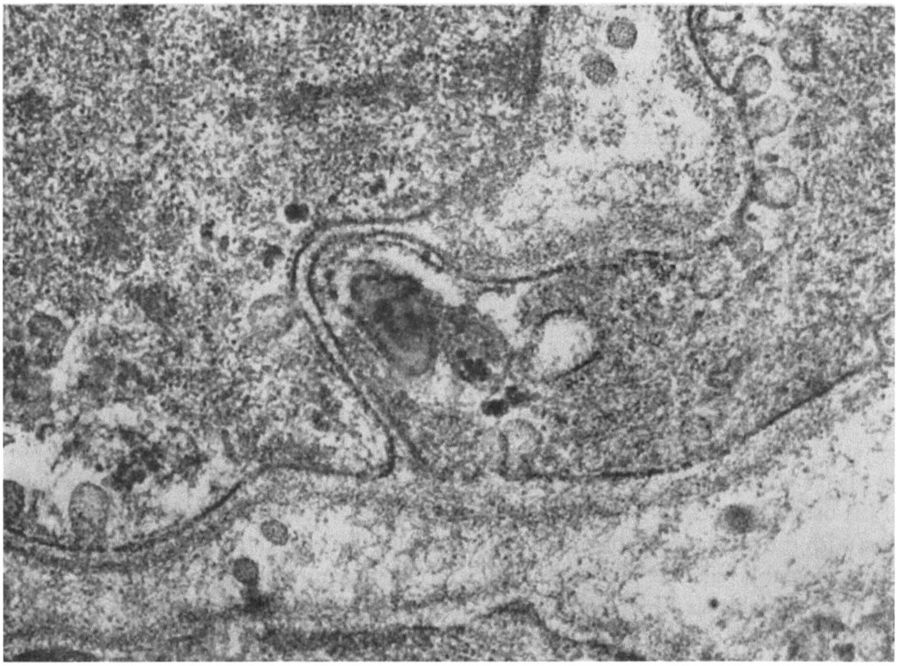

Fig. 1. Electronmicrograph showing close contact between two muscle cells in the wall of the human ureter. X 90,000

tion as other workers have recently demonstrated nexuses in the ureter by using a special technique which does not disrupt the membrane fusion (LIBERTINO, 1969).

Here then is an anatomical pathway by which the electrical excitation necessary to produce muscular contraction may be propagated along the ureter without involvement of nerve fibres. Peristalsis normally proceeds from the renal end of the ureter to the bladder, although its direction may be reversed by stimulation of the lower end of the ureter. This is to be expected if this concept of spread of electrical activity from cell to cell is correct as the cells have no directional conductivity.

In this context the possibility of a ureteric pacemaker requires examination. This is not a new concept (BOZLER, 1942), but more recent experimental evidence has made a ureteric pacemaker a more convincing proposition (SHIRATORI and KINOSHITA, 1961). WEISS, WAGNER and HOFFMAN (1967) adapted the techniques classically employed to define the sinus node of the heart for investigation of the ureter. The electrograms they recorded from the pelvi-ureteric segment with the unipolar electrode revealed what they suggest is an underlying pacemaker site. All the electrical activity was found to pass away from and none towards this region, which has been termed the primary pacemaker site. In addition the ureter, like the heart, appears to have muscle cells in other areas which possess the ability to exhibit pacemaker activity. These have been termed latent pacemaker sites; peristalsis arises in these regions only after suppression of the primary pacemaker site or interruption of conduction. As a parallel, if the sinus node of the heart is excised other areas become the site of paceamker acti-

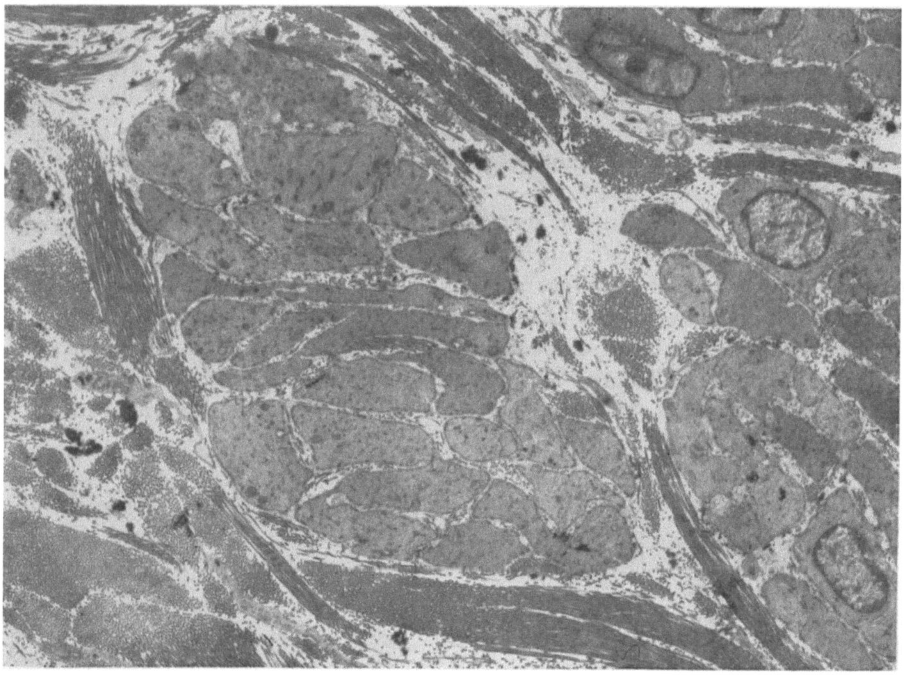

Fig. 2. Electronmicrograph showing excess amounts of collagen between the bundles of smooth muscle cells in the pelvi-ureteric segment of idiopathic hydronephrosis. X 4,300

vity. MURNAGHAN's observations (1957) that under various circumstances other parts of the ureter may initiate peristalsis support this view.

It must be emphasised at this point that these pacemakers, both primary and latent, are abstract concepts. There are no specialised muscle fibres which can be identified as the pacemaker fibres. Muscle contractions occur in the pelvis, but because of its funnel shape it is not an efficient pump; the contractions which occur in the pelvi-ureteric segment as a response to distension are the first effective propulsive contractions which send urine on its way to the bladder. It is in this sense, rather than a physical one, that this region acts as a pacemaker.

By combining these functional observations the following hypothesis describing the mechanism of ureteric peristalsis can be proposed and it can be supported by structural evidence revealed by the electron microscope: peristalsis appears to be initiated in the pelvi-ureteric segment, perhaps associated with pacemaker activity triggered off by distension of this region as urine is produced by the kidney. The electrical excitation arising in this region spreads from muscle cell to muscle cell along the ureter by way of the close contacts, producing the coordinated contraction wave of peristalsis.

This investigation into the detailed structure of the ureter was extended to include ureters exhibiting abnormal peristalsis. The narrow pelvi-ureteric segment in idiopathic hydronephrosis and the narrow terminal segment of the obstructive megaureter were examined.

The ultramicroscopic appearance of the ureteric wall in both these abnormal segments of ureter was strikingly similar. There was no abnormality in the microscopical innervation of these regions, nor did the muscle cells show any intrinsic abnormality. The muscle cells contacts looked no different from those elsewhere in the ureter. There was, however, a considerable increase in the amount of fibrous tissue. The bundles of muscle cells were widely separated by very many collagen fibres (Fig. 2), but this increase was most striking in the submucosa, which was sometimes as much as one millimetre thick.

This increase in fibrous tissue in the ureteric wall, at the pelvi-ureteric segment in idiopathic hydronephrosis and in the terminal part of the obstructive megaureter provides a simple explanation for the proximal dilatation present in both conditions. An excess of inelastic collagen at these points will prevent distension of the ureter during diuresis so that urine cannot drain into the bladder at the rate at which it is being produced by the kidney. This will cause dilatation of the proximal ureter and pelvis and the structural changes typical of idiopathic hydronephrosis and obstructive megaureter (Fig. 3).

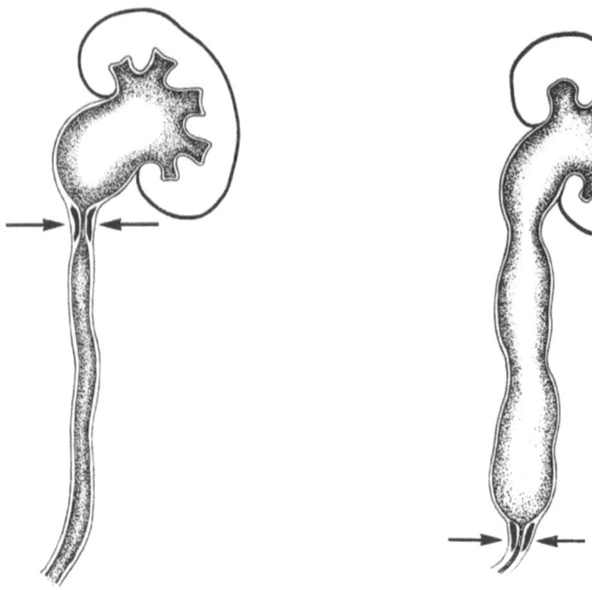

Fig. 3. Diagrams to show sites of abnormal amounts of fibrous tissue in the ureteric wall in idiopathic hydronephrosis (left) and obstructive megaureter (right)

Surgical treatment aimed at removing this strangling collar of fibrous tissue at either the pelvi-ureteric segment or the terminal ureter appropriately permits urine to drain away even during diuresis, preventing further progression of the obstructive changes.

References

BARR, L., BERGER, W., DEWEY, M. M.: Electrical transmission at the nexus between smooth muscle cells. J. gen. Physiol., **51**, 347–368 (1968).

BOZLER, E.: Electrical stimulation and conduction of excitation in smooth muscle. Amer. J. Physiol. **122**, 614–623 (1938).

BOZLER, E.: The action potentials accompanying conducted responses in visceral smooth muscles. Amer. J. Physiol. **136**, 553–560 (1942).

DEWEY, M. M., BARR, L.: Intercellular connections between smooth muscle cells: the nexus. Science, **137**, 670–672 (1962).

DEWEY, M. M., BARR, L.: A study of the structure and distribution of the nexus. J. cell. Biol. **23**, 553–585 (1964).

LIBERTINO, J.: Personal communication (1969).

MURNAGHAN, G. F.: Experimental investigation of the dynamics of the normal and dilated ureter. Brit. J. Urol. **29**, 403–409 (1957).

MURNAGHAN, G. F.: Experimental aspects of hydronephrosis. Brit. J. Urol. **31**, 370–376 (1959).

NOTLEY, R. G.: Electron microscopy of the upper ureter and the pelvi-ureteric junction. Brit. J. Urol. **40**, 37–52 (1968).

NOTLEY, R. G.: The innervation of the upper ureter in man and in the rat: an ultrastructural study. J. Anat. (Lond.) **105**, 393–402 (1969).

NOTLEY, R. G.: Electron microscopy of the lower ureter in man. Brit. J. Urol. **42**, 439–445 (1970).

PROSSER, C. L., SMITH, C. E., MELTON, C. E.: Conduction of action potentials in the ureter of the rat. Amer. J. Physiol. **181**, 651–660 (1955).

SHIRATORI, T., KINOSHITA, H.: Electromyographic studies on the urinary tract. II. Electromyographic study on the genesis of peristaltic movement of the dog's ureter. Tôhoku J. exp. Med. **73**, 103–107 (1961).

WEISS, R. M., WAGNER, M. L., HOFFMAN, B. F.: Localisation of the pacemaker for peristalsis in the intact canine ureter. Invest. Urol. **5**, 42–48 (1967).

The Autonomic Innervation of the Ureter and Ureterovesical Junction

C. C. Schulman*, O. Duarte-Escalante**, and S. Boyarsky**

The classic concept of urinary tract innervation maintains that the ureter is supplied exclusively by sympathetic nerves while the bladder receives essentially parasympathetic nerves.

The study of the pattern of innervation of the urinary tract has been limited for many years by light microscopic techniques which could not distinguish between sympathetic and parasympathetic nerves. New histochemical methods have allowed more extensive surveys. Parasympathetic components were studied by acetylcholinesterase staining methods (Elbadawi and Schenk), sympathetic components by localization of fluorescent catecholamines (Spriggs et al.), and monoamine oxidase reaction (Glenner et al.). In addition, electron microscopy has made it possibile to study important ultrastructural aspects of ureterovesical ganglia[1].

The ureter and bladder of mammals (dog and rabbit) are characterized by a dual sympathetic (adrenergic) and parasympathetic (cholinergic) innervation (Duarte-Escalante et al., Elbadawi and Schenk).

The nerves enter the organ, branching around the blood vessels. They ramify into different layers of the ureter and the bladder. In both organs the nervous pattern is characterized by a perivascular plexus supplying the muscular and subepithelial layers as well as the accompanying vessels.

In the ureteral musculature, the nerve fibers are related to muscle fascicles or small bundles of muscles. The nerve-muscle ratio is low (1/50) in both sympathetic and parasympathetic components. In the bladder, the nervous plexus is related to individual muscle cells by a 1/1 nerve-muscle ratio in the parasympathtic components and a rather lower ratio in the sympathetic elements. The sympathetic nerves are far less abundant in the dome of the bladder than in the base of the organ.

---

* Department of Urology, Brugmann University Hospital, University of Brussels, Belgium
** Division of Urology, Department of Surgery, Washington University Medical Center and Department of Pathology, Laboratory of Electron Microscopy, Cochran's Veterans Administration, St. Louis, USA

[1] Tissue for electron microscopy was obtained as follow: The ureterovesical junction was processed for histochemical examination using frozen sections; serial sections were prepared and stained with Hematoxylin-Eosin. When a large ureterovesical ganglion was observed externally to the ureter, it was dissected in the cryostat from the entire specimen and this fragment of frozen tissue was processed for electron microscopy. This technique made it possible to obtain ganglia with no fragments of ureter or bladder which might interfere with the electron microscopic study. The same specimens were used in the histochemical study

In the ureter, there is a progressive increase in the density of the nervous network towards the distal end. The terminal ureter is richly innervated, while only a few nerve fibers are observed in the upper segment (Fig. 1, 2).

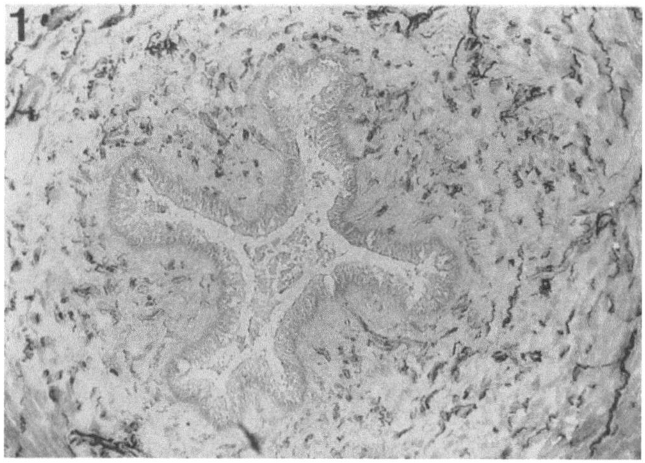

Fig. 1. Terminal ureter – AChE preparation (X 80). Rich in adventitial, muscular and submucosal cholinergic nerves. Adventitial nerve trunks give off muscular branches and terminate in the submucosal layer

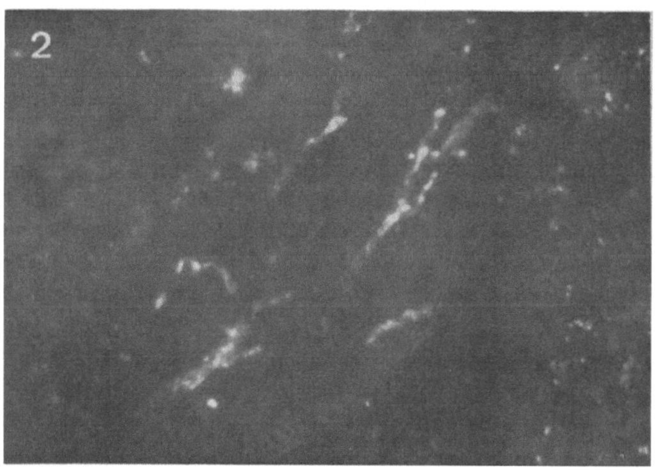

Fig. 2. Middle ureter – Fluorescent catecholamine (X 300). Typical fluorescent adrenergic nerve terminals in the muscular layer. The density of the adrenergic nerves is similar to the density in the cholinergic innvervation

The innervation of the periureteral muscular sheath, which is commonly designated Waldeyer's sheath, is identical to the innervation of the detrusor. This suggests that Waldeyer's sheath is of vesical origin (Fig. 3).

In the bladder, ganglia are most commonly found in the base of the organ and around the ureteral orifices. They are part of a ureterovesical ganglionic complex. This complex includes quite a large number of ganglion cells, located in the adventitia and surrounding adipose tissue of the terminal ureter and bladder base, the musculature of the bladder base surrounding the trigone, between the intramural ureter and the surrounding bladder musculature, and external to and around the periureteral musculature (Waldeyer's sheath).

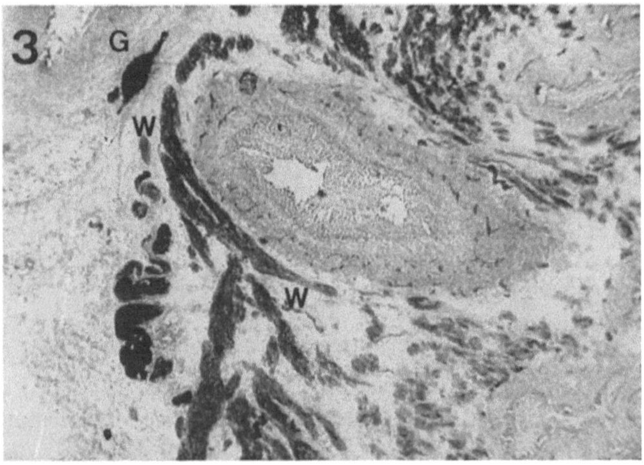

Fig. 3. Ureterovesical junction – AChE (X 60). The intramural ureter is surrounded by the vesical musculature. Striking contrast between the density of the detrusor innervation and that of the ureteral musculature. Innervation of the periureteral muscular sheath – Waldeyer sheath (*W*) is identical to that of the detrusor. Numerous ganglia *G* are observed at this level; they are part of a cholinergic – adrenergic ureterovesical ganglionic complex

Ganglia are extremely rare in the middle and upper ureter.

The ureterovesical ganglia are part of the pelvic plexus which includes branches of the pelvic and hypogastric nerves (Kuntz and Moseley). These ganglia are mainly parasympathetic (cholinergic) but must be considered as mixed structures, since sympathetic (adrenergic) elements are also demonstrable by fluorescent histochemical microscopy and by electron microscopy.

Histochemical techniques revealed two main elements in these ganglia. One of them consisted of fluorescent nerve cell bodies with fluorescent axonic processes corresponding to adrenergic neurons (Owman and Sjostrand). The other adrenergic element is represented by scattered fluorescent nerve terminals in the ganglia; some ganglion cell bodies are completely surrounded by a basket of fluorescent terminals which are considered to be synaptic structures (Hamberger et al., Hamberger and Norberg, Elfvin) (Fig. 4, 5).

Electron microscopy revealed the ultrastructure of these elements. The adrenergic nerve elements are detected by the presence of granular vesicles in their nerve terminals, which represent electron dense vesicles storing norepinephrine.

The adrenergic nerve endings are found less frequently than the cholinergic elements which contain numerous small agranular vesicles in their nerve endings (Richardson)

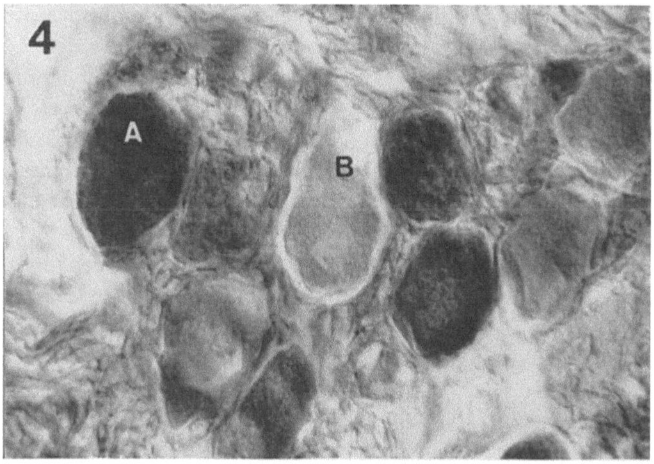

Fig. 4. Ureterovesical ganglion – AChE-Nomarski-differential phase microscopy-(X 500). Characteristic aspect of a mixed ureterovesical ganglion: some cells have cytoplasm richly filled with AChe granules *A* while others appear very clear and do not show any AChE activity *B*: these cells are considered to be adrenergic neurons

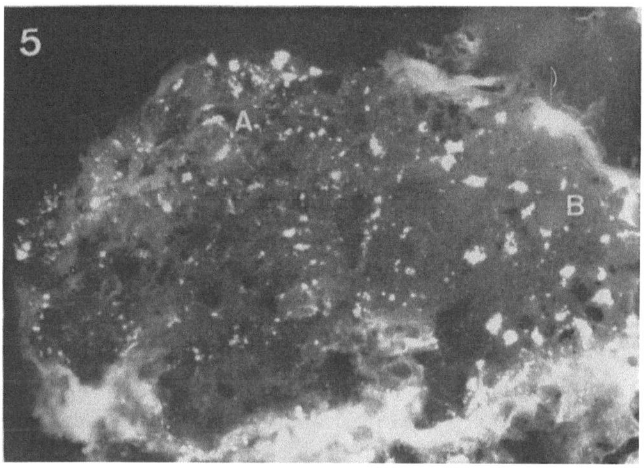

Fig. 5. Ureterovesical ganglion – Fluorescent catecholamine (X 125). Fluorescent adrenergic nerve endings surround ganglion cells (A): those fibers are considered to be synaptic structures. Fluorescent cells *B* correspond to adrenergic intraganglionic neurons

(Fig. 6). The detection of sympathetic nerve terminals in parasympathetic ganglia suggests the possibility of a new mechanism of adrenergic influence on ganglionic activity (HAMBERGER and NORBERG, NORBERG and SJOQVIST).
These elements are considered to have an inhibitory function (ECCLES and LIBET, FALCK et al., FERRY).
The ureterovesical ganglionic complex is not composed solely of intraganglionic neurons, but also includes extraganglionic cells represented by intramural cells in the lower ureter and bladder.

The existence of intramural ganglion cells in the ureter has long been a subject of controversy. In this study, according to morphological and histochemical criteria, a cell is considered to be a ganglion cell if the cell body stores either acetylcholinesterase or norepinephrine and if the cell body gives birth to an axon on serial sections.
Intramural ganglion cells, whether adrenergic or cholinergic, are usually associated with blood vessels. In the muscular and subepithelial layer of the ureter and bladder, Isolated cells are also present (Fig. 7, 8). This system is especially well developed in the ureterovesical junction.
The existence of sympathetic (adrenergic) intramural ganglion cells is focused since adrenergic neurons usually arise in the sympathetic chain or the prevertebral ganglia.

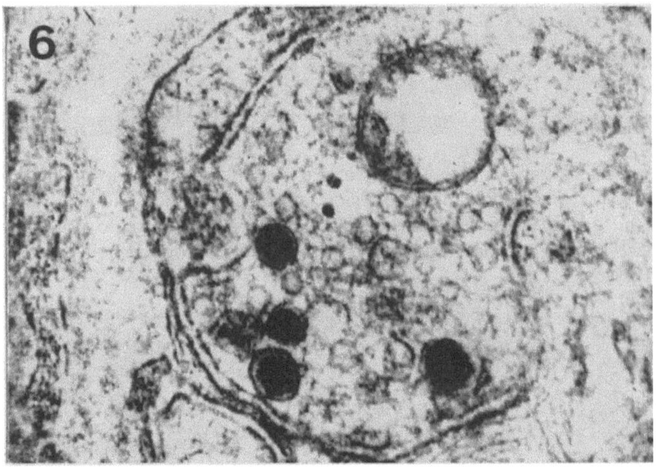

Fig. 6. Ureterovesical ganglion – Electron microscopy (X 135,000). Ultrasturctural aspect of an adrenergic nerve terminal in a parasympathetic ganglion: characteristic electron-dense granular vesicles storing norepinephrine

These elements are considered to be "short adrenergic neurons" since they derive from neurons located close to the innervated organ (Owman and Sjostrand). This pattern of sympathetic innervation resembles the pattern already seen in the parasympathetic system.
The nerves associated with ureterovesical ganglia innervate the terminal ureter, ureterovesical junction, trigone and bladder base. Hence, these structures are neurologically continuous to one another and their functions seems to be coordinated by both components of the autonomic nervous system.
The ureterovesical ganglionic system may play an important part in coordinating the behaviour of the bladder and terminal ureter, even during micturition. Therefore, it is suggested that the structures composing the ureterovesical junction must be considered as a single functional unit.
The existence of a short neuronal system is a fundamental feature of the peripheral autonomic innervation (Norberg and Sjoqvist). It introduces a new concept of innervation of the urinary tract, since morphological structures are present for an interaction of the two components of the autonomic nervous system and for a peripheral mechanism for mutual integration and modulation of the activity of the ureter and bladder.

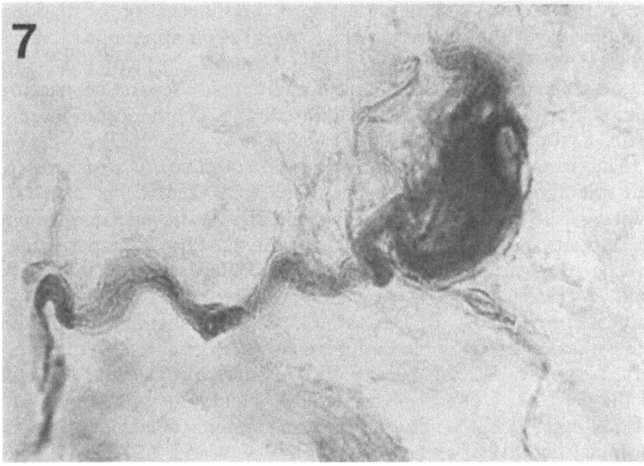

Fig. 7. Terminal ureter – AChE (X 1.250). Isolated intramural ganglion cell: excentric nucleus-rich AChE granular cytoplasm and axon running between muscle bundles

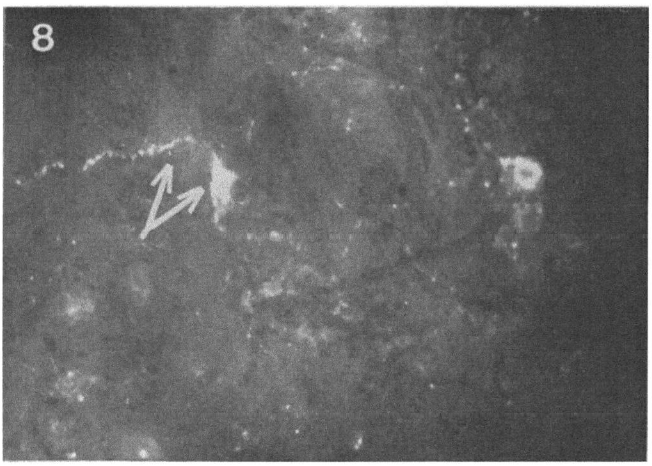

Fig. 8. Terminal ureter – Fluorescent catecholamine (X 500). Highly fluorescent sympathetic nerve cell with axonic process in the muscular layer (*arrow*)

References

Duarte-Escalante, O., Labay, P., Boyarsky, S.: The neurohistochemistry of mammalian ureter: a new combination of histochemical procedures to demonstrate adrenergic, cholinergic and chromaffin structures in ureter. J. Urol. (Baltimore) **101**, 803 (1969).

Eccles, R. M., Libet, B.: Origin and blockade of the synaptic response of curarized sympathetic ganglia. J. Physiol. (Lond.) **157**, 484 (1961).

Elbadawi, A., Schenk, E. A.: Intra and extraganglionic peripheral cholinergic neurons in the urogenital organs of the cat. Z. Zellforsch. **103**, 26 (1970).

Elbadawi, A., Schenk, E. A.: A new theory of the innervation of bladder musculature. Part I. Morphology of the intrinsic vesical innervation apparatus. J. Urol. (Baltimore) **99**, 585 (1968).

Elbadawi, A., Schenk, E. A.: Histochemical methods for separate consecutive and simultaneous demonstration of acetylcholinesterase and norepinephrine in cryostat sections. J. Histochem. Cytochem. **15**, 580 (1967).

Elfvin, L. G.: The electron-microscopic investigation of filaments structures in unmyelinated fibers of cat splenic nerve. J. Ultrastruct. Res. **5**, 51 (1961).

Falck, B., Owman, C., Sjostrand, N. O.: Peripherally located adrenergic neurons innervating the vas deferens and the seminal vesicle of the guinea-pig. Experientia, **21**, 98 (1965).

Ferry, C. B.: The post-ganglionic fibers of the vas deferens in the guinea pig. J. Physiol. (Lond.) **169**, 72 (1963).

Glenner, G. G., Burtner, H. J., Brown, G. W., Jr.: Histochemical demonstration of monoamine oxidase activity by tetrazolium salts. J. Histochem. Cytochem. **5**, 591 (1957).

Hamberger, B., Norberg, K. A.: Adrenergic synaptic terminals and nerve cells in bladder ganglia of the cat. Int. J. Neuropharmacol. **4**, 41 (1965).

Hamberger, B., Norberg, K. A.: Studies on some systems of adrenergic synaptic terminals in the abdominal ganglia of the cat. Acta physiol. scand. **65**, 235 (1965).

Hamberger, B., Norberg, K. A., Sjoqvist, F.: Evidence for adrenergic nerve terminals and synapses in sympathetic ganglia. Int. J. Neuropharmacol. **2**, 279 (1964).

Kuntz, A., Mosely, R. L.: An experimental analysis of the pelvic autonomic ganglia in the cat. J. comp. Neurol. **64**, 63 (1936).

Norberg, K. A., Sjoqvist, F.: New possibilities for adrenergic modulation of ganglionic transmission. Pharmacol. Rev. **18**, 743 (1966).

Owman, C., Sjostrand, N. O.: Short adrenergic neurons and catecholamine-containing cells in vas deferens and accessory male genital glands of different mammals. Z. Zellforsch. **66**, 300 (1965).

Richardson, K. C.: Electron microscopic identification of autonomic nerve endings. Nature **210**, 756 (1966).

Schulman, C. C., Duarte-Escalante, O., Boyarsky, S.: The ureterovesical innervation. Brit. J. Urol. **44**, 698 (1972).

Spriggs, T. L. B., Lever, J. D., Rees, P. M., Graham, J. D. P.: Controlled formaldehyde – catecholamine condensation in cryostat sections to show adrenergic nerves by fluorescence. Stain Technol. **41**, 323 (1966).

Discussion

Innervation of the Ureter

On the basis of histochemical investigations and in addition to the physiological and biochemical investigations of Boyarsky and co-workers, Schulman was able to demonstrate ganglion cells and nerve fibers in the wall of the ureter. While numerous ganglion cells were found in the terminal ureter, especially in the region of the ureterovesical junction, the upper and the middle segment of the ureter do not have ganglion cells. Allen suggested that the growth of the ganglion cells depends on the relative growth of the muscle fibers in the terminal ureter. Contrary to Allen, Schulman ascribed this observation to the fact that in this region there are numerous ramifications of the pelvic plexus, which includes branches of the hypogastric and pelvic nerves, and also many blood vessels.

Notley queried Schulman's argument because he himself was unable to demonstrate ganglion cells in or around any part of the ureter by electron microscopy. Notley pointed out that while nerve fibers in both axons and nerve cells were demonstrated individually for the first time by the electron microscope, with electron microscopy, as with optical microscopy, one must adhere to the defined morphological rules. During his presentation SCHULMAN showed pictures which he claimed were ganglion cells, but by electron microscopical morphology they looked to NOTLEY more like muscle cells, the two having similarities when seen in electron micrographs of poor definition. NOTLEY, however, agreed with SCHULMAN's general theory, which is the same as his own, although he rejected the evidence presented. For NOTLEY the ureter is "mysterious: there are so many nerve fibers that besides the myogenic peristaltic contractions there must surely be some nervous influence on ureteric behavior".

ZIMSKIND and TANAGHO asked: where are the ganglion cells observed by SCHULMAN localized? Are they in the wall of the ureter or in the adventitia? If these ganglion cells are outside the wall of the ureter in the adventitia, they may in fact belong to the bladder. Such nervous centers would also guarantee coordination of the functions of terminal ureter and the bladder.

SCHULMAN emphasized that he found ganglia in the adventitia of the terminal ureter, in the area of the trigone and the base plate of the bladder. These ganglia are localized in the ureter, periureterally and between the peri-ureteral musculature and the wall of the ureter itself. These ganglia must be differentiated from intramural ganglionic cells which are isolated cells. Numerous such ganglionic cells are found in the terminal segment of the ureter as well, and always near blood vessels.

The terminal ureter, trigone and the base plate have to be regarded as a complete integrated system which is under sympathetic and parasympathetic nervous control. Such structures are not found in the middle and upper ureter (SCHULMAN).

Dynamics of the Ureter — Ureteral Physiology in the Monkey

J. A. Roberts*

Evidence suggests the origin of the class mammalia 200 to 250 million years ago, the first mammals being represented by a primitive rodent. The first member of the order of primates, a tree shrew, was first found to exist around 60 million years ago. Paleontologists suggest that man has evolved from this terrestrial type of animal through an arboreal group of animals represented by the lower orders of primate species. While New World monkeys became differentiated from Old World monkeys around 40 million years ago, it was not until around 25 million years ago that the order Hominoidea became differentiated from the Old World monkeys. Within this group are the gibbon, orangutan, chimpanzee, gorilla and of course man. While the evidence of the paleontologist, based mainly on morphologic similarity of skeletal structures, has been accepted, more recent methods of determining phylogenetic differences have almost entirely confirmed the previous morphologic evidence. These include biochemical profiles in which comparisons of amino acids in serum albumin, the electrophoretic pattern of serum proteins, hemoglobins, similarities in chromosomes and enzymes systems such as chromosome oxidase have been used. All have shown the marked similarity between man and other apes and a close similarity between man and the Old World monkey (Lerner, 1968).

It was of interest to me, then, to determine how similar man and monkey were physiologically as regards the urinary tract. It has previously been shown by Odell et al. (1968) and Selkurt and Wather (1967) that the renal physiology of monkey is quite similar to that of man, especially as regards the concentrating ability of the kidney. Both rat and dog concentrate their urine more markedly than do the primates. What then of natural disorders in the monkey in comparison to those of man? Hydronephrosis of pregnancy has been studied by many investigators but determination of its exact etiology is lacking. This is probably because of a lack of an ideal experimental animal. The primate offers such an experimental animal in that primates naturally show this disorder unlike all other mammals, Roberts and Wolf (1971). This natural occurrence then is further suggestive evidence that the monkey would be an ideal animal for the study of ureteral physiology.

At the Delta Primate Center, two species have been evaluated using the techniques of Kiil (1957) with manometric measurement of ureteral peristalsis by means of non-

* Urology Laboratory, Delta Regional Primate Research Center, and Section of Urology, Tulane University Medical Center, Covington, and New Orleans, USA

obstructive ureteral catheters introduced cystoscopically. Both the rhesus monkey (*Macaca mulatta*) and the stumptail (*Macaca arctoides*) have been used. All studies were done under the influence of a tranquilizer-analgesic, Innovar-vet[1]. The monkeys studied were in a hydropenic state of 12 to 16 hours of dehydration. A positive urine culture excluded the animal from the normal group. In addition excretory urograms were done at the termination of each study and those showing ureteral dilation were excluded from the normal.

In the female monkey after appropriate urethral dilation a 16 Fr. cystoscope could be introduced and a 3 FR. ureteral catheter used for ureteral pressure studies. Fluid-filled catheters were attached to P23BB Statham pressure transducers for recording of peristaltic activity. Normal values for peristalsis, baseline or resting pressure, and amplitude of contraction were obtained; and in addition the effect of filling of the bladder, and diuresis on these normal values determined. Cystograms were performed at voiding pressure to determine the presence or absence of vesicoureteral reflux in the chimpanzee, four species of Old World monkeys and one species of New World monkey, using suprapubic bladder puncture for filling rather than a urethral catheter.

The values from 16 normal *Macaca mulatta* monkeys were subjected to statistical comparison using non-parametric statistics (Siegel, 1956). It was found that there was no difference between the right and left ureter at any level of the ureter. However, peristaltic rate was significantly greater in the upper ureter in comparison to the lower segments, and amplitude of contraction was greatest in the lower ureter. Baseline or resting pressure remained the same throughout (Table 1).

Table 1. Ureteral Physiology at Various Levels of the Ureter

	Peristalsis			Baseline			Amplitude		
	U	M	L	U	M	L	U	M	L
n	39	37	37	35	35	35	39	39	39
Rank Sum	88	64.5	69.5	79	65	66	63.5	82.5	88
Rank Statistic	8.28			3.485			8.474		

The critical Chi Square value is 5.99 at the 5 % level of significance, thus peristalsis is significantly greater in the upper ureter, and amplitude is greatest in the lower ureter, while baseline remains constant throughout the ureter

In a group of 7 monkeys, pressure studies were repeated at least three times to determine whether repeating a ureteral pressure study in one animal would give a more valid mean than a simple mean of a group of normal animals. It was found that a mean value from repeated measurements in one monkey was no more valid, than the mean of a similar number of measurements in different monkeys (Table 2). However, it should be noted that in either case the variance is great and to produce a significant change experimentally would require a wide difference from the normal mean. Filling the bladder resulted in marked increases in peristaltic rate, baseline pressure and

[1] Each ml contains 0.5 mg fentanyl and 20 mg droperidol, McNeil Laboratories, Inc., Fort Washington, Pa

Table 2. Repeated Pressure Studies in one Monkey Versus a Compilation of Studies in Several Monkeys Right Upper Ureter

Pressure	Peristalsis 1	2	3	Baseline 1	2	3	Amplitude 1	2	3
Monkey									
73	3	3	3	6	4	3	2	8	4
171	3	3	3	8	5	6	4	7	9
845	3	2	2	5	5	4	3	3	3
851	3	4	2	5	5	2	5	4	2
853	3	3	1	6	5	5	6	3	5
1874	5	4	2	7	10	7	5	4	7
215	2	4	4	7	7	5	3	6	3
Rank Statistic	4.324			11.901			7.165		

Repeated measurements in one monkey lend no more validity to a mean value than a similar number of measurements in different monkeys since the Chi Square value of 12.59 was found to be greater than any rank statistic. The variance is great in either case however.

amplitude of contraction as the bladder was filled in increments (Table 3). Diuresis, however, increases peristaltic rate but baseline and amplitude vary in response to increased flow.

As can be seen from Table 4, none of the chimpanzees showed vesicoureteral reflux. Only 2 of 150 Old World monkeys showed vesicoureteral reflux while 2 of 11 New World Monkeys showed vesicoureteral reflux. It is of interest that in the experimental production of pyelonephritis it is almost as easy to produce ascending infection by bladder infection in a New World monkey, the marmoset, as it is in the refluxing rat. This strongly suggests a high incidence of vesicoureteral reflux in this species in contrast to the Old World monkeys studied. A comparison of some of the normal values published for intraureteral pressures in dog and man is shown in Table 5. Comparison with our values in the monkey shows the close similarity of values for man and monkey.

Table 3. The Effect of Bladder Filling

	Peristalsis (Contraction/min)				Baseline (mmHg)				Amplitude (mmHg)			
Bladder Volume in ml	0	35	70	100	0	35	70	100	0	35	70	100
Mode	3	2	3	7	5	5			5			
Range	2–4	1–9	1–9	1–12	4–9	3–15	4–25	4–30	2–9	4–26	2–42	5–32
Median	3.0	3.0	4.5	6.0	5.5	7.0	10.0	12.0	5.0	8.0	9.0	10.0
Rank Statistic	9.9167				41.55				14.53			

The Chi Square value is 7.8 at the 5 % level, thus peristalsis, baseline pressure, and amplitude all significantly increase as the bladder is filled.

The absence of ureteral reflux in the Old World monkey, the normal values obtained for ureteral pressures and in addition the natural occurrence of disorders similar to those of man, such as hydronephrosis of pregnancy are all indicative of the potential value of this animal for studies of ureteral physiology.

Table 4. Vesicoureteral Reflux in the Nonhuman Primate

Species	–	Reflux
Chimpanzee	18	0
Rhesus	100	2
Stumptail	34	0
Patas	14	0
Talapoin	6	0
Squirrel	11	2

Table 5. Comparative Ureteral Physiology

	Peristalsis per min	Amplitude mmHg	Baseline mmHg
Dog (Boyarsky)	2–20 (9)	10–15 (13)	0–10 (10)
(DeLuca)	6– 8	15–37	0–10
Man (Weinberg)	1– 8	2–14	1– 6
(Edmond)	2– 6	11–59	
(Sala and Rubi)	1– 9 (4)	1–30 (13)	1–20 (9)
Monkey			
Rhesus (M. mulatta)	1–10 (3)	1–21 (7)	1–10 (5)
Stumptail (M. arctoides)	1–11 (3)	1–21 (6)	2– 9 (5)

Data reported on ureteral physiology (Table 1, 2, 3, 5) are reprinted through the permission of Williams and Wilkins Co., Baltimore, Maryland. See Roberts, J. A., Clayton, J. D., Mather, F.: Ureteral physiology in the monkey. Invest Urol., Jan. (1972).

References

Kiil, F.: The function of the ureter and renal pelvis. Philadelphia: W. B. Saunders Co. 1957.

Lerner, I. M.: Heredity, Evolution and Society. San Francisco: W. H. Freeman & Co 1968.

Odell, R. M., Radwin, H. M., Bernstein, L. H., Schlegel, J. U.: Renal function in *Macaca speciosa*. J. appl. Physiol., **24**, 366–368 (1968).

Roberts, J. A., Wolf, R. H.: Hydronephrosis of pregnancy: a disorder of primates closely resembling that of man. Folia Primat. **15**, 143–147 (1971).

Selkurt, E. E., Wather, R. L.: Renal concentrating mechanism of the squirrel monkey. Amer. J. Physiol., **213**, 191–197 (1967).

Siegel, S.: Non-parametric statistics. New York: McGraw Hill 1956.

The Effect of Adrenergic Substances on the Ureter Dynamics

E. MALATINSKY*

Our knowledge of pharmacological effects on ureteral activity is still at an early stage of development. One exact method for examining the effect of various drugs on the ureter activity is the registration of electric potential changes (p.c.) in the syncytial ureter muscles. Several authors (BAKER and HUFFER, 1953; HANLEY, 1953; WEINBERG and MALETTA, 1961; SHIRATORI et al., 1961) report different results. In our opinion the varying and even contradictory records of p.c. are due to different techniques of p.c. take-off and to the neglect of factors exerting a direct effect on ureteral dynamics, such as changes of diuresis, the temperature of tissue, disturbed urinary transport resulting from the obstruction constituted by the electrodes and the state of the animal's circulation during the experiment. It has already been mentioned in our earlier papers (MALATINSKY, 1966; 1967) that the sympathomimetics (epinephrine, norepinephrine, ephedrine) may affect ureteral activity. These drugs have a positive chronotropic effect and raise the frequency of the p.c. in the ureter muscles. The effect with a duration of only a few minutes is constant and reproducible at any time. BOYARSKY et al. (1966) demonstrated the presence of catecholamines in the wall of the ureter. This suggests that the ureteral muscles have a sympathetic innervation.

Methods

We exposed the ureter through a pararectal incision in our experiments on dogs (20 to 25 kg in weight, male and female). Basal anesthesia was induced with barbiturate prior to inhalation anesthesia with Fluothane and a N_2O—O_2 mixture. We administered the drugs by intravenous infusion at a constant drip rate in Ringer's solution. The bladder was emptied by an indwelling catheter. The p.c. were recorded from the ureter muscle by means of two bipolar platinum electrodes approximately 5 cm apart following careful preparation of a section of the ureter. An intra-abdominal oilbath ensured isolation from the surrounding tissue and constant temperature.

We then examined the effect of several sympathomimetics and parasympathomimetics in 10 experiments on 10 dogs. Since the positive chronotropic effect of this group of drugs was constantly and repeatably reached only with sympathomimetics, we continued examination of the effects of this group of drugs. In previous experiments on cats (MALATINSKY, 1969), we had already found that the effect of epinephrine on the ureteral activity was statistically highly significant ($P < 0{,}01$).

* Department of Urology University of Bratislava, Czechoslovakia

The higher frequency of the p.c. and consequently of peristalsis following administration of the therapeutic doses of epinephrine did not diminish the effect of ureteral peristalsis. The higher frequncy of the peristaltic waves also increased the transport of the urine through the ureter (Tab. 1).

Table 1. Amount of Urine before and after Administration of Epinephrine to Cats

Experiment	Amount of urine/ 5 min before administration	Amount of urine/ 5 min after administration
I	1.1 ml	3.0 ml
II	1.3 ml	2.1 ml
III	1.05 ml	2.0 ml
IV	1.5 ml	2.0 ml

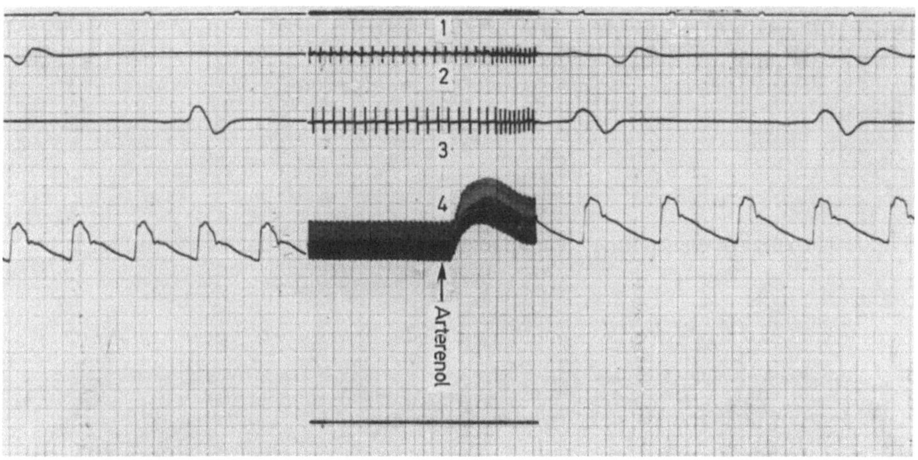

Fig. 1. The effect of epinephrine on the propagation of electric impulses along the ureter. *Line 1* = time in sec; *line 2* = upper electrode; *line 3* = lower electrode; *line 4* = blood pressure in arteria renalis

In our experiments the rate of conduction of the electrical impulses along the ureter was measured, while the distance between the electrodes remained constant. The time difference of the wave passage between the electrodes was the basis for the measurements. We can state as a preliminary result that *the propagation of the electrical impulses along the ureter after the administration of sympathomimetics is diminished due to the higher frequency or as a direct effect of epinephrine (Fig. 1).*

To acquire further insight into the mechanism of effect of epinephrine, we interrupted the bloodstream in the arteria renalis during the experiment, thereby also making it possible to study the effect of diuresis on the ureter activity. The electric p.c. gradually diminished and disappeared. Epinephrine was administered at this point and ureteral activity became normal within a few minutes. *Thus epinephrine has a direct effect on the smooth muscle of the ureter.*

All the observations described relate to the effect of sympathomimetics on the ureter activity of a ureter in isoperistalsis.

During the experiments we observed antiperistalsis in 3 dogs, in one case during a gradational overfilling of the bladder and in two cases for no known reason. Electro-

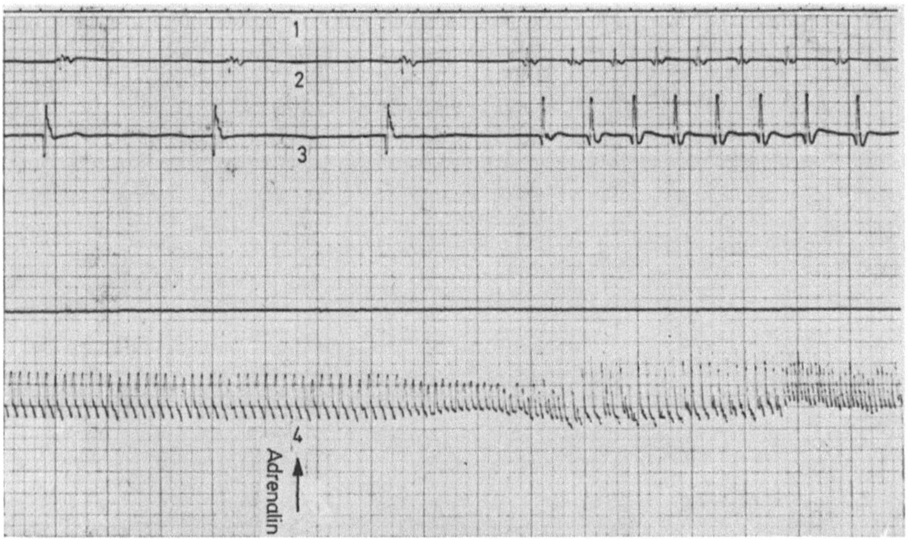

Fig. 2. The effect of epinephrine on antiperistalsis of the ureter (the lines as in Fig. 1)

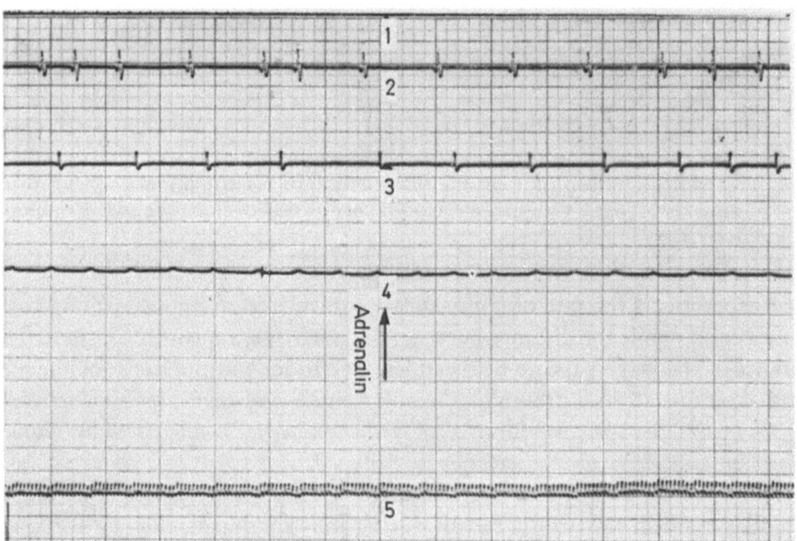

Fig. 3. The effect of epinephrine on the dissociation of the ureteral peristalsis. *Line 1* = time (1 sec); *line 2* = upper electrode; *line 3* = lower electrode; *line 4* = application of epinephrine; *line 5* = blood pressure in arteria femoralis

ureterography (EUG) offers the possibility of obtaining an excellent graphic record of the antiperistaltic waves (Fig. 2) by means of typical changes in the individual p.c. configuration. Since antiperistaltic waves have most often been observed on in cases of mega-ureter and of ureters with vesico-ureteral reflux, we focused our interest on the effect of epinephrine on the antiperistaltic ureter in our experiments. *We found that epinephrine (1 γ/kg i.v.) promptly regulates the course of peristalsis.* During these experiments we never observed recurrence of antiperistalsis after the administration of epinephrine.

During a less pronounced diuresis in three other experiments dissociation of the ureteral peristalsis between the upper and the lower parts of the ureter was observed for no apparent reason. P. c. changes passing rhythmically from the upper part of the ureter to the lower part were well recorded by the upper electrode, while the lower one showed no record. After the intravenous infusion of epinephrine (1 γ/kg) all p. c. passed regularly from the upper ureter down to the bladder (Fig. 3). Peristalsis of the ureter was normalized by an injection of epinephrine in a similar way to bigeminal pulse.

In spite of our knowledge about the positive effect of sympathomimetics on the peristalsis of the ureter and of our experience that no effect follows the administration of parasympathetic drugs, no therapeutic conclusions for clinical practice have yet been made. We must admit that since our results were obtained by electromyography in animal experimentation it is essential that they are verified in human beings. MELCHIOR and TERHORST (1969) observed the same positive chronotropic reaction of the ureter in man after intravenous administration of epinephrine.

We agree with MELCHIOR and TERHORST (1969) that sympathomimetics, epinephrine in particular, have a positive chronotropic effect on ureteral activity. The fact that the speed of the urine stream, measured by means of urorheography (LUTZEYER and MELCHIOR, 1969), increases after the administration of epinephrine, may correspond with our results concerning the increased urine transport after the same injection. Furthermore, our results confirm that *epinephrine in therapeutic doses does not only raise the frequency of p. c. but also corrects various experimental and spontaneous ureteral arrhythmias and irregularities in the conduction of the p.c. along the ureter.* While the administration of epinephrine immediately raises the frequency of peristalsis, the speed of the electric conduction of individual p.c. decreases along the ureter. This is due to the rise in frequency up to 28 impulses/min. We certainly cannot say that this is an effect of epinephrine, because a similar result was obtained by SLEATOR and BUTCHER (1955) by the application of pressure manipulations only to the ureter without administration of any drugs.

Although the effect of sympathomimetics on the ureteral muslces, epinephrine in particular, has also been shown in human beings, their therapeutic use is limited. Their effect does not last long, and the side effects make therapeutic trials on a larger scale difficult.

References

BAKER, R., HUFFER, J.: Ureteral electromyography. J. Urol. (Baltimore) **70**, 874–884 (1953).

BOYARSKY, S., KIRSCHNER, N., LABAY, P.: Catecholamine content of the normal ureter pelvis. Invest. Urol. **4**, 97–102 (1966).

HANLEY, H. G.: The electroureterogram. Brit. J. Urol. **25**, 358–365 (1953).

LUTZEYER, W., MELCHIOR, H.: Uro-Rheographie. Eine neue Methode der Harnflußmessung im Harnleiter mit Wärmeleitelementen. Urologe **8**, 208–214 (1969).
MALATINSKY, E.: Význam elektroureterografie v experimente a v klinickej praxi. Rozhl. Chir. **45**, 661–666 (1966).
MALATINSKY, E.: Farmakodynamický účinok na močovod in situ sledovaný EUG metodou. Rozhl. Chir. **46**, 796–803 (1967).
MALATINSKY, E.: Sledovanie fyziologickej a arteficiálne zmenenej činnosti močovodov pomocou elektroureterografie. Folia fac. med. Univ. Comenianae Bratislava **7**, 191–215 (1969).
MELCHIOR, H., TERHORST, B.: Die Wirkung sympathicotroper Substanzen auf die Ureterdynamik. Urologe **8**, 348–352 (1969).
SHIRATORY, T., KATAYOSE, K., SUETAKE, Y., KAWAMATA, S., KINOSHITA, H., TAKAKURA, K., ENOKIDO, K., SHIGA, R., NOMURA, T., USUBA, G., CHIBA, K.: Electromyographic Studies on Urinary Tract. I. Fundamental Studies on EMG of Animal and Human Ureters. Tohoku J. exp. Med. **69**, 103–117 (1961).
SLEATOR, W., Jr., BUTCHER, H. R., Jr.: Action potentials and pressure changes in ureteral peristaltic waves. Amer. J. Physiol. **180**, 261–276 (1955).
WEINBERG, S. R., MALETTA, T. J.: Measurement of peristalsis of the ureter and its relation to drugs. J. Amer. med. Ass. **175**, 15–18 (1961).

Discussion

Electromyography on the Ureter

The interpretation of electromyographic investigations on the ureter is very problematic. A change in potential in the case of bipolar mass derivation depends very much on the implantation of the electrodes, the mechanical alteration, and the artefact of loose contacts. No conclusions should be drawn from the individual types of potentials (BOYARSKY, HANNAPPEL).

Ureteral Peristalsis and Adrenergic Drugs

K. Haubensak, P. May, and E. Schindler*

We tested the effects of various sympathomimetics on ureteral peristalsis, blood flow, renal cortex and circulation in 16 dogs under inhalation anesthesia. The experiments were carried out according to the Ahlquist scheme of adrenergic drugs. The muscle summation potential in the middle of the ureter was transmitted by platinum wire electrodes (0,3 mm in diameter) which were implanted intramurally and

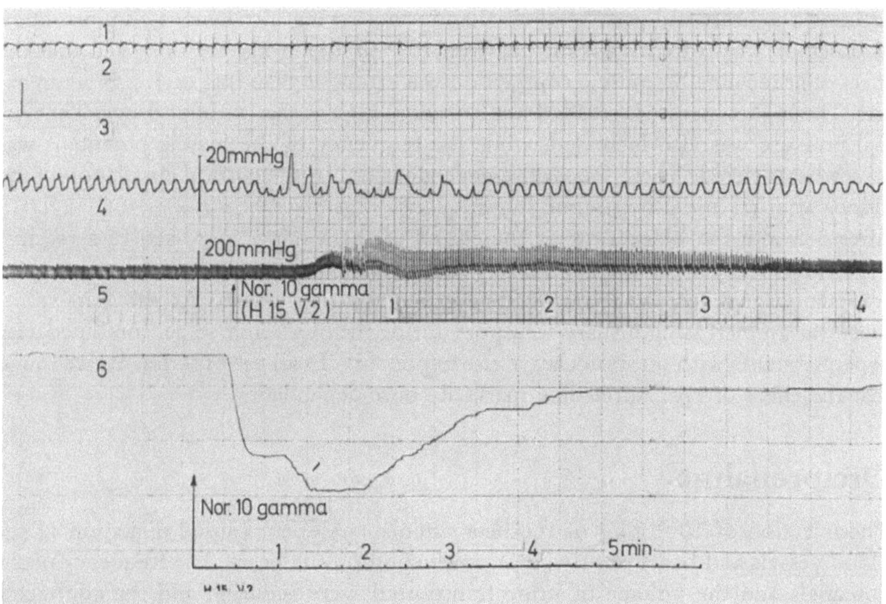

Fig. 1. Effect of noradrenaline: 2. bipolar electroureterogram of left ureter. 3. drop-counter (oliguric phase). 4. pressure profile in left ureter. 5. blood pressure in art. fem. sin. 6. cortical blood flow of the left kindey

retroperitoneally. Arterial blood pressure was recorded at the same time and we were able to determine the urinary volume transported by means of an electric drop-counter. A rubber T-tube connected to a Statham element was used to determine the internal pressure of the ureter directly below the electrodes.

* Urologische Universitätsklinik (Director: Prof. Dr. C. E. Alken), Homburg/Saar, Germany

Noradrenaline

Arterenol in doses from 10^{-6} to 10^{-7} g/kg caused a rise in arterial and venous blood pressure. Within the first minute there was a decline in frequency of the muscle potentials. Correspondingly the mechanical pressure waves became weaker but the individual muscle contractions stronger. The urinary output diminished during this period. This stage was typical of the action of 1. c. Noradrenaline and was accompanied by deficient blood flow to the renal cortex and the skin, resulting from vasoconstriction for 1–3 min. After this, the cortical blood flow to the kidneys rose to the original level and the ureteral potentials became more abundant. At the same time, normal mechanical contraction waves of the ureter transported more urine. After the decrease of the reaction, we observed interference in the transmission of the electrical potential along the ureter.

Adrenaline

The administration of Suprarenin caused a complex circulatory reaction by a direct effect on the heart and peripheral circulation. Following intravenous administration of doses of 1.c. Adrenaline ranging from 10^{-6} to 10^{-7} g/kg, we observed that the effect of adrenaline on muscle contraction was stronger than that of 1.c. Noradrenaline. The highest pressure reached was 3 times as high as that without drugs. The normal pressure was also increased, while the frequency of the muscle potentials was somewhat lowered. There was a dose-dependent fall in the cortical circulation of the kidney after an initial hyperemic stage.
After 3–5 min the blood flow to the cortex of the kidneys increased to a reactive hyperemic stage. At first, the urinary volume transported due to strong mechanical contraction and decreased during the vasoconstricotry phase. At the same time, there was an increasing urinary transport during the hyperemic stage, combined with hyperperistalsis with no appreciative rise in pressure. In all the experiments we found that the effect of 1.c. Adrenaline markedly dose-dependent.

Orciprenaline

The injection of 10^{-5} g/kg of the beta-stimulant Alupent caused dilatation of the blood vessels and had a positive inotropic action on the heart. The frequency of the potentials and the volume of urine transported were reduced, and the contractile force declined. The duration of effect was longer than had been anticipated on the basis of the published pharmacological data. The blood flow to the renal cortex showed a minimal decline, and the electromyogram revealed changes in the transmission of electrical stimulation.

Phentolamine

The alpha-blocker Regitin showed a circulatory reaction characterized by a relative predominance of the beta-receptors when given in a dosage of 10^{-4} g/kg. The blood flow to the cortex of the kidney was not essentially changed. It was observed in the ureter that the frequency of contractions was diminished and the mechanical contractions were weaker, while the resting tonus appeared unchanged. The electromyogram revealed interference in the transition of the electrical stimulation, as in the case of Alupent. Following the administration of 1.c. Adrenaline during alpha-blockade we

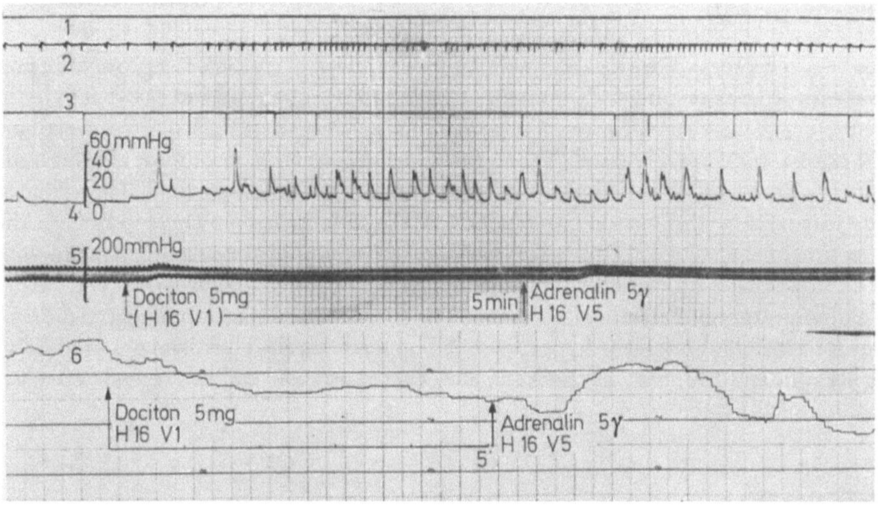

Fig. 2. Effects of Dociton and adrenaline on: 2. bipolar electroureterogram left ureter. 3. urinary volume transported. 4. pressure profile left ureter. 6. cortical blood flow left kidney

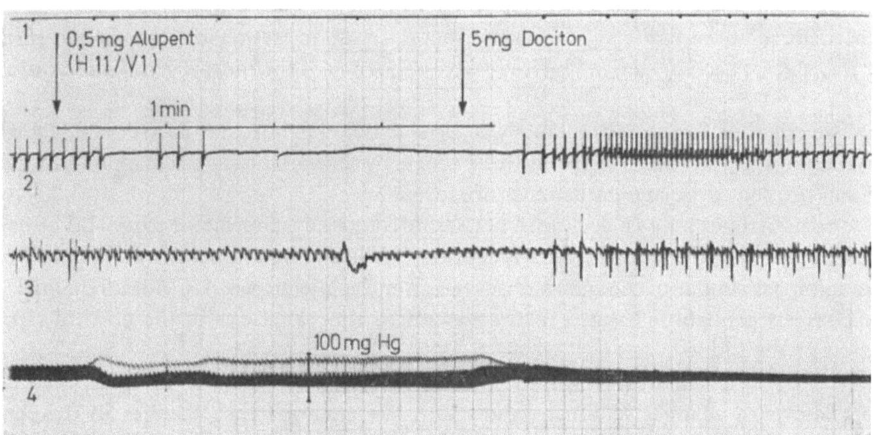

Fig. 3. Administration of orciprenalin (Alupent) followed by propranolol (Dociton): 1. drop-counter. 2. bipolar electroureterogram of left ureter. 3. bipolar electroureterogram of right ureter. 4. arterial blood pressure

observed a circulatory reaction which was described as a reserve reaction of 1.c. Adrenaline and was characterized by beta-stimulation like the effect of Alupent. The transport of urine, frequency in electromyogram and pressure of urine were diminished. The effect of noradrenaline, as previously described, was suppressed by Regitin.

Competitive antagonism in the ureter was evident because it was possible to abolish thus blockade with high doses of 1.c. noradrenaline. Moreover low doses of 1.c. noradrenaline eliminated the transmission disturbances caused by alpha-blockade.

Propranolol

The beta-receptors were blocked by 10^{-4} g/kg Docition. The alpha-receptors were predominant. In our results, we found a rise in blood pressure and increased activity in the ureter, in the electromyogram and also in the strength of muscular contractions. The normal pressure increased. There was also a greater urinary output, although the cortical circulation of the kindney decreased due to vasoconstriction. The effect of 1.c. adrenaline with Dociton corresponded to stimulation of alpha-receptors. At the same time, the effect of Alupent was annulled. Blocking of the receptor system with Dociton and Regitin caused slight diuresis to transmission disturbances and finally to a disappearance of electrical potential. The diuresis did not stop and forced diuresis was not possible though the hyperperistalsis apparent in the electromyogram following Regitin and Alupent was checked. The effect of Alupent on the ureteral peristalsis was annulled by the action of Dociton.

Summary

1. Peristalsis of the ureter is transmitted through diuresis which depends upon blood flow to the kidney and can be influenced by exogenous adrenergic drugs.
2. Injection of 1.c. adrenaline resulted in a distinct positive inotropic effect on the musculature in the first stage, apparent in a decrease in frequency; the second phase of the effect was characterized by increased frequency of the muscle summation potentials.
3. The effect of 1.c. noradrenaline was similar but the muscular inotropic effect was weaker. It was possible to suppress these effects by the administration of alpha-blockers which proved to be antagonistic to noradrenaline.
4. Orciprenaline caused a decline in peristaltic frequency, contractile force and urinary transport by its effect on the beta-receptors, while the beta-blockade following propranolol was similar to the action observed after the injection of 1.c. noradrenaline.
5. Diuresis probably changed in proportion to the variations in the cortical circulation of the kidney.
6. The transported urinary volume during the administration of the drugs did not correlate with changes in renal circulation. We assumed that changes in the renal pelvis and the ureteral content resulted from increased or decreased muscular tone in the upper part of the urinary tract.
7. The Ahlquist scheme was also applicable to the ureter. The excitability of the alpha-receptors had a stimulatory effect while the beta-receptors had an inhibitory effect on the ureter.
8. Blockade of alpha-receptors and stimulation by means of beta-receptors caused disturbances in the transmission of electrical stimulation along the ureter, so that administration of adrenergic substances was assumed to have a direct effect on the electrical activity of the ureter.

References

AHLQUIST, R. P.: Arch. int. Pharmacodyn. **139**, 38 (1962).
BÄCKLUND, L.: Acta physiol. scand. **59**, Suppl. 212 (1963).
BORGSTEDT, H. H., BENJAMIN, J. H., EMMEL, V. M.: J. Pharmacol. exp. Ther. **136**, 386 (1962).
BOYARSKY, S., LABAY, P., KIRCHNER, N., GERBER, C.: J. Urol. (Baltimore) **97**, 627 (1967).
BOYARSKY, S., KIRSHNER, N., LABAY, P., Invest. Urol. **4**, 97 (1966).
DUARTE-ESCALANTE, O., LABAY, P., BOYARSKY, S.: J. Urol. (Baltimore) **101**, 803 (1969).
ENGELMANN, T. W.: Pflügers Arch. ges. Physiol. **2**, 243 (1869).
GRUBER, C. M.: J. Urol. (Baltimore) **20**, 27 (1928).
KUSCHINSKY, G.: Lehrbuch er Pharmacologie. Stuttgart: Georg Thieme 1966.
LABAY, P., BOYARSKY, S., DUARTE-ESCALANTE, J. A.: J. Urol. (Baltimore) **103**, 38 (1970).
LAPIDES, J.: J. Urol. (Baltimore) **59**, 501 (1948).
MACHT, D. I.: J. Urol. (Baltimore) **1**, 97 (1917).
ORBELL, L.: Pflügers Arch. ges. Physiol. **133**, 341 (1910).
SATANI, Y.: Amer. J. Physiool. **49**, 474 (1919).
TSUCHIDA, S.: Us. Japan. Seminar: Clinical Appl. of smooth muscle electromyogramm, Honolulu 1968.

Spasmolysis by β-Adrenergic Agents

H. Melchior and A. Kettner*

The treatment of acute ureteral colic and of asymptomatic calculi in the ureter falls in the domain of conservative therapy. The therapeutic aims are analgesia and expulsion of the concrement. At present our usual treatment schedule is as follows (Bandhauer, 1970):

1. In cases of acute ureteral colic we give "spasmoanalgesics" as combination preparations of a pyrazolone derivative and a parasympatholytic agent or papaverine derivative.
2. To expel a calculus we use continuous spasmoanalgesic therapy with diuretic effects.

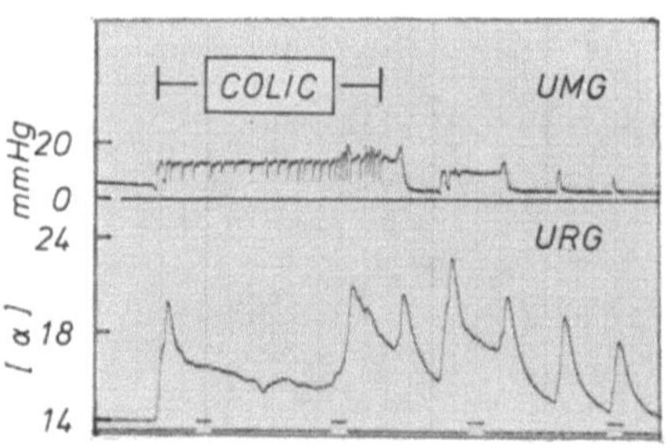

Fig. 1. Ureteral dynamics in colic. (UMG = uretero-manometrogram, URG = urorheogram) (s. Urologe A **10**, 1971)

We now have to assume that the parasympathiolytic agents and their derivatives have no direct influence on ureteral peristalsis (Lapides, 1948; Rutishauser, 1965; Melchior, 1969). We rather have to take it for granted that the ureter reacts mainly to sympathetic influences. The clinical success of the "spasmoanalgesic" combination treatment described above is therefore based solely on the effects of the analgesic components (Melchior et al., 1971). Experimental studies have shown that corpuscular particles in the ureteral lumen disturb normal urodynamics if they do not swim freely in the central flow of the urine spindle without touching the wall. Mechanical alterations in the ureteral wall induce local contractions of the ureter which as "extrasystoles" disturb the normal peristaltic stimulation process. In extreme cases the heterotopic polarisation is so extreme that it causes a continuous contraction (ureteral spasm) resulting in colic (Fig. 1).

* Abteilung Urologie der Medizinischen Fakultät der Rheinisch-Westfälischen Technischen Hochschule (Director: Prof. Dr. W. Lutzeyer) Aachen, Germany

It is possible to achieve both direct myogenic and indirect neurogenic drug effects on this continuous contraction of the ureter. Papaverine and its derivatives reduce the tonus of the smooth musculature independently of its vegetative innervation. Neurogenic "spasmolytics" exert their effect by way of the autonomic nervous system including the carrier substances and the receptors.
The theory of the sympathotropic sites of action in the upper urinary tract suggests that a neurogenic effect on the urodynamics is only possible by way of the sympathetic nervous system, and that the ureter is supplied, corresponding to the "receptors theory" of AHLQUIST (1948), by α-receptors with stimulant effects and by β-receptors with inhibitory influences (MELCHIOR et al., 1971). We performed animal experiments and clinical investigations in an attempt to determine whether β-adrenergic drugs have a spasmolytic effect on the urodynamics. Our conclusions were mainly based on the results of uro-rheomanometry (MELCHIOR 1969).

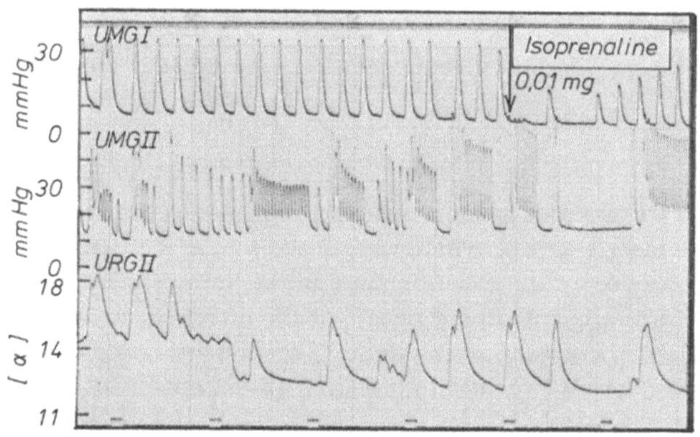

Fig. 2. The influence of isoprenaline on the dynamics of the mechanically altered ureter. (UMG I = uretero-manometrogram of the normal ureter, UMG II = uretero-manometrogram of the mechanically alterated ureter, URG II = uro-rheogram of the mechanically altered ureter)

Intravenous administration of β-adrenergic agents in animal experiments (dogs) does not only reduce normal ureteral activity (MELCHIOR, 1971) but also influences a pathologic ureteral peristalsis:

1. The contraction frequency of the constantly perfused ureter decreases following intravenous administration of 1.c. isoprenaline.
2. The ureteral contractions caused by mechanical alteration of the ureteral wall can be suppressed by 1.c. isoprenaline (Fig. 2).

The inhibitory influence of 1.c. isoprenaline on ureteral peristalsis cannot, however, be used clinically because of the considerable side effects on the coronary circulation with tachycardia. These side effects are substantially miligated by slow administration of the β-adrenergic agent 1.c. orciprenaline, which makes clinical application possible. Intravenous administration of 1.c. orciprenaline (0,01 mg/min) results in acceleration of the pulse rate but not to more than 100/min. When this level is reached the systolic and diastolic pressure in the ureter decrease and the contraction frequency increases.

In contrast to the animal experiment, we observed in clinical trials increased diuresis following the administration of β-adrenergic agents. We were unable to find any satisfactory explanation for this in the literature. We therefore suppose that it had something to do with the dose and the specific sensitivity. In our animal experiments we usually found diuresis was reduced by the administration of β-adrenergic agents, as did THURAU and WOBER (1962). HEIDENREICH et al. (1969) and FÜLLGRAF et al. (1969) think the direct tubular sites of action of the β-adrenergic agents in dog are responsible for this reduced diuresis. It causes an increase in the absorptive capacity in the proximal tubulus.

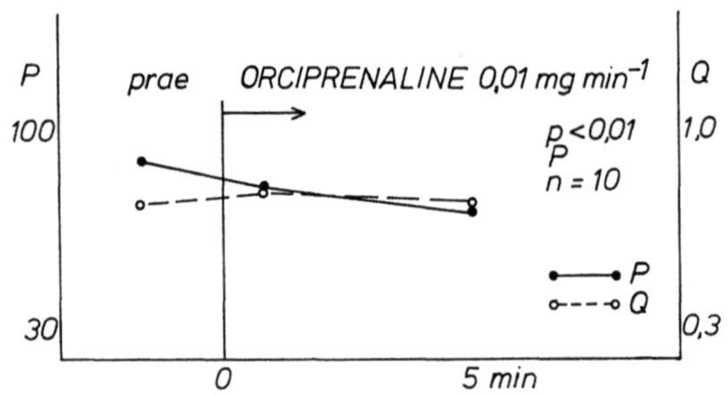

Fig. 3. The dependence of the pressure-velocity products (P) and the urodynamic quotients (Q) on intravenous application of orciprenaline. (Clinical studies, $n = 10$)

In clinical work we observed an increase in the frequency of contractions after administration of β-adrenergic agents, which was caused by the increased diuresis. Though a significant reduction of the mean pressure in the flow velocity was measured, it is impossible to form a quantitative estimate of the direct effect on the ureteral tonus and activity from these parameters alone. The pressure-velocity product and the urodynamic quotient give further information (MELCHIOR and LUTZEYER).

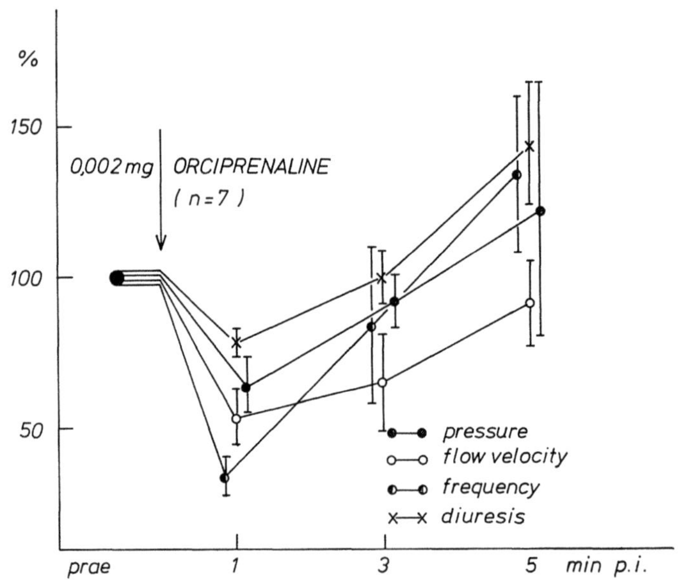

Fig. 4. The dependence of urodynamic parameters on intravenous application of orciprenaline. (Clinical studies, $n = 7$)

Under the influence of β-adrenergic agents the pressure-velocity product becomes definitely lower, though the amount of liquid increases, while the quotient changes only slightly (Fig. 3). This means that a larger volume of urine is transported through the ureter, while the endoureteral transport mechanism can only function according to Hagen-Poiseuille's law; this means has the cross section of the ureter become larger reduced wall tonus is necessary for the transport mechanism to function.

We used the inhibitory influence of β-adrenergic agents on ureteral peristalsis for medicinal treatment of urolithiasis in about 70 patients with a sound circulatory system. During an attack of colic we slowly gave 0.2 mg of 1.c. orciprenaline intravenously in 20 ml iso-osmotic NaCl-solution while the pulse rate was continuously recorded. The typical reports given by the patients stated that the spasmodic colicky pains abated, while the dull pain of acute hydronephrosis in the renal bed persisted. If these symptoms last for any length of time a burning inflammatory pain is felt in the region of the ureter. We also used β-adrenergic agents to induce expulsion of the calculi: intravenous infusion of 0.5 mg 1.c. orciprenaline in 500 ml of 10 percent mannitol solution. The velocity selected for the infusion is such that the pulse rate does not exceed 100 min. This treatment did result in an accumulated discharge of calculi, but the number of patients treated is still too low to justify a definitive statement on this subject.

It is important to avoid the administration of β-adrenergic agents in combination with parasympatholytic agents. This would cause dangerous tachycardias. Careful examination of the previous medical history is essential before treatment is instituted.

Summary

In our investigations in clinical practice and animal experiments we were able to demonstrate the inhibitory effect of the β-adrenergic agents on urodynamics.

Use of the β-adrenergic agents as spasmolytics for the medicinal treatment of urolithiasis is not recommended because of serious cardiogenic side effects. Widespread use of these drugs will only be possible if the pharmaceutical industry succeeds in modifying them in such a way that the cardiogenic sites of action are suppressed farther than the myogenic sites of action on the ureter.

References

AHLQUIST, R. P.: A study of adrenergic receptors. Amer. J. Physiol. **153**, 586 (1948).

BANDHAUER, K.: Therapie des Uretersteines. Acta Urol., 42–47 (1970).

FÜLLGRAF, G., HEIDENREICH, O., HEINTZE, K., OSSWALD, H.: Die Wirkung von α- und β-Sympathikomimetika und Sympathikolytika auf die renale Exkretion und Resorption von Flüssigkeit und Elektrolyten in Ausscheidungs- und Mikropunktionsversuchen an Ratten. Naunyn-Schmiedeberg's Arch. exp. Path. Pharmak. **262**, 295 (1969).

HEIDENREICH, O., FÜLLGRAF, G., LAAFF, H., BALSHÜSEMANN, E.: Die Wirkung von β-Sympathikomimetika und Sympathikolytika auf die Nierenfunktion von Hunden. Naunyn-Schmiedeberg's Arch. exp. Path. Pharmak. **263** (1969).

LAPIDES, J.: The physiology of the intact human ureter. J. Urol. (Baltimore) **59**, 501 (1948)

MELCHIOR, H.: Uro-Rheomanometrie (Simultane Druck- und Strömungsmessungen im Ureter). Urodyn. Symposion, Aachen 1969. Publ. in: LUTZEYER, W., H. MELCHIOR (Ed.): Ureterdynamik. Stuttgart: Georg Thieme 1971.

Melchior, H.: Pharmakologische Beeinflussung der Ureterdynamik. 10. Tag. Südwestdtsch. Ges. Urol. 1969.
Melchior, H., Diemer, K. F., Simhan, K., Lutzeyer, K., Lutzeyer, W.: Ureterdynamik II. Der Einfluß des autonomen Nervensystems. Z. Urol. **64**, 93–101 (1971).
Melchior, H., Lutzeyer, W.: Functional parameters of ureteral peristalsis. s. p. 71–79.
Melchior, H., Terhorst, B., Kettner, A.: Die medikamentöse Behandlung des Harnsteinleidens. I. Spasmolyse und Analgesie. Urol. int. (Basel) **26**, 367–378 (1971).
Rutishauser, G.: Kann die pyeloureterale Dynamik unter klinischen Bedingungen durch Medikamente beeinflußt werden? Helv. chir. Acta **32**, 447 (1969).
Thurau, K., Wober, E.: Zur Lokalisation der autoregulativen Widerstandsänderungen in der Niere. Pflügers Arch. ges. Physiol. **274**, 553 (1962).

Discussion

Pharmacodynamic Investigations on the Ureter

Pharmacodynamic investigations on the ureter in situ are often performed without adequate controls. Certain facts should be kept in mind, then much of the present confusion could be avoided. First there are great fluctuations in spontaneous activity. Secondly, ureteral function can be influenced by changes in urine flow and by the possible effect of drugs on the urinary bladder. On account of the inconstant spontaneous activity, the function of both ureters should be observed before and after intravenous application of a drug. For constant and reproducible conditions, the drug should be infused over a long period. It is important for the interpretation of the results to know whether the effect is a transient or a persistent one (Kiil).

It is not enough to interpret the pharmacodynamic effects of observations of ureter contraction frequency alone. It may even be disputed whether a rise in contraction frequency indeed implies improvement in the function of the ureter as a transport organ. If urine flow is high, the passage is actually disturbed and delayed by high contraction frequency. Although recording of the frequency of ureteral contractions is simple, it should not be forgotten that other parameters such as force of contraction and velocity of propagation of contraction waves have more to do with the function of the ureteral muscle (Kiil).

Concerning pharmacodynamic investigations, it should further be remembered that ureteral function can be influenced by urine flow and by variations in bladder tone. Drugs which increase bladder tone are likely to produce a retention pattern in the ureter, although the ureteral function is not directly influenced. The simplest measure is to keep the urinary bladder empty during studies of ureteral function (Kiil).

In case of pharmacodynamic investigations, the secondary effects of the respective pharmaca have also to be considered. β-adrenergics, for instance, stimulate the heart. In the first place such substances should be classified according to their molecular biological or pharmacodynamic properties: substance, structure, enzyme system, specific receptor and energy system. Such a classification is much better than a global classification of α-and β-receptors (Boyarsky). Neuro-humoral substances like epinephrine and noradrenaline (these are no vasopressor substances!) have an effect on the smooth muscles of the ureter and change the peristalsis. The level of the circulating catecholamines in connection with the local stimulating moments produces the pattern which regulates the activity of the ureter (Boyarsky).

The ureteric peristalsis is a very complex process: vascular elements like glomerular arteries, peritubular capillaries and main arteries succumb to the same influences as the muscular elements of renal pelvis and ureter. You have to look at all these effects as a whole. It is very difficult to differentiate all the individual effects and to decide which valency is suitable for the urinary flow and which is primarily suitable for the ureteric peristalsis.

Functional Changes in Peristalsis: Hydronephrosis of Pregnancy

J. A. ROBERTS*

The etiology of hydronephrosis of pregnancy has not been completely elucidated. The rhesus monkey has been used as the test animal since hydronephrosis of pregnancy occurs naturally in the primate as opposed to other mammals (ROBERTS and

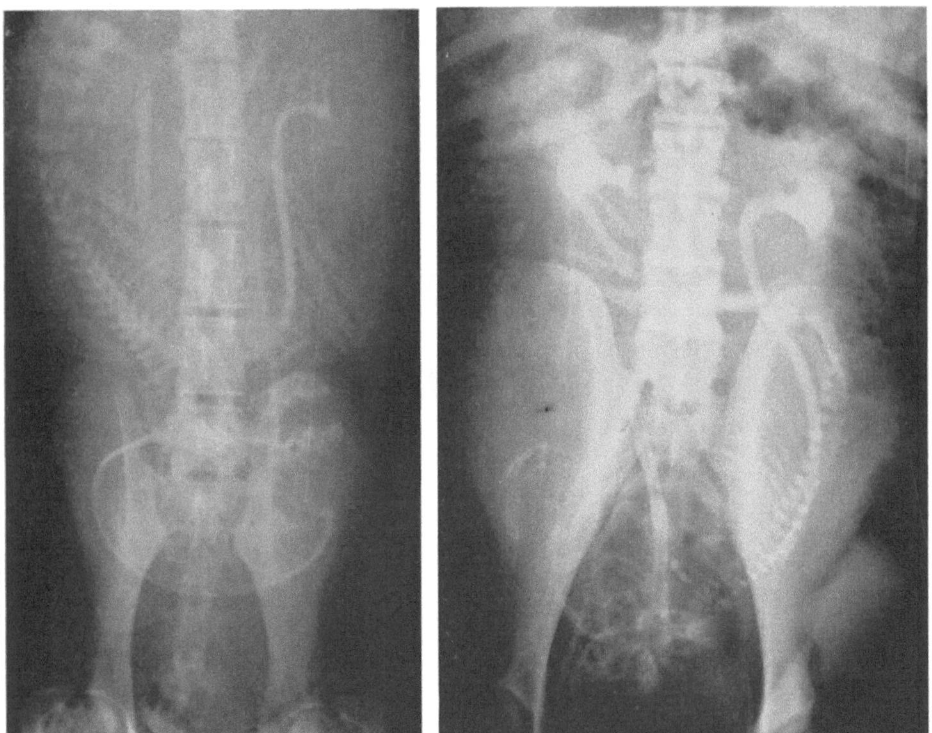

Fig. 1. a Excretory urogram of female rhesus monkey, when pregnant, showing hydronephrosis of pregnancy. b Excretory urogram of chimpanzee when pregnant shows a normal pelvic ureter with marked ureterectasis, caliectasis and ureteral tortuosity. (Reproducted by permission of the publishers of Folia Primatologica Basel, S. Karger.)

* Urology Laboratory, Delta Regional Primate Research Center, and Section of Urology, Tulane University Medical Center, Covington, and New Orleans, USA

WOLF, 1971). The fact that the pelvic ureter is usually normal suggests that hydronephrosis is due to obstruction at the pelvic brim by the gravid uterus.

In addition, the right ureter is more commonly affected than the left in both woman and the monkey (VAN WAGENEN and JENKINS, 1943). Fig. 1 shows the appearance of this disorder not only in the rhesus monkey, but also in the chimpanzee, which shows changes which are almost identical with those in man.

It is suggested, then, that hydronephrosis in pregnancy is an obstructive disorder. If this is so, it represents a very common cause of chronic ureteral obstruction. The obstruction in woman lasts for six to nine months and in monkey for three to six months with no change in renal function. Studies of renal function including inulin and PAH clearance in five monkeys have shown no decreased renal function after multiple pregnancies, and the same has been shown to be true in man (SMITH, 1951). Thus, in this disorder the ureter must act as a hydrodynamic cushion between the site of obstruction and the renal parenchyma. During a normal pregnancy, ureteral pressure studies were performed and baseline ureteral pressures were found to be elevated (Tab. 1), while there was little change in amplitude of contraction or peristaltic rate. As can be seen, removal of the uterus from the abdomen decreases the elevated pressure immediately, and pressure remains low during this "placental pregnancy" until almost term.

Table 1. Ureteral pressure at hysterotomy and excision of the fetus of rhesus monkey[a]

	R	L
Pregnant initial	9 ± 3.6	9 ± 2.5
Uterus lifted out	6 ± 1.7	4 ± 0.6
Fetus out – wound closed	10 ± 3.5	7 ± 2.6
1 week later	6 ± 0	7 ± 1.5
(Normal)	(4 ± 2.6)	(4 ± 1.0)

[a] Baseline pressure of 3 monkeys in the upper ureter in mmHg, mean ± 1 s tandard deviation

This study also included an evaluation of whether the ovarian vein might be responsible for hydronephrosis of pregnancy (ROBERTS, 1971). It is known that the ovarian veins dilate markedly during human pregnancy and this is also true in the monkey. Since the veins cross the ureter at the pelvic brim and since the anatomical course is different between the right and the left and the right is more commonly dilated than the left, it was hypothesized that the dilated vein might obstruct the ureter. Excision of the right ovarian vein during a pregnancy in the monkey did not however affect the ureteral dilation in this or a subsequent pregnancy.

The fetus was removed while the placenta was left intact, and ureteral diameter was followed on excretory urograms along with ureteral pressure throughout the pregnancy. When the fetus was removed ureteral diameter and pressure immediately decreased, but, late in the pregnancy ureteral pressure and ureteral diameter both rose. The ureteral diameter appeared to increase before ureteral pressure, but since excretory urograms were recorded during a different week from the ureteral pressure studies this could not be definitely determined. If ureteral diameter did increase before ureteral pressure, it might be due to a hormonal effect to alter compliance of the ureter and make it more easily dilated.

Several studies were then performed to evaluate the effect of hormones upon the ureter. A synthetic progestin melengestrol acetate was added to powdered silastic and a catalyst was then added to form the silastic into a solid tube. These tubes of silastic were implanted next to the ureter in rabbits and later the rhesus monkey. A similar amount of silastic without the active hormone was implanted next to the opposite ureter as a control and the animals followed with excretory urograms. There was no significant difference between the ureter before surgery and that after silastic implant but further no statistically significant difference between the silastic implant ureter and

Table 2. The *in-vivo* effect of implanted melengestrol aceteate on the ureter. Ureteral diameter ± I S.D. (mm)

I Non-operated ureter	0.7 ± 0.12
II Silastic implant	0.7 ± 0.34
III Silastic implant with MGA	1.0 ± 0.72
Comparing groups I and II	$P > 0.8$, $t = 0.21235$
Comparing groups II and III	$P > 0.05$, $t = 1.70097$

Table 3. The effect of implanted periureteral MGA on the ureter of Macaca mulatta

	mm	
I Control ureter	1.51 ± 0.5	$P > 0.3$
II Implant with MGA	2.26 ± 0.9	
III Change in ureteral diameter		
a. From pregnancy	0.3 mm	
b. Pregnancy plus MGA	0.4 mm	

that containing the hormone, although preliminary studies in the rabbits had suggested an effect (Table 2). The convincing evidence was that the administration of melengestrol acetate by parenteral injection at a dose expected to produce serum levels of progestins much higher than during pregnancy had no effect on the ureter of the rabbit.

Similar studies in the rhesus monkey showed no effect of the implanted MGA. The hormone was then implanted in silastic next to the upper ureter of the pregnant monkey to determine whether a combination of implanted hormone plus obstruction of the ureter by the uterus might produce more dilation of the ureter on the side of the hormone. However, the increase in diameter in the normal pregnant ureter was the same as that in the ureter next to implanted melengestrol acetate (Roberts and Auger, 1971) (Table 3).

Pure progesterone, since it has previously been implicated as a causative agent in hydronephrosis of pregnancy, has been studied more recently using techniques designed to show changes in compliance as well as in ureteral peristalsis. Progesterone was given at a dose expected to produce progesterone levels 4 to 5 times that found during pregnancy (CLAYTON and ROBERTS, 1972). Excretory urograms were performed, followed by a second series with the bladder full. It was noted that filling the bladder produced a marked increase in the size of the ureter with some constancy (Fig. 2).

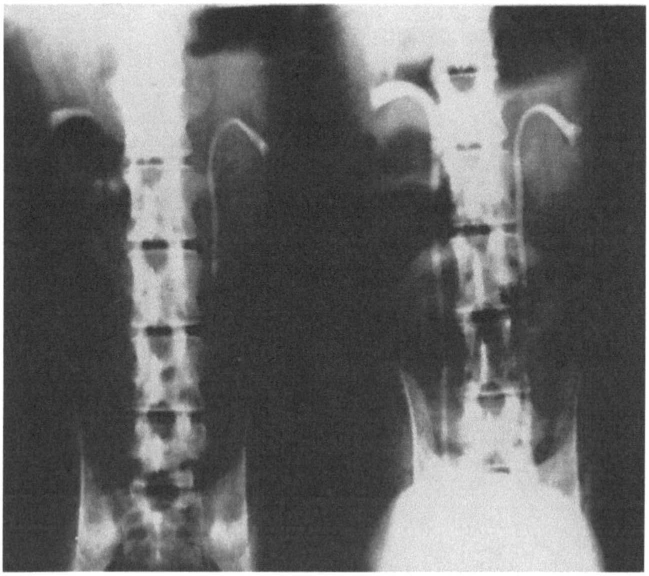

Fig. 2. Illustration of the increase in ureteral diameter from normal (on left) to that seen with bladder full (on right)

Control ureteral pressure studies were performed and all studies were repeated over the 90-day period of treatment. No change in ureteral physiology as noted on pressure studies occurred. Moreover, the diameter of the ureter on excretory urograms or on full-bladder excretory urograms did not differ from the control values. There was a definite progestational effect as noted by: the presence of an enlarged uterus (to about the size of an eight weeks pregnancy), the absence of active ovarian follicles, the presence of a bright sex skin and the absence of menstruation throughout the period of administration of the hormone. Thus, progesterone had no effect upon the ureter of the rhesus monkey when given parenterally in its pure form.

I feel that hydronephrosis of pregnancy is due to obstruction of the ureter by the gravid uterus alone. The importance of the studies lies in the fact that an understanding of hydronephrosis of pregnancy may decrease the incidence of pyelonephritis of pregnancy. This occurs in 2 to 6% of all normal pregnancies and may be one of the major initial factors in the pathogenesis of chronic pyelonephritis.

References

Clayton, J. D., Roberts, J. A.: The effect of progesterone on ureteral physiology in a primate model. J. Urol. (Baltimore) **107**, 945 – 947 (1972).

Roberts, J. A.: The ovarian vein and hydronephrosis of pregnancy: experimental studies in the rhesus monkey (*Macaca mulatta*). Invest. Urol. **8**, 610–621 (1971).

Roberts, J. A., Wolf, R. H.: Hydronephrosis of pregnancy: a disorder of primates closely resembling that of man. Folia Primat. **15**, 143–147 (1971).

Roberts, J. A., Auger, M.: The effect of a progestational agent on ureteral physiology. Invest. Urol. **9**, 206 – 210 (1971).

Smith, H.: The kidney, pp. 636–637. New York: Oxford Univ. Press 1951.

Van Wagenen, G., Jenkins, R. H.: Pyelo-ureteral dilation in successive pregnancies. J. Urol. (Baltimore) **49**, 228–235 (1943).

Histochemical Studies on the Ureter after Ligation

S. LYMBEROPOULOS, H. BREINING, and B. KLINKHAMMER*

Diagnosis and plastic reconstruction of a hydro- or megalo-ureter do not solve the main problem of our therapeutic efforts, the conservation of the kindney, at all. An irreversibly damaged ureter with no function cannot transport the urine even after surgical correction. Therefore this leads to the progression of urinary retention and to the loss of the kindney.

The experimental model of acute and chronic, complete and incomplete ureteral obstruction gives us information on the reaction of the kindney to urinary retention, and on its capacity for recovery and to regeneration after removal of the obstruction. About the ureter, in contrast, we know comparatively little. The investigations hitherto have mainly been based on the morphological changes in the wall and the pathogenesis.

In our animal experiments we used ligation of the ureter and histochemical methods of examination in an attempt to determine the moment when regeneration of the peristaltic wall function of the ureter still seems possible and when damage to the contractile elements of the ureter must be considered irreversible.

Methods

In 50 rabbits the middle segment of the left ureter was ligated. The animals were sacrificed 1 day to 12 weeks postoperatively and cryostat sections (12 μ thick) were prepared. For the sections we took ureter specimens of the obstructed as well as the contralateral non-obstructed ureter. In these sections we examined lactate dehydrogenase, alkaline phosphatase and nonspecific esterase using the methods of PEARCE. Ureters affected by inflammation were not used.

Evaluation was based on the intensity of the color reaction. We expressed the activity as low, medium and high, corresponding to the graduation of "+" to "+++" with results in between. We added the single results and calculated the arithmetic mean and deviation.

Results and Discussion

Like other authors, we were able to demonstrate that nonspecific esterase was localized mainly in the epithelial cells and in the stroma of the normal ureter while this enzyme only showed a slight reaction in the musculature. Alkaline phosphatase was

* Abteilung Urologie (Director: Prof. Dr. W. LUTZEYER) and Abteilung Pathologie (Director: Prof. Dr. J. SCHOENMACKERS) der Medizinischen Fakultät der Rheinisch-Westfälischen Technischen Hochschule, Aachen, Germany.

confined mainly to the subepithelial stroma. The lactate dehydrogenase was detected in the musculature of the ureter. Lactate dehydrogenase is an enzyme of glycolytic metabolism and sited in the mitochondria of the smooth muscle cells.

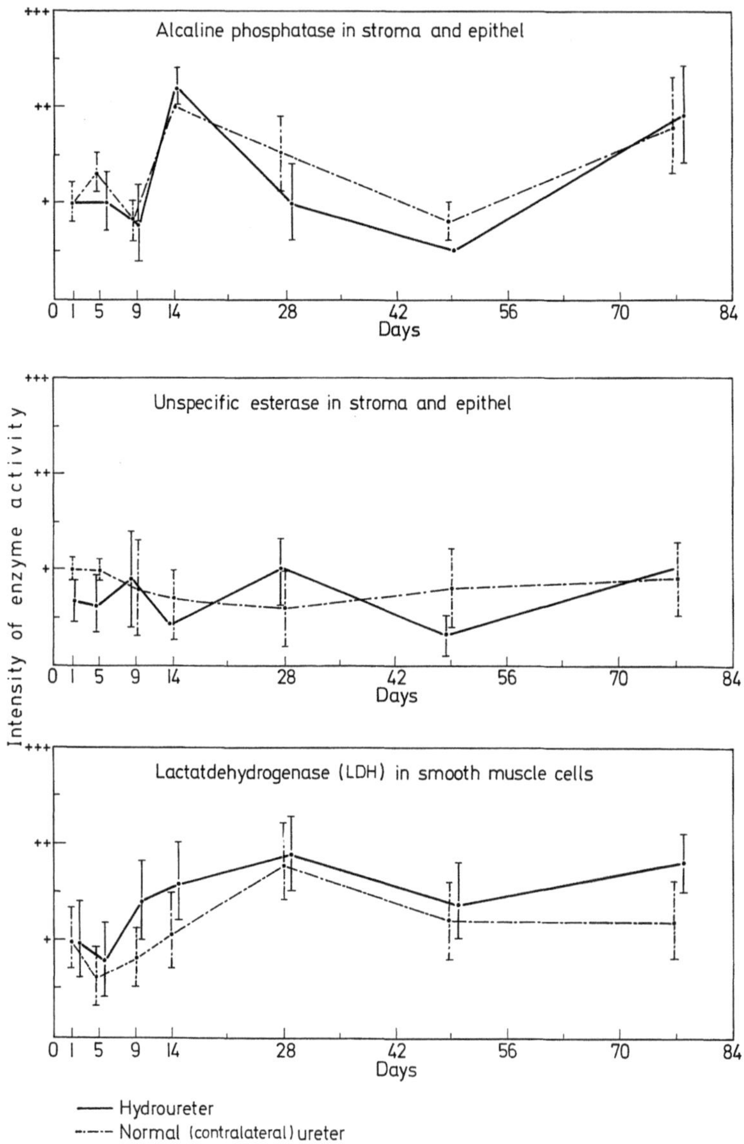

Fig. 1 graphically represents the arithmetic means and deviations for nonspecific esterase and alkaline phosphatase in the transitional epithelium and in the subephithelial stroma of the obstructed and non-obstructed contralateral ureter in relation to the time of ligation. In the case of nonspecific esterase, we found that the enzyme activity for both sides remained almost unchanged during the experiment, nor did alkaline phosphatase show a significant difference between the two sides. The relative increase of the enzymes in the second postoperative week, which we found in both ureters, could be explained by the increased water reabsorption (LESSON, MARTIN, VACED and SCHÜCK) but later results and the greater differences measured make this theory very dubious.

While alkaline phosphatase and nonspecific esterase did not support the hypothesis of histochemical changes in the contractile elements of the ureteral wall due to their mainly epithelial and subepithelial localization, examination of lactate dehydrogenase revealed interesting aspects.

In the obstructed ureter there is a pronounced increase in activity which starts on the first day and has its maximum during the second postoperative week. The same level of enzyme activity was demonstrable until the end of the experiment with only very slight differences.

We also found raised enzyme activity in the non-obstructed contralateral ureter, starting a little later, at the end of the first prostoperative week. It subsequently declined only slightly.

We suggest the following interpretation for the reaction of lactate dehydrogenase:

1. The immediate increase of activity in the obstructed ureter might be due to the initially increased functional capacity of the smooth muscle cells to overcome the mechanical obstruction. The increasing intraluminal pressure keeps the cells in an almost continuous state of irrigation. This causes an increased energy metabolism from the fourth up to the twelfth postoperative week. This enzyme activity suggests to us that the cells are still able to work and therefore to recover after twelve weeks.
2. In the non-obstructed, contralateral ureter the initial enzyme increase up to the fourth postoperative week could be interpreted as an increased functional capacity resulting from a compensatory increase of urine. The subsequent slight fall may correspond to the functional adaptation of the cells.

Summary

1. We used histochemical methods of examination to demonstrate the localization of nonspecific esterase and alkaline phosphatase in the epithelial cells and the subepithelial stroma of the ureter.
2. Because of their localization in the hydroureter neither enzyme showed any significant changes dependent on the functional degeneration of the ureter wall.
3. The initial increase in the alkaline phosphatase in the hydroureter during the second postoperative week could be due to increased water reabsorption.
4. Therefore we believe that our histochemical demonstration of the activity changes of lactate dehydrogenase in the smooth muscle fibres of the experimental hydroureter as well as the contralateral ureter has shown the functionally increased stress and the increased functional activity.
5. After twelve weeks, we still found lactate dehydrogenase activity in the hydroureter. This shows that the smooth muscle cells still maintained biological activity and were able to recover.

These results are to be considered provisonal as our experiments should be supplemented by investigations of other enzymes, especially of adenosine triphosphatase (ATPase) and of the nucleoproteids for up to one year after complete ureteral occlusion.

References

Gomori, G.: The distribution of phosphatase in normal organs and tissues. J. cell. comp. Physiol. **17**, 71 (1941).

Lesson, C. R.: Histology, histochemistry and electron microscopy of the transitional epithelium of the rat urinary bladder in response to induced physiological changes. Acta anat. (Basel) **48**, 297 (1962).

Martin, B. F.: Histological and histochemical studies on the bladder and ureter, with particular reference to alcaline phosphatase and golgi material. J. Anat. (Lond.) **92**, 286 (1958).

Pearse, A. G. E.: Histochemistry, theoretical and applied. Vol. I, Third Edition. London: J. &. A. Churchill 1968.

Vaced, Z., Schück, O.: Histology and histochemistry of the transitional epithelium of the rat bladder in response to experimental filling. Anat. Rec. **136**, 87 (1960).

Changes in the Electrophysiologic Activity of the Ureter in Irregular Renal and Ureteral Hemodynamics

N. KULOV, TZ. TZEKOV, and K. VELIKOV*

The bioelectric activity of the ureters in normal and in pathologic conditions is a widely discussed problem (BOZLER, 1941; LAPIDES, 1948; BAKER et al., 1953a, 1953b; WEINBERG, 1964; SUGAVARA, 1967).

Most authors think the ureters have a clearly expressed automatism, though there are complicated neurohumoral connections between the ureters and a number of organs and structures inside and outside the abdominal cavity. In spite of this, the ureters are dependent on a number of temporal and lasting influences. The aim of the present work was the study of the changes in the electrophysiological activity of the ureters in all kinds of irregularities of renal and ureteral hemodynamics. The organs in the human body undergo different disturbances in their organic hemodynamics, and written data on these problems are inadequate.

Materials and Methods

The experiments were performed in the central electrophysiologic laboratory of the Departement of Pathophysiology of the High Medical Institute, Varna. In 12 animals (cats) a mixture of ink and gelatine was injected into the arteries to demonstrate the separate contribution of each artery to the blood supply of the ureter. In 6 of the animals the injection was given through the renal artery and in the remaining 6 through the internal hip-artery. Histological preparations were obtained from the ureters when the intramural vessels were fully dituted. The thickness of the capillary network in the mucous membrane, i.e. the degree of vascularization yielded information on the contribution of each of the arteries mentioned to the blood supply to the ureter. It could be shown that the internal renal artery contributes considerably less to the blood supply to the ureter than the interior hip-artery.

Biophysiological tests were carried out on 40 cats to find out how a disturbed blood supply affects the bioelectric activity of the ureter. Anesthesia was induced and maintained with a 10 percent urethane solution (10 cm^3/kg). The electrical activity (E.U.G.) was recorded by a cardioscript (Schwarzer Co.). The animals were divided into 4 groups. In the first group (11 cats) one of the 2 renal arteries was ligated. In the

* Department of Propedeutics of Surgical Diseases (Prof. Dr. ANRDEEV). High Medical Institute, Varna, Rumania.

second group (10 cats) the renal vein was ligated on one side. In the third group (9 cats) the whole renal pedicle was ligated on one side, artery and vein together. In the fourth group (10 cats) one of the two internal hip-arteries was ligated at a site above the point where the artery branched for the ureter. The number of the bioelectric complexes of the four groups was measured at 30-minute intervals, as well as the average absolute quantity of the amplitudes.

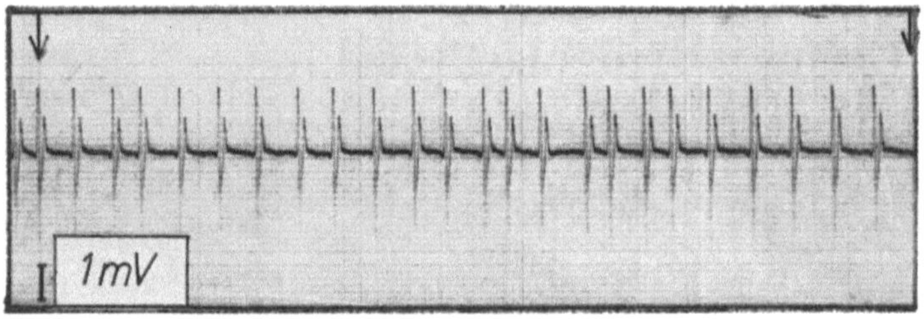

Fig. 1. E.U.G. of the ureter before ligation of the renal vein

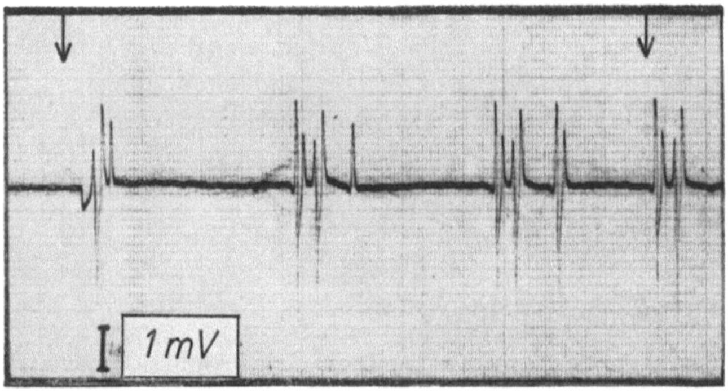

Fig. 2. E.U.G. of the ureter immediately after ligation of the renal vein

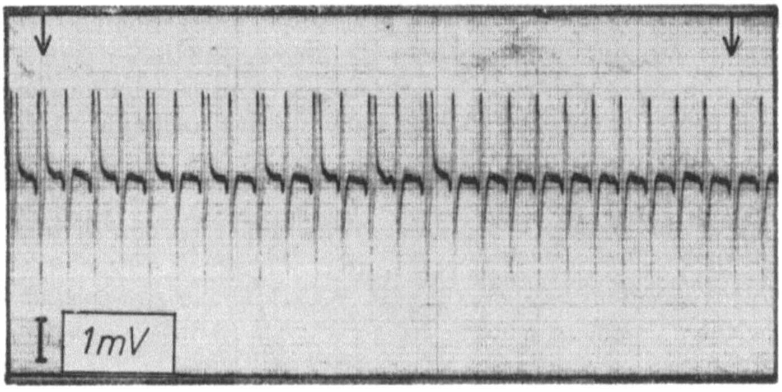

Fig. 3. E.U.G. of the contralateral ureter immediately after ligation of the renal vein of the opposite kidney

Results and Discussion

The results are given in Table 1 and Table 2.

Table 1. Average number of bioelectrical complexes in the ureter in case of disturbed haemodynamics

MEAN FREGUENCY OF BIOELECTRICAL ACTIVITY IN DISTURBED HAEMODYNAMICS													
Ligature of:	before ligature	time in minutes after the ligature 0 30 60 90 120 150 180 210 240 270 300 330 360											
ART. RENALIS	3,2	4,3	4,5	4,7	4,7	4,6	4,9	4,8	4,8	4,9	4,5	5	4,5
VENA RENALIS	2,5	3	2,6	2,5	2,4	2,3	1,6	2,4	3	2,2	2,4	2,5	2,8
PEDICULUS RENALIS	3,1	2,7	3,9	3,4	3,8	3,9	4,2	4,7	2,9	3,5	3,3	3,2	2,1
ART. ILIACA INTERNA	1,6	-	1,8	-	1,8	-	1,8	-	2	-	2	-	2,1

Table 2. Average value of the magnitude of the amplitude of the bioelectric complexes of the ureter in case of disturbed haemodynamics

MEAN VALUES OF THE AMPLITUDES OF THE COMPLEXES IN DISTURBED HAEMODYNAMICS													
Ligature of:	before ligature	time in minutes after the ligature 0 30 60 90 120 150 180 210 240 270 300 330 360											
ART. RENALIS		4320	4490	4700	4680	4580	4870	4820	4870	4870	4500	5000	4470
VENA RENALIS		3000	2600	2530	2360	2320	1570	2400	3000	2200	2420	2480	2760
PEDICULUS RENALIS		2680	3900	3360	3780	3900	4160	4660	2900	3550	3320	3250	2100
ART. ILIACA INTERNA		-	1760	-	1805	-	1850	-	1970	-	2055	-	2070

Table 1 clearly shows that occlusion of the renal artery leads to a rapid decrease in the number of bioelectric complexes of the corresponding ureter. The average absolute amplitude undergoes a gradual increase (Table 2). Following ligation of the renal vein, the changes in the bioelectric activity show the same tendency as in the case of a ligated renal pedicle. The changes in the number of the bioelectric complexes, and the average amplitude following occlusion of the internal hip-artery above the point where it branches for the ureter are not significant. The results demonstrate that ligation of the branches of the internal hip-artery, which makes the largest contribution to the blood supply of the ureter does not lead to significant changes. On the contrary, ligation of the artery which supplies mainly the pyeloureteral segment leads to rapid deviations in the bioelectric activity. We assume the renal artery supplies the vein with blood where the excitatory impulses are born and scattered. After ligation of the renal vein, we found no increase in the ureteral activity as described by BAKUNZ in 1870.

Conclusions

1. Experimental ischemia, obtained by occluding the renal artery, leads to certain changes in the bioelectric activity of the ureter but not to complete disappearance of the activity.
2. Ligation of the internal hip-artery which makes the largest contribution to the blood supply of the ureter does not lead to any significant changes in the bioelectric activity of the ureter.

Summary

Macroscopic and histological experimental findings show that the internal hip-artery relatively is the one most concerned in the blood supply of the ureter. In contrast, the ligation of the renal artery leads to substantially more changes, as could be demonstrated by recording the bioelectrical activity. It is supposed that ligation of the renal artery disrupp the physiology of the impulsogenic zone or pace-maker.

References

Baker, R. *et al.*: Ureteral electromyography. J. Urol. (Baltimore) **70**, 874–884 (1953).
Baker, K. *et al.*: Electromyography in the normal, dilated, transected and transplanted ureter. Amer. J. Physiol. **174**, 381–390 (1953).
Bozler, E.: Action potentials and conduction of excitation to muscles. Biological Symposia **3**, 95–109 (1941).
Lapides, J.: The physiology of the intact human ureter. J. Urol. (Baltimore) **59**, 501–537 (1948).
Sugavara, H.: The effect of various drugs on the human ureter electromyogram. Jap. J. Urol. **58**, 566–581 (1967).
Weinberg, S. R.: Electro-Physiology of the ureter. J. Urol. (Baltimore) **91**, 482 (1964).

The Effect of Acute and Chronic Urinary Obstruction on Renal Function

P. May*

Quantitative tests of renal function in the presence of urinary obstruction have hitherto been seriously hampered by the interference with urine flow.
However, the measurement of isotope clearance in each kidney separately with ^{131}I-hippuran now makes it possible to follow the changes in effective renal plasma flow despite urinary obstruction, as the collection of urine samples is not required.

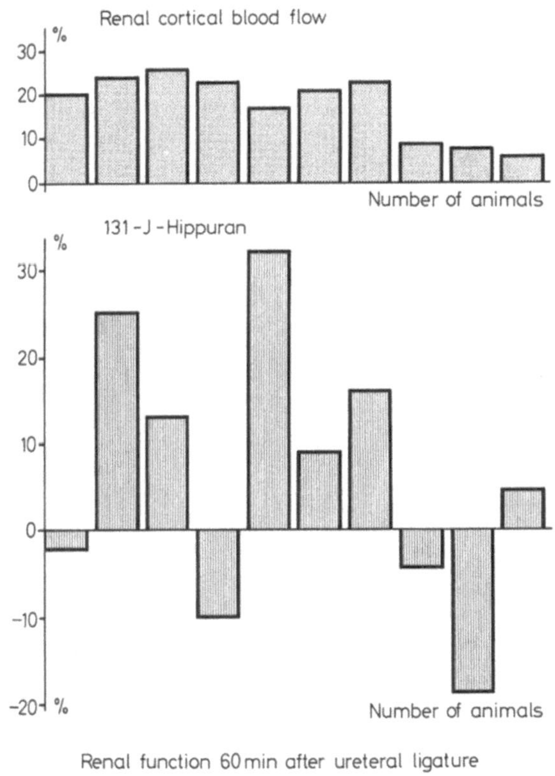

Fig. 1. Changes in effective renal plasma flow and renal cortical blood flow 60 minutes after prevesical ureteral occlusion in dogs

* Urologische Universitätsklinik (Director: Prof. Dr. C. E. Alken), Homburg/Saar, Germany

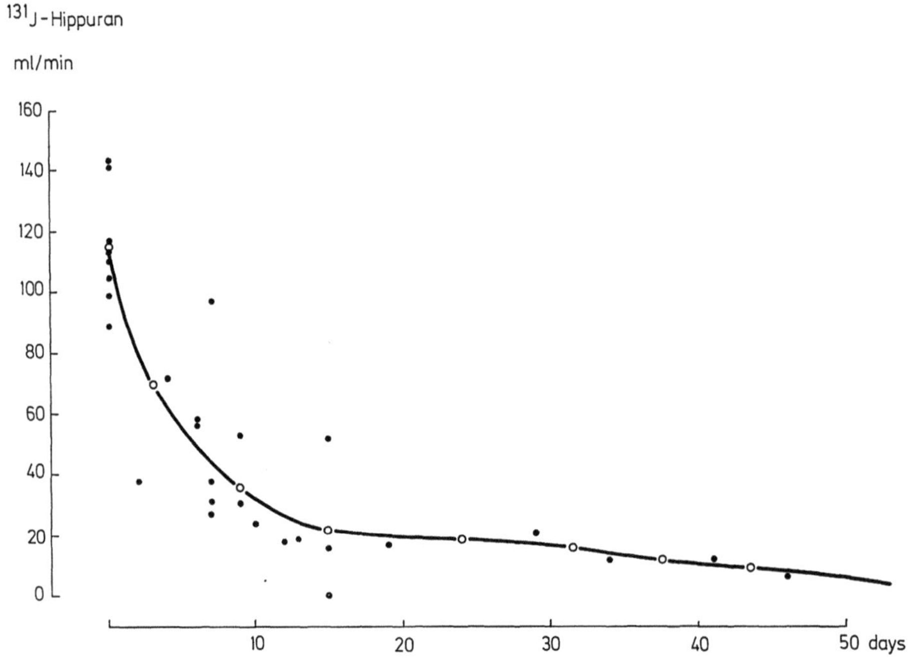

Fig. 2. Curve showing average fall in tubular secretory capacity after unilateral ureteral ligation in dogs

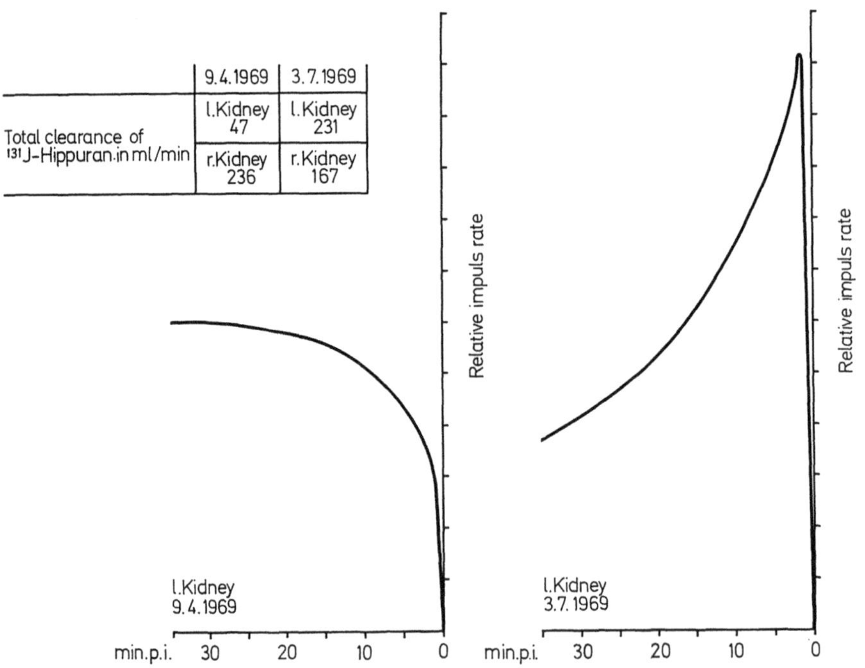

	9.4.1969	3.7.1969
Total clearance of 131J-Hippuran in ml/min	l.Kidney 47	l.Kidney 231
	r.Kidney 236	r.Kidney 167

Fig. 3. R. R., 45 years. Isotope renograms and clearance tests before and after removal of a stone obstructing the renal pelvis

This method is based on a combination of the so-called whole–body clearance with a synchronously recorded renogram.
Comparisons with clearances determined by conventional methods have shown good agreement.
A total of 126 bilateral ^{131}I-hippuran clearance estimations were carried out in 10 dogs up to 46 days after ligation of one ureter.
60 minutes after prevesical ureteric occlusion in 6 dogs the average increase in effective renal plasma flow amounted to 16% of the original value (Fig. 1). In the same experimental conditions I had previously found that acute urinary obstruction was followed by a rise in renal cortical blood flow up to a maximum of 26%. The initial rise in renal plasma flow corresponds to this. The changes in tubular secretion rate are not as consistent as the rise in blood flow. THURAU believes that this may be due to the fact that flow within the tubules stagnates because of the rise in pressure in the renal pelvis, though renal blood flow increases due to the dilatation of the afferent arterioles.
Fig. 2 shows the average reduction in tubular secretion which follows ligation of the ureter. At the beginning there is a steep fall and after 3 weeks the residual capacity is approximately 20% of the initial value. After a further 3 weeks it is no longer possible to detect any measurable readings over the obstructed kidney.
Many workers have reported that chronic urinary obstruction is followed by a reduction in renal blood flow and in tubular secretion. Though there duction in blood flow may progress more slowly at first, after 5 weeks both values are extremely low. The present pathological observations appear to confirm this.
46 days after ureteric ligation the main features are intimal oedema in the small arteries, separation of the endothelial layer and disaggregation of the muscle.
On the basis of various independent investigations in dogs, HOLDER and WIDEN postulate that the period of complete ureteric occlusion beyond which no functional recovery is possible is approximately 40 days.
The technique of bilateral measurement of effective renal plasma flow with ^{131}I-hippuran has been used for routine diagnostic purposes in this hospital for over 2 years. I can therefore show you clinical examples of the changes in tubular secretory capacity caused by urinary obstruction, together with the corresponding radiological findings.
A patient had an obstruction caused by a stone in the left renal pelvis. The pyelogram showed only a faint shadow. Attacks of colic began 18 days before admission to hospital. Three months after removal of the stone another pyelogram showed prompt excretion of dye and normal outflow (Fig. 3).
The clearance, though at first seriously impaired – 47 ml – had also returned to normal.
This sequence of events is familiar to urologists and shows that after complete ureteric occlusion the tubular secretion rate in the human kidney falls rapidly, as it does in experimental animals. However, from extensive clinical experience we have learned that more or less complete recovery can be expected if the obstruction is removed within about 30 days.
When obstruction to urine outflow is incomplete the decline in tubular secretory capacity seems to be a good deal slower, providing secondary infection does not occur.
We have now measured effective renal plasma flow before and after successful plastic repair in some 60 patients with incomplete ureteric stenosis below the brim or just above the bladder.

Fig. 4 shows clearance measurements before and after pyeloplasty in patients aged from 4 to 56 years.
There was no definite evidence of any correlation between age and residual function or between age and recovery of renal functional after relief of the obstruction. However, there was still surprisingly good secretory capacity except in a few cases. The patients with values over 300 ml/min each had only one kidney. The difference be-

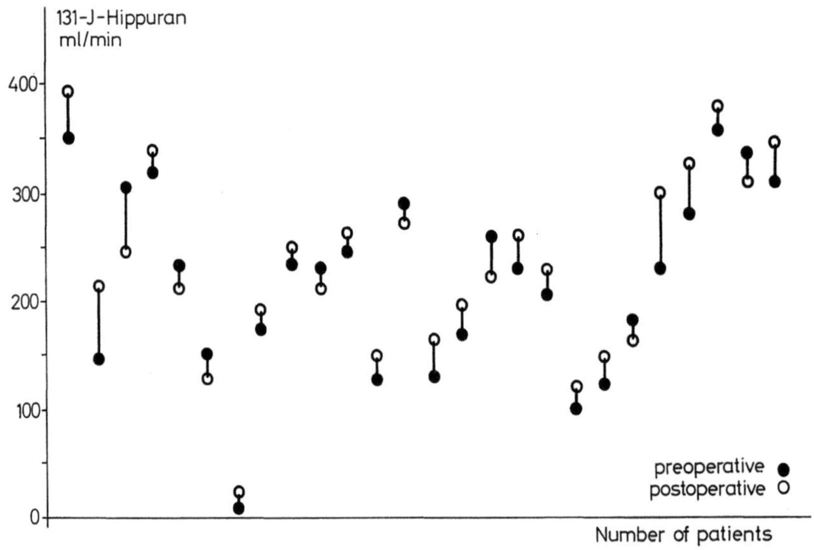

Fig. 4. Unilateral clearance measurements before and after pyeloplasty in patients of ages 4 to 56 years

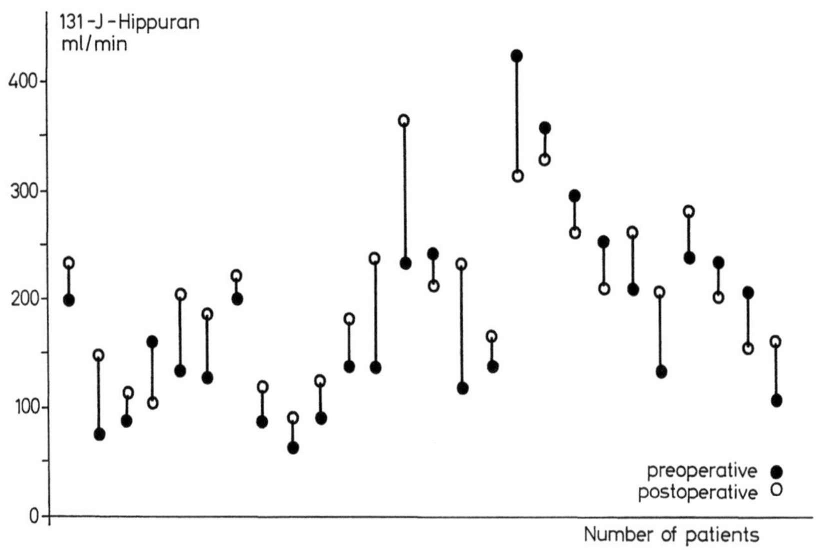

Fig. 5. Clearances before and after reimplantation of the ureter for congenital ureteral stenosis with or without reflux

tween the reading obtained before and after surgical removal of the obstruction shows that an operation which may appear successful when judged by morphological criteria will not necessarily succeed in improving residual tubular function; in cases of chronic obstruction, as a general rule, the lower the initial secretory capacity, the poorer are the changes of recovery.

Comparison of our clearance results before and after reimplantation of the ureter for congenital ureteric stenosis with or without presents a similar picture (Fig. 5). If we take the lower limit of normal clearance as roughly 250 ml/min, we see that even under these conditions half the patients had normal values even before operation. On the basis of our follow-up checks, we do not normally regard a kidney with a hippuran clearance below 80 ml/min due to chronic urinary obstruction as worth preserving, providing the contralateral kidney is healthy. The decision to advise nephrectomy should not, however, be based on the clearance alone; the pyelogram must also be taken into account, and if necessary the angiographic and operative findings.

Numerous clinical cases have taught us that tubular secretory capacity may be surprisingly well preserved despite advanced urinary stasis, if the obstruction to urinary flow is incomplete.

Prolonged infusion urography with films taken over up to 48 hours, perhaps with the use of Conray, often reveals that ureteric stenosis at first thought to be complete is in fact incomplete.

In patients with extreme urinary stasis and still relatively well preserved secondary capacity the surgeon has to decide during the operation whether is it worth while attempting to save the kidney by means of a plastic operation on the dilated renal pelvis, which is sometimes fibrotic and scarred.

In borderline cases of this kind arteriography is of special value because by comparing the renal arteriogram with the isotope clearance it is possible to obtain further information on the condition of the kidney.

Comparative measurements have shown that the parenchyma of nonfunctioning kidneys from patients with urinary stasis, though extremely thinned and apparently stretched out over the dilated renal pelvis and calyces, nevertheless still has a total weight comparable with that of the normal renal parenchyma.

In summary, it may be said that all the animal experiments showed a short-lasting rise in blood flow after urinary obstruction. On the other hand, the effect of one hour's acute ureteral obstruction on effective renal plasma flow is inconstant, an observation possibly attributable to the increasing stagnation of tubular flow which follows urinary obstruction.

^{131}I-hippuran clearance measurements on each kidney in dogs after ligation of one ureter show that residual function persists for only about three weeks after complete urinary obstruction, and that after about 40 days there is no longer any measurable tubular secretion.

Follow-up isotope clearance measurements on each kidney after the injection of ^{131}I-hippuran in some 60 patients with congenital ureteric stenosis below the pelvic brim or just above the bladder before and after plastic operations have confirmed that in the presence of incomplete urinary obstruction renal function may persist for long periods. Under these conditions, it appears that an equilibrium is established between the remaining secretory capacity and the rate at which urine can be removed, either through the stenosed ureteral segment or via the so-called backflow mechanisms.

References

BARTHA, J., HARZA, T., HARSING, L.: Die intrarenale Verteilung der durch die Niere strömenden Blutmenge in osmotischer Diurese und unter Stopflow-Bedingungen. Arch. ges. Physiol. **288**, 315 (1966).

DROM, G.: Pathologie der sekundären Abflußstörungen im Bereich der Niere und der oberen Harnwege. 86. Tag. dtsch. Ges. Chir. (1969).

ERD, W., FRITZLER, W., HÖFER, R., OEL, H. J., PFEIFFER, G.: Vergleichende Untersuchungen der renalen Clearance mit Inulin, PAH, Cr.51-EDTA und J-125-Hippuran. Wien. Z. inn Med. **50**, 106 (1969).

HARSING, L., SZANTO, G., BARTHA, J.: Renal circulation during stop flow in the dog. Amer. J. Physiol. **213**, 935 (1967).

HOLDER, E.: Die mechanische Hydronephrose und ihre Fähigkeit zur Rückbildung im Experiment. Erg. Chir. und Orthop. **40**, 266 (1956).

IDBOHRN, H., MUREN, A.: Renal blood flow in experimental hydronephrosis. Acta physiol. scand. **38**, 200 (1956).

KERR, W. S.: Effect of complete ureteral obstruction for one week on kidney function. J. appl. Physiol. **6**, 762 (1954).

KIIL, F., AUKLAND, K.: Renal concentration mechanism and hemodynamics at increased ureteral pressure during osmotic and saline diuresis. Scand. J. clin. Lab. Invest. **13**, 276 (1961).

KIRSCH, W., MAY, P., OBERHAUSEN, E.: Bestimmung des effektiven Plasmastromes der Einzelniere mit 131-J-Hippuran – Tierexperimentelle und klinische Ergebnisse. Urologe **9**, 135 (1970).

MAATZ, R., KRÜGER, E.: Das Verhalten der Nierendurchblutung in der experimentellen Hydronephrose. Z. Urol. **31**, 736 (1967).

MALVIN, R. L., WILDE, W. S., SULLIVAN, L. P.: Localisation of nephron transport by stop flow analysis. Amer. J. Physiol. **194**, 135 (1958).

MAY, P.: Lokale Nierenrindendurchblutung und Nierenfunktion bei akuter und chronischer Harnstauung. (Habilitationsschrift) Homburg 1969.

MAY, P., LAUBENBERGER, TH., OBERHAUSEN, E.: Renovasographie und Radioisotopendiagnostik bei Harnstauungsnieren. Urologe **7**, 360 (1968).

MAY, P., SÖKELAND, J., SCHMIDT, P.: Zur Erholungsfähigkeit der Harnstauungsniere. Internist **9**, 174 (1968).

MOTZKUS, F., BACHMANN, D. Entstehung und Deutung des Radio-Renogramms. Nuclearmed. **7**, 46 (1968).

MURPHY, G. P., SCOTT, W. W.: The renal hemodynamic response to acute and chronic ureteral occlusion. J. Urol. (Baltimore) **95**, 636 (1968).

NASH, F. D., SELKURT, E. E.: Effects of elevated ureteral pressure on renal blood flow. Circ. Res. (Suppl. 1) **14**, 142 (1964).

OBERHAUSEN, E.: Bestimmung der Nierenclearance durch externe Gammastrahlenmessung. 5. Jahrestag. Ges. Nuclearmed. Wien (1967).

OESTER, A., MADSEN, P.: Determination of the renal plasma flow and tubular mass by means of radioisotopes. Invest. Urol. **6**, 322 (1968).

RAM, M. D., EVANS, K., CHISHOLM, G. D.: A single injection method for measurement of effective renal plasma flow. Brit. J. Urol. **40**, 425 (1968).

RAZZAK, M. A., BOTTI, R. E., MACINTYRE, W. J., PRITCHARD, W. H.: Determination of renal blood flow by extrarenal monitoring of radiohippuran disappearance. J. Urol. (Baltimore) **100**, 209 (1968).

SCHIRMEISTER, J., SCHMIDT, L., SÖLING, H. D.: Die renale Extraction verschiedener Clearance-Substanzen beim Hund während maximal erhöhtem Ureterdruck. Arch. exp. Path. u. Pharmakol. **237**, 473 (1959).

Congenital Ureteral Strictures

T. D. Allen*

There are a number of diseases, evidently congenital in origin, which produce a localized narrowing in the ureter with obstruction to the passage of urine yet which strangely offer little or no resistance to the passage of catheters or probes. Depending upon their location, they are known by any of a wide variety of names such as ureteropelvic junction obstruction (Shopfner), congenital hydronephrosis (Murnaghan), functional stenosis (Bäcklund), achalasia of the ureter (Hepler), megaureter (Williams), atonic distal segment (Creevy), megaloureter (Caulk), and congenital ureteral strictures (Cambell), but seldom have they been considered together as a single entity. As this study proposes to show however there are valid reasons for believing that they are all examples of the same basic pathological process. For the purposes of this report, they will be referred to as congenital ureteral strictures regardless of their location.

Little is known of the basic pathogenesis and histopathology of congenital ureteral strictures and what has been reported has been largely conflicting. An association with overlying vessels was noted early in the case of obstructions at the ureteropelvic junction (Young, H. H.) and more recently in instances of lower ureteral strictures (Young, J. D. Jr.), but it appears that the more basic lesion is to be found intrinsically within the ureter itself. Williams attributes the obstruction to a collarette of hypertrophied circular muscle fibers and Tanagho has demonstrated such fibers on cross sections of the ureter, but Murnaghan has reported a preponderance of longitudinal fibers with a decrease in muscle bulk at the site of the stricture, a finding which has been largely confirmed by MacKinnon et al. To cloud the issue further, Swenson proposed that these segments might be functionally obstructive because of absent autonomic ganglia while Shopfner denied the existence of primary ureteropelvic junction disease altogether, attributing it instead to the effects of vesicoureteral reflux.

Some insight into the nature of congenital ureteral strictures might be gained by a review of the microanatomy of the normal ureter if one assumes the former to be an altered development of the latter, but unfortunately this information appears equally unsettled. Satani stated that the renal pelvis consisted chiefly of a circular coat of smooth muscle, that an inner longitudinal layer became evident in the upper ureter, accompanied by an outer longitudinal layer in the mid ureter, and that finally the middle

---

* Division of Urology, University of Texas Southwestern Medical School, Dallas, Tex., USA

circular layer disappeared in the lower ureter at which point the two longitudinal coats fused into one. KOELLIKER, however described the ureter as having an inner circular and an outer longitudinal layer and HENLE described an inner longitudinal and an outer circular layer. DISSELHORST felt that the muscle bundles intertwined and ran in all directions while SCHNEIDER concluded that the muscle fibers began as an outer longitudinal layer which, as they descended, passed internally to follow a more circular course then finally terminated as an inner layer once more directed longitudinally.

From these introductory remarks it is evident that not only is the basic nature of congenital ureteral strictures uncertain, but there exists considerable confusion as to the structure of the normal ureter as well. Therefore the following study (ALLEN) was undertaken to seek clarification of these points through microdissection and longitudinal histological section of normal and abnormal specimens, supplementing this information with a clinical analysis of 95 patients with congenital ureteral strictures of various types.

Materials and Methods

Thirty normal collecting systems from the minor calyces to the bladder were obtained from patients dying of non-urological disease. These were formalin fixed and tubular segments were taken from 1. the infundibulo-calyceal area, 2. the ureteropelvic area including a large segment of renal pelvis and upper ureter, 3. the mid ureter, and 4. the lower ureter including the intravesical segment. Each was split longitudinally and pinned out flat without tension and subjected to microdissection, histological examination or both. Microdissection was carried out under a dissecting microscope using fine tweezers and picks; histological examination was of longitudinal sections stained with Masson's trichrome stain.

In addition, the involved ureteral segment was obtained from 27 patients undergoing surgery because of clinically significant congenital ureteral strictures. Twenty-nine such segments were obtained, including 5 from the lower ureter, 1 from the mid ureter and 23 from the ureteropelvic area, 8 of which were also associated with overlying accessory renal vessels. Only tubular segments containing adequate tissue for examination were accepted in this study. Each was fixed in formalin, split longitudinally, and pinned out flat without tension, then subjected to microdissection, longitudinal histological section or both, using the techniques previously described.

Finally, the records of all patients with congenital ureteral strictures seen at Parkland Memorial Hospital, Baylor University Medical Center, and Children's Medical Center over the past several years were reviewed. There were 95 individuals who had undergone surgical procedures yielding adequate anatomical and pathological data for evaluation. The efforts of some 21 surgeons were represented by the series which included 85 ureteropelvic junction obstructions, 7 lower ureteral strictures and 3 strictures of the mid ureter.

Results

Normal ureter : The study showed that contrary to widely accepted opinion, the ureter is composed of but a single layer of smooth muscle whose fibers, while intertwining extensively, are nonetheless all oriented predominantly in a longitudinal direction

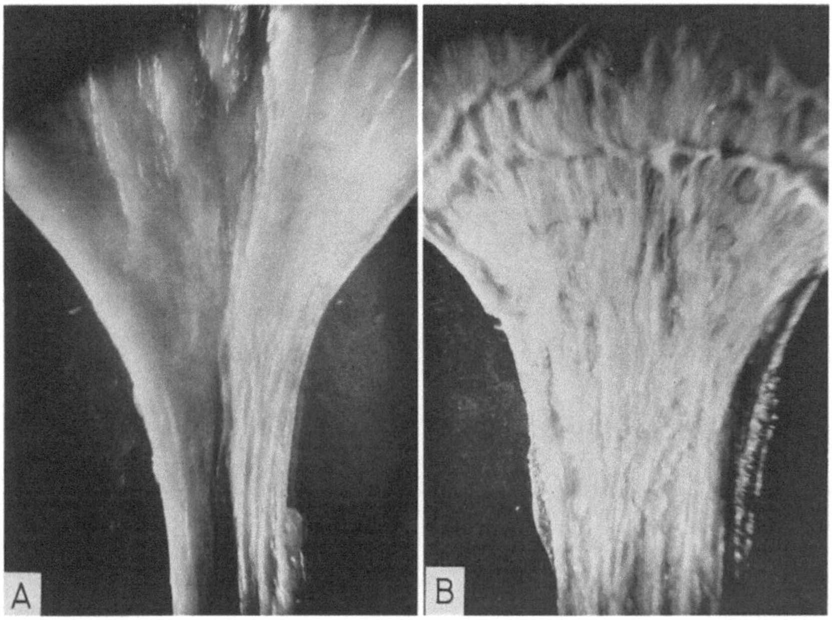

Fig. 1. Microdissection of normal ureteropelvic area. A. Segment opened longitudinally and pinned out flat. B. Epithelium removed to expose muscular layer showing predominantly longitudinal orientation of fibers.
(Reproduced from The Journal of Urology [Allen] by permission of The Williams and Wilkins Co.)

(Fig. 1). Beginning at the minor calyces (Fig. 2A), these fibers are seen as wispy filaments within a collagenous matrix, but as they descent into the pelvis and upper ureter (Fig. 2B), they increase in bulk at the expense of collagen until in the middle and lower ureter (Fig. 2C), they become quite prominent indeed. In the intramural ureter (Fig. 2D) the smooth muscle bundles are so rigidly longitudinal in their orientation that individual fibers can be stripped out with ease for a considerable distance.

It appeared then that the musculature of the ureter was basically a continuous matte of interwoven longitudinally directed fibers flowing imperceptibly from one part of the collecting system to the next, increasing in bulk as it descended and terminating ultimately in the parallel longitudinal fibers of the intramural ureter (Fig. 3). At no point were separate muscular layers identified nor were any circular fibers or sphincters encountered in any of the specimens.

Congenital ureteral strictures: Congenital ureteral strictures, regardless of their location in the collecting system were all quite similar histologically and not appreciably altered in a qualitative sense from the normal, i.e. they exhibited the same single layer of longitudinally oriented smooth muscle fibers as had been noted in the normal collecting system (Fig. 4). Quantitatively however, there were important differences. In the first place, despite considerable variation in diameter from one specimen to another, they all tended to be narrowed when compared to the adjacent normal ureter. This was often more striking when the specimen was opened than it appeared from external examination. Secondly, the musculature appeared deficient within the strictured area,

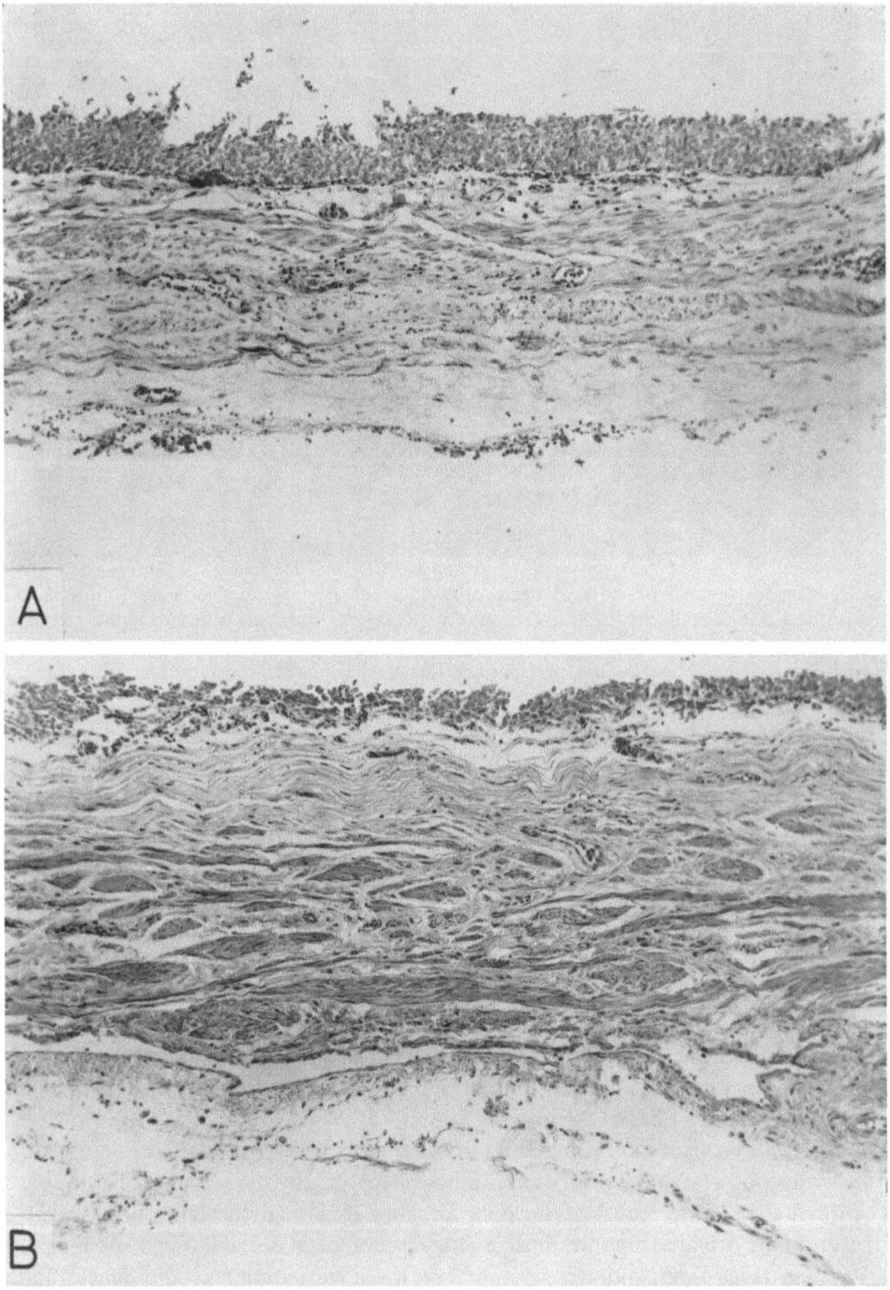

Fig. 2 A–B. Longitudinal sections through the normal ureter at different levels. A Section through the infundibulocalyceal area showing sparsity of smooth muscle. B Section through the ureteropelvic area. Smooth muscle is now more prominent

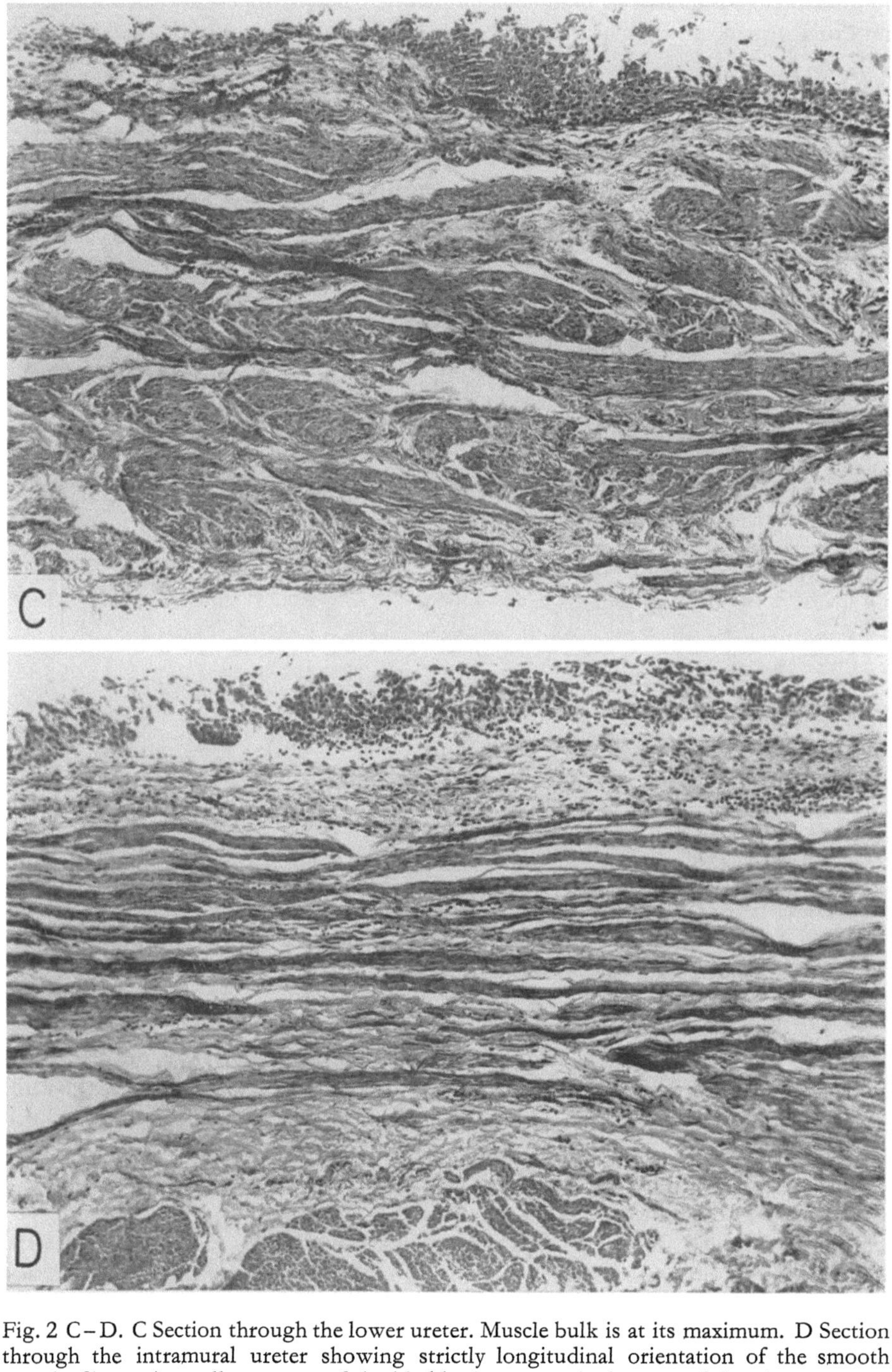

Fig. 2 C–D. C Section through the lower ureter. Muscle bulk is at its maximum. D Section through the intramural ureter showing strictly longitudinal orientation of the smooth muscle fibers. A small segment of the bladder is seen at the bottom of the picture. All pictures X 125 (Reproduced from The Journal of Urology [Allen] by permission of The Williams and Wilkens Co.)

both in total and relative amounts (Fig. 5), while collagen appeared concommitantly increased. Again, no circular fibers or sphincters were seen in any of the specimens (Fig. 6).

Proximal to the stricture, the collecting system was dilated and its hypertrophied muscle fibers, while still oriented fundamentally in a longitudinal direction, were now stretched laterally so that they tended to approach the stricture more obliquely (Fig. 7), It is easy to see that cross sections through the ureter at this point could be misinterpreted as circular fibers, a fact which may account for the reports suggesting such an arrangement (Fig. 8).

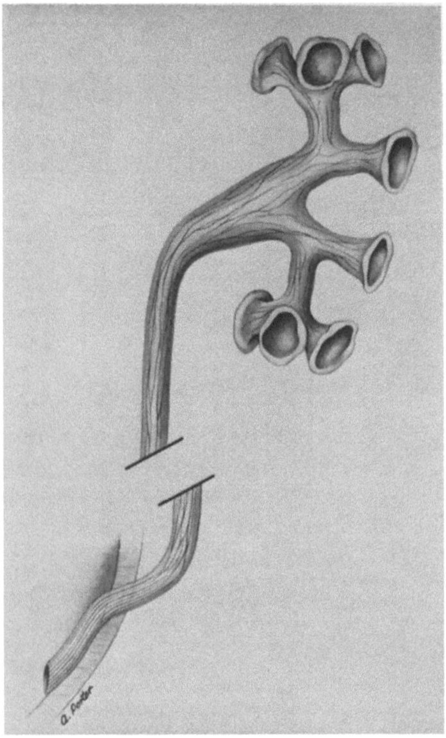

Fig. 3. Diagram of the arrangement of the smooth muscle of the normal collecting system indicating that it exists as a continuous sheet of intertwining longitudinally oriented fibers with rigidly parallel fibers in the intramural segment.
(Reproduced from The Journal of Urology [ALLEN] by permission of The Williams and Wilkins Co.)

Strictures of the ureteropelvic area were of particular interest because of the high incidence of associated aberrant renal vessels crossing the ureter in the region of the stricture. An effort to correlate the stricture with the overlying vessel suggested that the more severe the stricture intrinsically, the less likely there was to be an associated vessel and vica-versa, but there was too much overlap to permit considering the two processes as distinct and unrelated phenomena.

Clinical observations: From a clinical analysis of 95 patients with congenital ureteral strictures it was apparent that the disease afflicts the ureter at one of three anatomical sites: The ureteropelvic area, the mid ureter at the pelvic brim, and the lower ureter at or just proximal to the ureterovesical area. Other regions of the ureter seem to remain remarkably free from such a complication. Accessory renal vessels to the lower pole of the kidney were found in 35% of the patients with strictures of the ureteropelvic junction and these generally passed anterior to the ureter as they coursed to the kidney. Again, it was noted that such vessels were usually associated with the milder strictures, and that severe intrinsic strictures of the ureteropelvic area only rarely exhibited this association.

It was characteristic that the urinary tract proximal to the stricture was dilated more than it was elongated so that it appeared somewhat tubular in outline. The kidney itself, particularly in instances of strictures of the lower ureter often seemed relatively unaffected. Repeatedly the comment was heard that despite good peristaltic waves, little urine passed through the stricture even when it was detached distally; yet instruments of sizable caliber could be inserted through the involved segment without apparent obstruction.

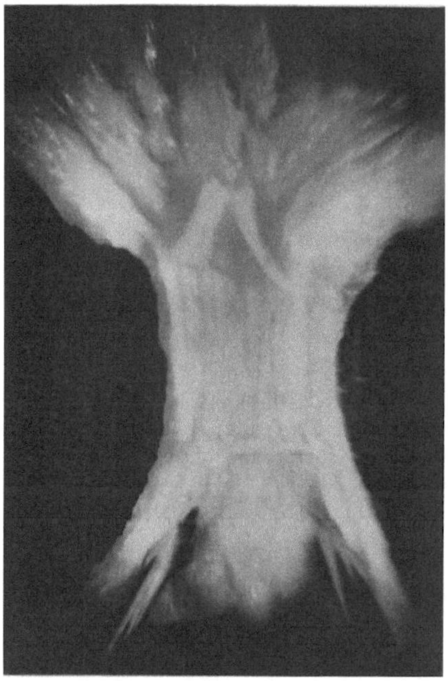

Fig. 4. Microdissection of a ureteropelvic junction stricture. Epithelium removed to show underlying muscle to be longitudinally oriented

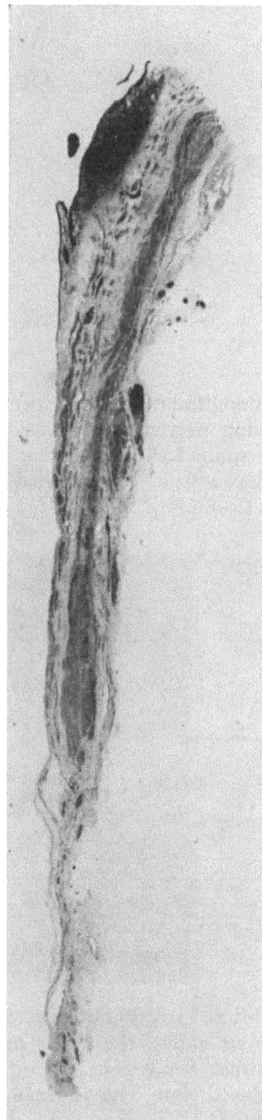

Fig. 5. Longitudinal section through one wall of a lower ureteral stricture showing almost complete absence of smooth muscle distally at the site of the stricture. X8

Discussion

From this study it is evident that the muscular wall of the normal collecting system is not divided into distinct layers but exists rather as a single-layered matte-like network of smooth muscle whose individual fibers are oriented primarily in a longitudinal direction. This muscular sheet flows without interruption from the minor calyces to

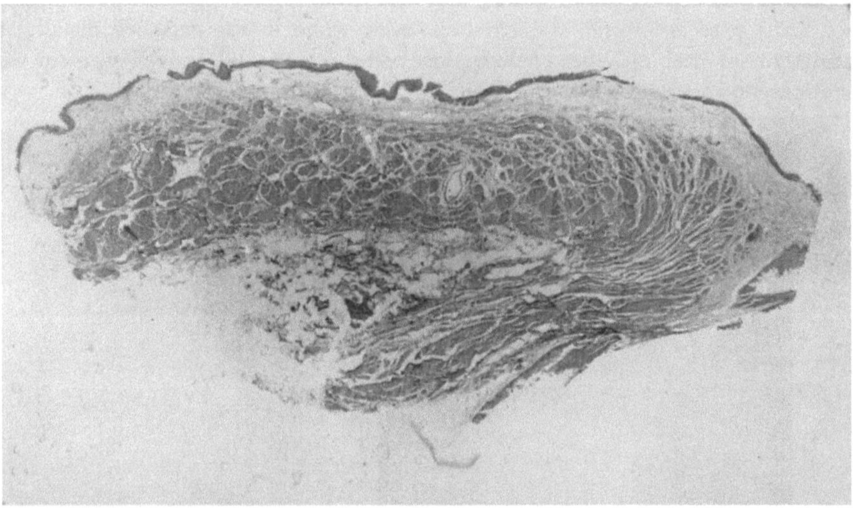

Fig. 6. Longitudinal section through one wall of a ureteropelvic stricture showing the pelvis lying transversely above with the ureter extending obliquely below. No circular fibers or sphincters are apparent in the section. X 20.
(Reproduced from The Journal of Urology [Allen] by permission of The Williams and Wilkens Co.)

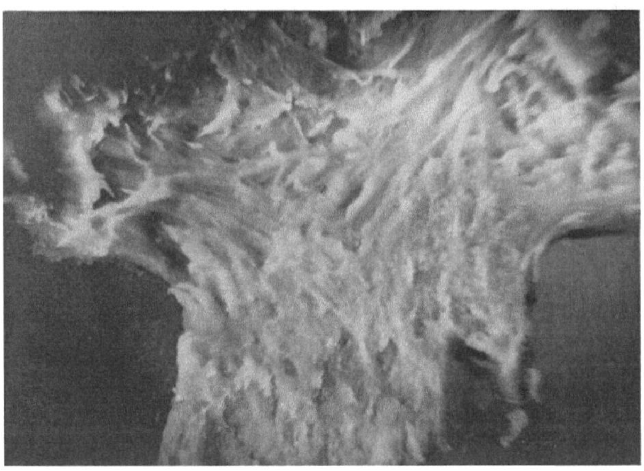

Fig. 7. Microdissection of a ureteropelvic junction stricture. Because of the dilatation of the renal pelvis above, the pelvic muscle fibers approach the upper ureter at a greater angle than normal. See Fig. 8.
(Reproduced from The Journal or Urology [Allen] by permission of The Williams and Wilkens Co.)

the lower ureter, its fibers intertwining and spiraling lazily in the process until the bladder wall is reached, whereupon they become strictly longitudinal in orientation and remain so throughout the length of the intravesical ureter. Smooth muscle fibers are rather sparse in the minor calyces and infundibula suggesting that peristaltic contractions in this area may be relatively inefficient, an impression which compares favorably with the conclusions reached by BECKER and POLLACK during cinefluorographic studies. The increase in muscle bulk as the ureter descends suggests a progressive rise in intraluminal pressure as the more distal segments of the ureter are approached and this impression too is consistent with established physiological data (KIIL). The

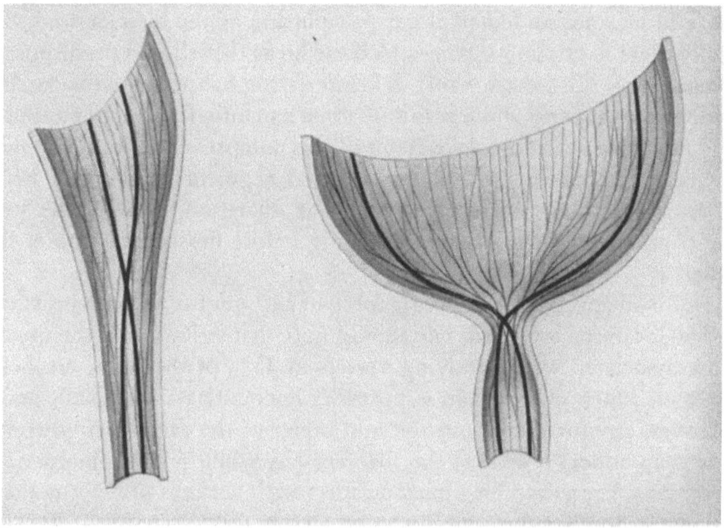

Fig. 8. Diagram illustrating the effect of dilatation of the renal pelvis on the orientation of the longitudinal muscle fibers. Compare with Fig. 1b and Fig. 7.
(Reproduced from The Journal of Urology [ALLEN] by permission of The Williams and Wilkens Co.)

absence of a circular layer of smooth muscle seems to preclude the possibility that the peristaltic wave is propagated by the progressive constriction of annular muscular rings as it is in the bowel, pushing its contents ahead of it. Instead it appears that as each segment is distended it contracts, shortens and draws itself over the bolus of urine, advancing it much as one might advance his foot in a sock by pulling the sock up over it. Evidently there is sufficient spiralling of the muscle fibers to occlude the ureteral lumen after the bolus has passed, thus assuring a unidirectional movement of the fluid.

The histological picture presented by congenital ureteral strictures does not represent a radical departure from this basic pattern but rather the changes encountered are those of quantitative alterations from the normal. It is as though the segment suffered from a maturation arrest in which not only is the intraluminal diameter reduced but there is also an absolute and relative deficiency of smooth muscle with a corresponding increase in collagen. From a physiological standpoint their propulsive efficiency might be impaired on this basis alone although it is possible that the reduced muscle to collagen ratio could mean an inadequate muscle to muscle contact with resultant defective myogenic conduction (NOTLEY). MurnagHAN has shown that these segments are able to transmit a peristaltic wave when the flow rate is low, but that they decompensate when the load surpasses certain critical levels. In any event, the picture is one of an effective, even if relative, mechanical obstruction with dilatation and hypertrophy of the ureter proximally. Nothing was found in the study to suggest that there might be any merit in seeking an explanation for the obstruction in some elusive dyskinetic neuromuscular phenomena.

What is particularly intriguing about these strictures is the fact that they seem to be so sharply localized to three areas of the ureter; the ureteropelvic area, the mid ureter at the pelvic brim, and the lower ureter just outside of the bladder wall. Since embryologically there is nothing intrinsic to these areas that should predispose them to such a complication (OSATHANONDH), it is interesting to speculate that perhaps their failure to develop properly is the result of some extrinsic force applied during a critical period in muscularization. As a result of this compression, both ureteral growth and muscular development within the affected segment might well be retarded (MEHTA). In addition, the characteristic tubular dilatation of the ureter proximally could be a consequence of dilatation occurring before muscularization of the ureter was complete.

In searching for an appropriate extrinsic force to account for this compression as well as the specific locations involved, one should note that strictures of the ureteropelvic junction are associated with overlying vessels in 35% of the cases, an incidence of aberrant vessels much higher than is normally encountered (MERKLIN and NICHOLAS). Moreover, strictures occur in the mid ureter at the exact part where they pass over (and rarely under) (CORBUS) the iliac vessels, while similar obstructions in the lower ureter have been described in association with vestiges of the umbilical artery. This relationship appears too striking to be merely fortuitous and it does not seem illogical therefore to propose that these strictures do, in fact, represent ureteral "blight" resulting from pressure by fetal vessels during a critical period in ureteral development. Nor does it seem imperative that these vessels should be identifiable at the time of operation. To effect muscularization, the extrinsic compression must occur relatively early in embryonic life (at about the 12th week) (CUSSEN) and subsequent atrophy with disapparance of the offending vessel is not unreasonable in view of the rapidly changing events of intra-uterine life. The fact that the more severe strictures are less likely and the least severe more likely, to be associated with overlying vessels, could indicate that the former represents an earlier insult with a more profound effect upon the muscularization process.

Summary

To characterize the pattern of musculature in the normal ureter and the manner in which this pattern is altered in congenital ureteral strictures, 30 normal collecting systems and 29 congenital ureteral strictures from the ureteropelvic area, the mid ureters at the pelvic brim, and the ureterovesical area were subjected to microdissection and longitudinal histological section. It was determined that the normal ureter is composed of but a single layer of smooth muscle whose fibers, though intertwining extensively, are all oriented predominantly in a longitudinal direction.

Congenital ureteral strictures were similar qualtitatively, but were characterized by a reduction in intraluminal diameter and an absolute and relative deficiency of smooth muscle fibers with a corresponding increase in collagen within the involved segment. The picture was suggestive of a localized area of developmental arrest, possibly brought about by extrinsic compression by fetal vessels during intrauterine life. In any event, the obstruction which results appears to be purely mechanical in nature rather than secondary to some neuromuscular dyskinetic phenomena.

References

Allen, T. D.: Congenital Ureteral Strictures. J. Urol. (Baltimore) **104**, 196–204 (1970).

Bäcklund, L., Grotte, G., Reuterskiöld, A.: Functional Stenosis as a Cause of Pelvi-ureteric Obstruction and Hydronephrosis. Arch. Dis. Childh. **40**, 203–206 (1965).

Becker, J. A., Pollack, H.: Cinefluorographic Studies of the Normal Upper Urinary Tract. Radiology **84**, 886–893 (1965).

Campbell, M. F.: The Dilated Ureter in Children. Amer. J. Surg. **39**, 438–451 (1938).

Caulk, J. R.: Megaloureter. The Importance of the Ureterovesical Valve. J. Urol. (Baltimore) **9**, 315–330 (1923).

Corbus, B. C., Estrem, Robert D., Hunt, W.: Retroiliac Ureter. J. Urol. (Baltimore) **84** 67–68 (1960).

Creevy, C. D.: The Atonic Distal Ureteral Segment (Ureteral Achalasia). J. Urol. (Baltimore) **97**, 457–463 (1967).

Cussen, L. J.: The Structure of the Normal Ureter in Infancy and Childhood. Invest. Urol. **5**, 179–194 (1967).

Disselhorst: Cited by Schneider.

Henle: Cited by Schneider.

Hepler, B.: Non Obstructive Dilatations of Upper Urinary Tract in Children. J. Amer. med. Ass. **109**, 1602–1606 (1937).

Kiil, F.: The Function of the Ureter and Renal Pelvis, Philadelphia: W. B. Saunders Co. 1957.

Koelliker, A.: Cited by Schneider.

MacKinnon, K. J., Foote, J. W., Wiglesworth, F. W., Blennerhassett, J. B.: The Pathology of the Adynamic Distal Ureteral Segment. J. Urol. (Baltimore) **103**, 134–137 (1970).

Metha, H. J.: Development of the Ureter. In: Bergman, Harry (Ed.): The Ureter, pp. 1–21. New York: Hoeber Medical Division, Harper and Row Publishers Inc. 1967.

Merklin, R. J., Michels, N. A.: The Variant Renal and Suprarenal Blood Supply with Data on the Inferior Phrenic, Ureteral, and Gonadal Arteries. J. int. Coll. Surg. **29**, 41–76 (1958).

Murnaghan, G. F.: The Mechanisms of Congenital Hydronephrosis with Reference to the Factors Influencing Surgical Treatment. Ann. roy. Coll. Surg. Engl. **23**, 25–46 (1958).

Notley, R. G.: The Musculature of the Human Ureter. Brit. J. Urol. **42**, 724–727 (1970).

Osathanondh, V., Potter, E. L.: Development of Human Kidney as Shown by Microdissection. Arch. Path. **76**, 271–302 (1963).

Satani, Y.: Histological Study of the Ureter. J. Urol. (Baltimore) **3**, 247 (1919).

Schneider, W.: Die Muskulatur der oberen harnableitenden Wege. Z. ges. Anat. **109**, 187–196 (1938).

Shopfner, C. E.: Ureteropelvic Junction Obstruction. Amer. J. Roentgenol. **98**, 148–159 (1956).

Swenson, O., Fisher, J. H., Cendron, J.: Megaloureter: Investigation as to Cause and Report on Results of Newer Forms of Treatment. Surgery **40**, 223–233 (1956).

Tanagho, E. A., Smith, D. R., Guthrie, T. H.: Pathophysiology of Functional Ureteral Obstruction. J. Urol. (Baltimore) **104**, 73–78 (1970).

WILLIAMS, D. I.: "Megaureter in Children", in RICHES, Sir ERIC (Ed.): Modern Trends in Urology, 2nd Series, pp. 147–158. New York: Paul B. Hoeber 1960.
YOUNG, H. H.: Obstruction to the Ureter Produced by Aberrant Blood Vessels. SGO **54**, 26–38 (1932).
YOUNG, J. D. Jr., KISER, W. S.: Obstruction of the Lower Ureter by Aberrant Blood Vessels. J. Urol. (Baltimore) **94**, 101–106 (1965).

Discussion

ALLEN's theory that the longitudinal muscles are interlaced like a net is contradictory to the view of SCHNEIDER, STEIGLEDER and KÖRNER. They believe that the muscle fibers of the ureter have a spiral arrangement in connection with different angles of inclination (LUTZEYER).

ALLEN's theory is only based on the dissection of the ureter including the surface muscle fibers. SCHNEIDER was the first to stain the muscle fibers successfully. From the serial sections he built up a phantom (HINMAN). These conceptions are identical to that of MURNAGHAN, who, with the help of serial sections, shows that in the ureter the spiral arrangement of muscle fibres predominates except in the area of the ureterovesical junction where only longitudinal muscles are to be found. MURNAGHAN was able to demonstrate that the longitudinal muscles are only dominant in cases of functional stenoses in the region of the pyelo-ureteral junction (LUTZEYER). Where such changes existed MURNAGHAN could find no circular course of the fibres. This is generally thought to be functionally important (BOYARSKY). ALLEN does not, however, believe that MURNAGHAN holds the view that the dominance of longitudinal muscle fibres is pathological.

A critical discussion of ALLEN's results raises the following question: can the longitudinal muscles be affected by the preparation of the ureter when it is extended and dilated? In order to prevent this possibility ALLEN fixes the muscles in formalin, then spreads the specimen by minimal tension.

In embryological studies TANAGHO observed that, at a rather early stage of development of the ureter, a coating of mesenchymal cells lies close to the ureter. While the kidneys ascend into the lumbar region, the mesenchymal cells in their circular arrangement are extended and arranged spirally. By surgical intervention, however, he was able to influence the orientation of the mesenchymal cells in such a way that they either remained arranged circularly or exclusively longitudinally, or became fibrous tissue. Thus, one may conclude that different arrangements of the muscle fibres may be favored depends on the extension and dilation: stagnation of the development in connection with a circular wall structure, or total extension in connection with collagen formation of the tissue.

Such a relative increase in the collagen tissue would seem to be correlated with the development of functional stenosis in this area (ALLEN). Not only in the case of functional stenosis but in almost every pathologically changed ureter, it is possible to an enormous increase of the collagenous tissue (NOTLEY).

Functional Ureteral Stenosis

P. RATHERT and H. MELCHIOR*

Functional ureteral stenosis is characterized by the following clinical triad: 1. a roentgenologically demonstrable proximal ureteral ectasia, 2. absence of mechanical obstruction, i.e. free passage for at least a No. 7 French ureteral catheter and 3. absence of a clear pathological-anatomical substrate.
Since 1901, when Israel discussed this clinical picture for the first time in connection with the genesis of hydronephrosis, very different definitions and ideas have been proposed from pathophysiological and morphological points of view, for example atonic segment, aperistaltic segment, segmental ureteral atonia, non-obstructive hydronephrosis, ureteral achalasia.

Methods and Results

Over the last 3 years functional ureteral stenosis has been diagnosed by the mentioned clinical triad and our urodynamic examinations and treated in 6 patients (4 female, 2 male) between 19 and 40 years of age.
The following symptoms were clinically dominant:

1. recurrent urinary infections
2. pain of unknown origin and changing intensity in the lumbar region.

On i. v. urograms a slight to medium-grade dilatation of the proximal ureter and some times of the renal pelvis and the calyces was demonstrable in all cases. Localization of the stenosis showed no predisposition for any particular part of the ureter. The free passage of a No. 7 French ureteral catheter was defined decisive for the final diagnosis of a functional ureteral stenosis.
For further differentiation of the syndrome a functional diagnosis was made by urorheomanometry in all patients (MELCHIOR 1971).
The results varied widely. In spite of the same clinical signs and roentgenological findings urodynamically there were principally two different types.
In group I proximal to the stenosis there was a high basal pressure, no pressure amplitude, irregular contraction complexes, and a relatively slow velocity. These are the urodynamic signs of peristalsis against resistance.

* Abteilung Urologie der Medizinischen Fakultät der Rheinisch-Westfälischen Technischen Hochschule (Director: Prof. Dr. W. LUTZEYER) Aachen, Germany

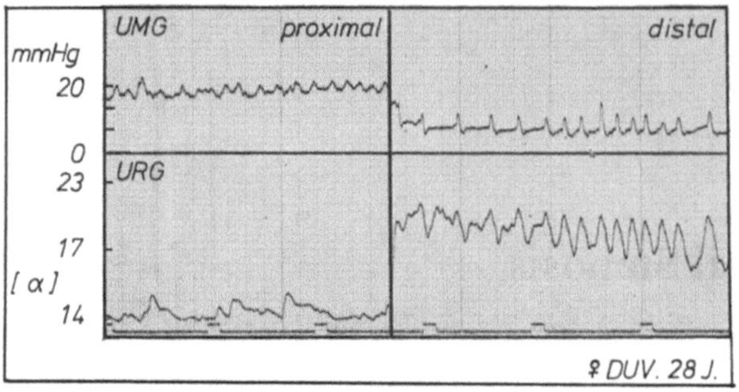

Fig. 1. Pat. Duv. ♀, 28 years. Typical *UMG* and *URG* of group I patients

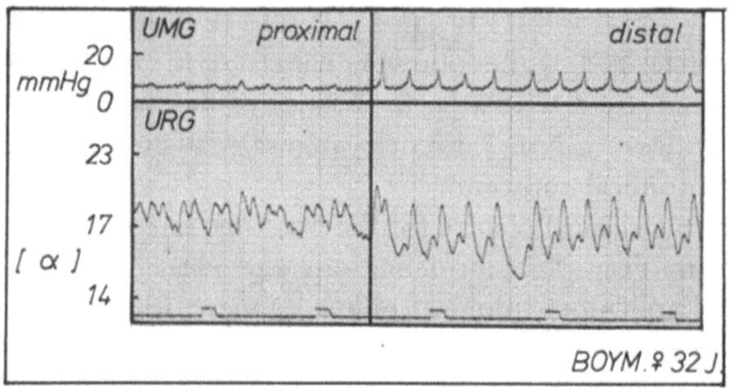

Fig. 2. Pat. Boym. ♀, 32 years. Typical uromanometrogram (*UMG*) and urorheogram (*URG*) of group II patients proximal and distal of a functional ureteral stenosis

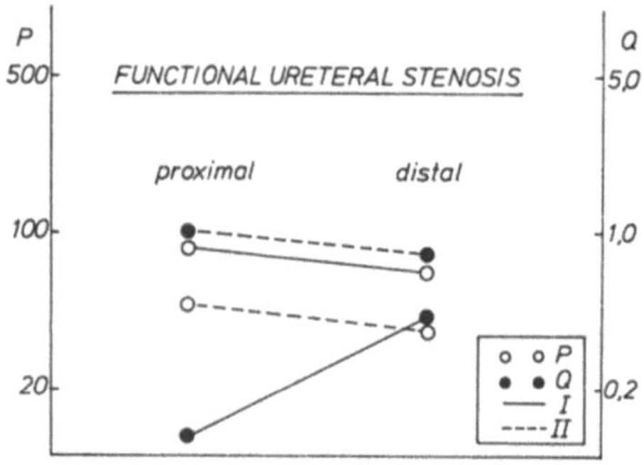

Fig. 3. Pressure-velocity product (*P*) and urodynamic quotient (*Q*) of group I —————) and group II (--------) proximal and distal of the stenosis

When the tip of the probe is drawn back several centimeters, the urodynamic situation is found to regulate directly distal to the stenosis; basal pressure declines, pressure amplitude increases and flow velocity rises substantially (Fig. 1).

In group II (Fig. 2) basal pressure was normal in the dilated segment but peristaltic contractions were slight, with minimal amplitude; in the rheogram an almost normal flow velocity was apparent, showing only slight changes between systole and diastole. Below the dilated segment there was normal ureteral peristalsis with regular and efficient contractions; basal pressure was the same as in the dilated segment (Fig. 2).

The differences between the two groups were even more obvious in the pressure-velocity product (*P*) and the urodynamic quotient (*Q*) (Fig. 3).

The pressure-velocity product shows no significant difference in either group between the dilated and non-dilated ureteral segment. On the contrary, the urodynamic quotient of group I is low in the dilated segment thereby proving an increased flow resistance.

In group II the urodynamic quotient is relatively high, judging by the minimal peristaltic activity in this segment.
Histological examination usually revealed slight hypertrophy of the musculature and sometimes an increase of elastic fibers in group I. In group II the muscle fibers were often diminished and the elastic fibers showed a marked increase. The differences in the histological results were not, however, significant.

Discussion

No adequate explanation for the different urodynamic results was found at the site of the operation or at the histologic examinations. This bears out WILLIAM'S statement (1958) that the pathology of hydroureter and -nephrosis cannot be discussed solely in terms of anatomy.

Table 1. Characteristics of Type I and II stenosis

Type I *Functional ureteral stenosis*	Type II *Hypoperistaltic segment*
Peristaltic against resistance	Hypoperistaltic
Prognosis: Favourable	Prognosis: Unfavourable

The urodynamic investigations performed hitherto are inadequate for a decision as to whether the two groups represent two different clinical syndromes (BOEMINGHAUS, 1960, LUDWIG, 1971). They may be two stages of a single clinical syndrome as discussed by BISCHOFF (1961) on the basis of clinical and embryological studies and by ZIMSKIND (1968), who used an experimental model.
On the other hand the prognosis of a surgical correction is definitely correlated with the type of functional stenosis (Table 1).
If the ureter is able to produce peristaltis against resistance (Type I), it has an adequate functional reserve and offers promising conditions for a plastic correction.
If, on the other hand, the ureter shows signs of decompensation (Type II), a favourable result can be expected only from resection or a large plastic operation of the whole segment with diminished peristalsis. The ureter no longer has sufficient adaptation.

Summary

Urodynamic studies were performed in 6 patients with functional ureteral stenosis. The clinical diagnosis was made from the general symptoms, the roentgenological findings and the free passage of a No. 7 French catheter. Two types were differentiated on the basis of the functional studies. Type I is characterized by peristalsis against resistance in the dilated segment. Type II, however, has a markedly reduced ureteral activity. The significance of these results lies in the finding of objective criteria for the therapy and prognosis of functional ureteral stenosis.

References

Bischoff, P.: Betrachtungen zur Genese des Megaureters. Urol. int. (Basel) **11**, 257–286 (1961).
Boeminghaus, H.: Urologie. Operative Therapie. Klinik-Indikation. München: Werk-Verlag Dr. Edmund Banaschewski 1960.
Israel, J.: Chirurgische Klinik der Nierenkrankheiten. Berlin: August Hirschwald 1901.
Ludwig, K. S.: Funktionelle Anatomie und Embryologie der oberen Harnwege, speziell des Ureters. In: Ureterdynamik. Stuttgart: Georg Thieme 1971.
Melchior, H.: Uro-Rheomanometrie (Simultane Uro-Rheographie und Elektromanometrie). In: Ureterdynamik. Stuttgart: Georg Thieme 1971.
Williams, D. Ines: Urology in childhood. Handbuch der Urologie, Band XV, Berlin-Göttingen-Heidelberg: Springer 1958.
Zimskind, P. D., Davis, D. M., Lewis, P. L.: The aperistaltic ureteral segment: experimental production and urodynamic significance. J. Urol. (Baltimore) **100**, 436–400 (1968).

Discussion

Functional Ureteral Obstruction

Functional ureteral obstructions exist in opposition to mechanical narrowing of the human. Their etiology, however, is mainly unknown. Notley thinks that the decisive etiological factors are the collagen fibers and Allen the relatively small mass of muscles in the region of the obstruction. Tanagho examined more than 500 serial sections of obstructions. He thinks the condition is due to an abnormal arrangement of muscles, both in the area of the obstruction and in the area proximal to the obstruction.
Clinical investigations by Zimskind failed to find any, or only much reduced, ureteral contractions in the ureteral segment directly below the dilatation. This segment looks perfectly normal at first sight. Further proximal, in the dilated segment, there are only the slowly increasing pressure complexes with small amplitudes and broad bases, which are typical for incomplete ureteral obstructions. Both proximal and distal to this segment ureteral peristalsis is in general completely normal. A functional obstruction of this kind may not be diagnosed by an ordinary excretory urogram. It can only be demonstrated as a manifest obstruction by infusion or hydration urography.
In animal experiments a functional obstruction can be reproduced as an aperistaltic segment. A balloon catheter is inserted into the ureter, intermittently for 1 to 2 weeks, inflating the ballon with each insertion. In this case all animals develop a dilatation proximal to the injured area and sooner or later a moderate to severe hydronephrosis. In the area of the traumatized segment there are virtually no contraction complexes (Zimskind). The histology of such a segment shows that the muscle fibers are torn and partly replaced by fibrous tissue. The segment itself remains widely patent, confirming its role as a functional obstruction.
Zimskind's observations do not explain functional ureterostenosis in connection with typical resistance peristalsis. This fact has been observed by Rathert and Melchior apart from the aperistaltic segment. Both types seem to be different stages of the same clinical syndrome: the resistance peristalsis as a symptom of compensation, the hypo- or aperistalsis as a stage of decompensation.

What is an Obstruction?

The word "obstruction" has three meanings (Hinman):

1. stenosis
2. increased pressure in front of a stenosis and
3. urinary stasis: that is, a larger urine volume in front of the stenosis within the unit of time.

"Obstruction" can also be an expression of infections and of calculi formation. Instrumental or radiological examinations disclose the volume of obstruction and the effectiveness of the voiding mechanism, thus enabling the clinician to diagnose hydronephrosis, hydroureter or disturbed micturition. Functional examintation prepares the way for a differential diagnosis (the reason for the disturbed function).

The Influence of Bladder Dynamics on Ureteral Dynamics

P. D. ZIMSKIND*

Recent studies in the pathophysiology of the ureter have demonstrated disorders of ureteral motility which can contribute to the development of urinary tract infection or hydronephrosis. Significantly, some of these disorders were demonstrated by pressure studies of the ureter and bladder, while by intravenous urography the urinary tract appeared normal (ZIMSKIND, et al., 1968; DAVIS and ZIMSKIND, 1963; DAVIS, et al., 1963).

The ureter propels a small bolus of urine from the renal pelvis to the bladder. The rhythm of these propulsions is fairly regular. The ureter is empty between boluses and the columns of urine in the ureter are widely separated (KIIL, 1957; NARATH, 1951). With resistance to flow or with a physiological overload, widening and lengthening of the bolus occurs. This may result in dilatation with filling of the entire lumen with urine and suppression of peristaltic contractions (LAPIDES, 1948; MURNAGHAN, 1957). Similarly, overt ureteral obstruction causes a rise in ureteral resting pressure, leading to overdistension and hypotonia (DAVIS and ZIMSKIND, 1962).

Ureteral activity primarily depends on the volume rate of urine flow (DAVIS and ZIMSKIND, 1963; BOYARSKY and Martinez, 1962). While autonomic nerve fibres reach the ureter, their role is apparently not essential since ureteral function continues after denervation (MELICK et al., 1961). It has been previously demonstrated that the ureteral resting pressure and contractility are profoundly affected by changes in bladder pressure and volume in patients with reflux (DAVIS and ZIMSKIND, 1963). It was the purpose of the present study to compare the effects of altered intravesical pressure and volume on ureteral dynamics in the presence and absence of reflux.

Materials and Methods

Pressure recordings of the ureter and bladder were made in 30 female patients with a history of recurrent urinary tract infection. No abnormalities of the ureter or bladder had been demonstrated radiographically. None of these patients exhibited reflux. Patients were uninfected at the time of examination. A specially constructed thin-walled

* NATHAN LEWIS HATFIELD Professor of Urology and Chairman of the Department, Jefferson Medical College, Philadelphia, USA

4F ureteral catheter was passed into the ureter cystoscopically and a second catheter was inserted into the bladder. Each catheter was connected to a Statham P 23AA strain gauge pressure transducer and pressures were recorded continuously on a Gilson minipolygraph[1].

Following baseline recordings with the bladder empty, the pressures in the ureter and bladder were monitored simultaneously as the bladder was gradually filled with sterile water. In 9 patients the bladder was filled with a solution of Retrografin and normal saline solution in equal volumes. While the catheters were in place, cystograms were taken at low and high intravesical pressures. These included A–P and left and right oblique views. No complications occurred in any patients following these studies.

Ten patients with vesicoureteral reflux also had simultaneous ureteral and bladder pressure studies during bladder filling and emptying.

Results

With the bladder empty, the ureteral resting pressure varied from 0–10 mmHg ranging from between 2–4 mmHg in 21 of the 30 patients. The peristaltic contraction frequency ranged from 2–6 per minute. Contraction amplitudes varied irregularly.

A ureteral response to a gradual increase in bladder pressure during bladder filling was noted in 29 of the 30 patients without reflux. While filling the bladder, the average vesical pressure at which ureteral resting pressure and peristalsis showed early change was 8 mmHg.

In 12 patients the ureteral resting pressure rose and the frequency of ureteral peristaltic contractions increased. Ureteral contractions eventually disappeared and a high resting pressure remained (Fig. 1A). In 7 other patients, the ureteral resting pressure rose, but peristaltic contractions persisted at an increased frequency (Fig. 1B). In 6 patients, the ureteral resting pressure rose but the frequency of contractions remained unchanged (Fig. 1C). In 4 patients there was no increase in ureteral resting pressure or ureteral peristalsis during the study. Contraction amplitudes exhibited no consistent response to bladder filling.

During bladder filling the average rise in ureteral resting pressure was 8 mmHg. This was calculated on the basis of highest ureteral resting pressure during filling minus ureteral resting pressure before filling. Cystograms obtained with the ureteral catheter in place revealed no reflux. Upon emptying of the bladder and release of bladder pressure, the ureter reverted to its normal pattern. In these patients without reflux, the ureteral responses to bladder filling and emptying did not appear immediately, but occurred after certain levels of bladder filling or emptying had been reached. These levels were inconstant.

In the 10 patients with reflux, even small degrees of filling invariably produced a rise in ureteral resting pressure which occurred simultaneously with the increase in bladder pressure. Fig. 2A shows an intravenous urogram of a 14-year old girl with a spastic neurogenic bladder following excision of a congenital meningomyelocele 24 hours after birth. Bilateral pyeloureterectasis was evident. A cystogram revealed bilateral vesicoureteral reflux (Fig. 2B) with fair emptying of the upper tracts after

[1] Gilson Medical Electronics

20 minutes of ambulation (Fig. 2C). The pressure study performed with a recording catheter in the right ureter and another in the bladder showed rapid increase in intravesical pressure as spasms occur subsequent to fluid instillation (Fig. 3). After each

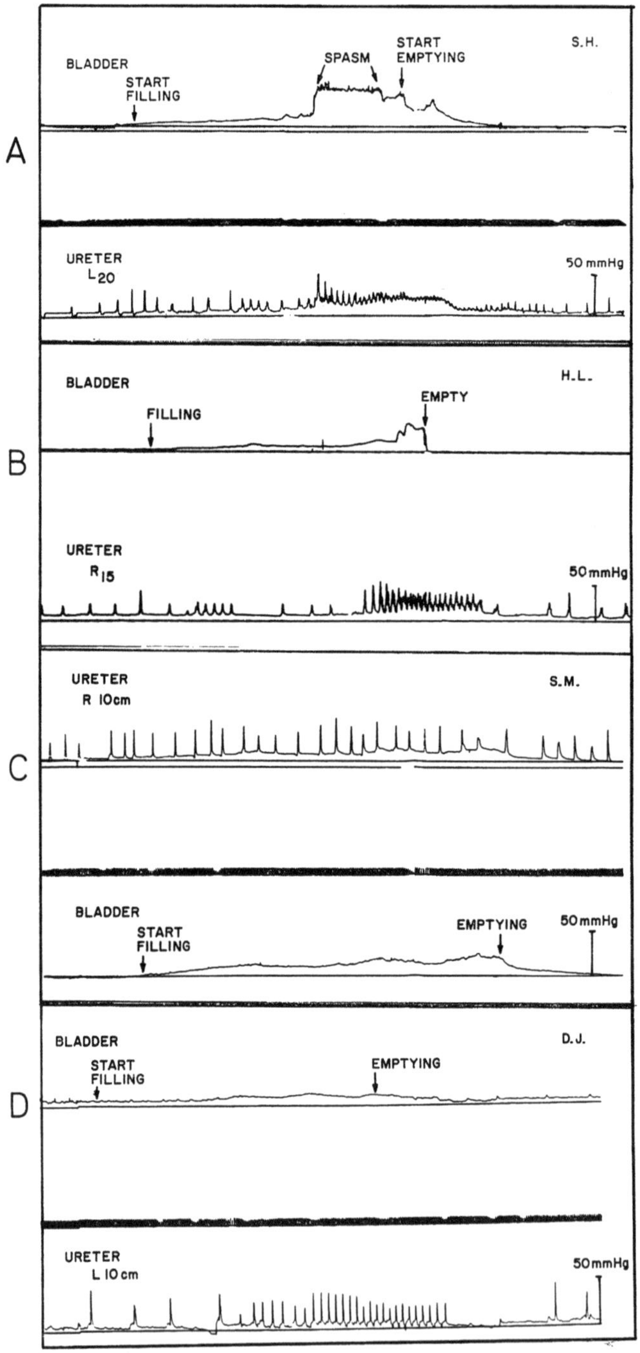

Fig. 1. Ureteral response to bladder filling. A Ureteral resting pressure rises and contraction frequency increases. Ureteral contractions are eventually abolished. With bladder emptying, ureteral peristalsis returns. B Ureteral resting pressure rise is accompanied by increased frequency of peristalsis. After emptying of bladder, peristaltic pattern returns to normal. C Ureteral resting pressure rises during bladder filling. Rate of ureteral contractions remains unchanged. Upon emptying of bladder, ureteral resting pressure returns to normal. D During bladder filling, no change in ureteral resting pressure occurs, but frequency of ureteral peristalsis increases. With bladder emptying, ureteral peristalsis returns to normal (*L* indicates left ureter, *R* indicates right ureter. Number indicates distance from ureteral orifice in centimeters. Time marker indicates 2-second interval.) [From Zimskind *et al.*, 1969]

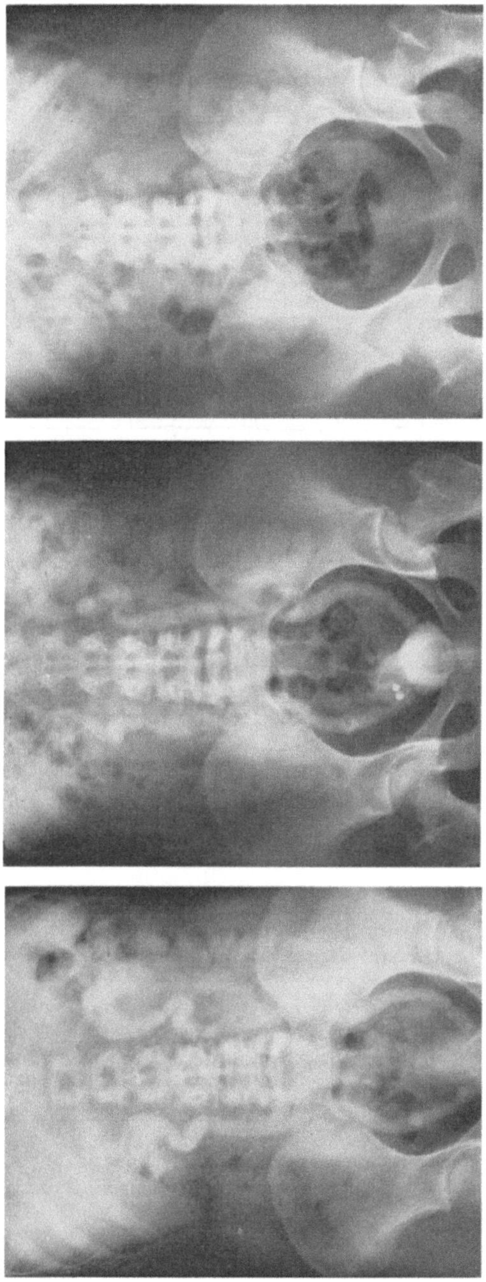

A B C

Fig. 2. A Intravenous urogram shows bilateral dilatation of upper tracts. B Cystogram reveals bilateral reflux. C Following ambulation, upper tracts empty fairly well. [From DAVIS and ZIMSKIND]

spasm, an immediate, but slight rise in ureteral resting pressure was seen in the upper, middle and lower ureter, suggesting reflux into flaccid atonic ureters. Discrete peristaltic contractions in the right ureter were not discernible. It is likely that contraction complexes, if present, would be damped by the distribution of the pressure changes in the ureter to the fluid within the bladder and left ureter vie the widely patent and freely communicating ureterovesical ostia. It is interesting to note that after the bladder was emptied, some degree of ureteral emptying did occur despite the comparative atonicity of the ureters.

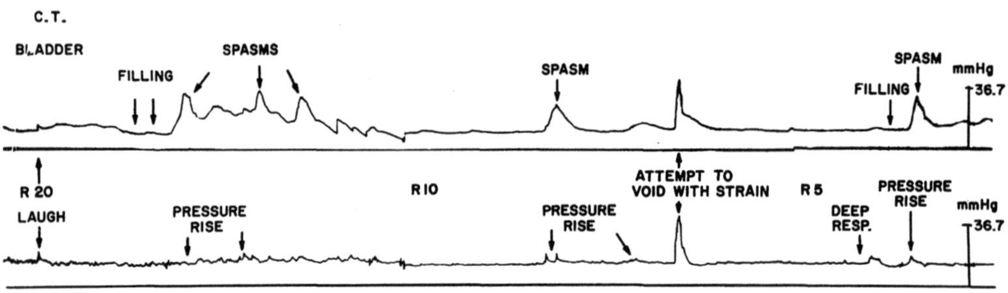

Fig. 3. Same patient as in Fig. 2. Pressure study shows pressure rise in refluxing right ureter (bottom) as bladder is filled. Peristaltic contractions are absent. [From Davis and Zimskind, 1963]

Summary

The ureteral response to changes in bladder pressure was studied in 40 patients, 10 of whom exhibited reflux. An intrinsic dynamic relationship between intravesical pressure and volume and ureteral peristalsis was demonstrated even in the absence of reflux.

Physiologic mechanisms were considered and certain clinical implications were cited.

References

Boyarsky, S., Martinez, J.: Ureteral peristaltic pressures in dogs with changing urine flows. J. Urol. (Baltimore) **87**, 25–32 (1962).

Davis, D. M., Zimskind, P. D.: Progress in urodynamics. J. Urol. (Baltimore) **87**, 243–248 (1962).

Davis, D. M., Zimskind, P. D.: Pathologic types of ureteral pressure graphs with remarks on their relation to lower tract obstruction. J. Urol. (Baltimore) **90**, 677–690 (1963).

Davis, D. M., Zimskind, P. D., Paquet, J. P.: Studies on urodynamics: New light on ureteral function. J. Urol. (Baltimore) **90**, 150–159 (1963).

Kiil, F.: The function of the ureter and renal pelvis. Philadelphia: Saunders 1957.

Lapides, J.: The physiology of the intact human ureter. J. Urol. (Baltimore) **59**, 501–533 (1948). Discussion 534–537.

Melick, W. F., Naryka, J. J., Schmidt, J. H.: Experimental studies of ureteral peristaltic patterns in the pig: Myogenic activity of the pig ureter. J. Urol. (Baltimore) **86**, 46–50 (1961).

Murnaghan, G. F.: Experimental investigation on the dynamics of the normal and dilated ureter. Brit. J. Urol. **29**, 403–409 (1957).

Narath, P. A.: Renal pelvis and ureter. New York: Grune-Stratton 1951.

NOGRADY, M., DUNBAR, J., MACEWAN, D.: The effect of bladder distention on the intravenous pyelogram – an experimental study. Amer. J. Roentgenol. **90**, 37–54 (1963).
PIERCE, J. M., Jr., BRAUN, E.: Ureteral response to elevated intravesical pressures in humans. Surg. Forum **11**, 482–484 (1960).
TANAGHO, E. A., HUTCH, J. A., MEYERS, F. H., RAMBO, O. N.: Primary vesicoureteral reflux: Experimental studies of its etiology. J. Urol. (Baltimore) **93**, 165–176 (1965).
ZIMSKIND, P. D., DAVIS, D. M., LEWIS, P. L.: The aperistaltic ureteral segment: Urodynamic aspects and clinical significance. J. Urol. (Baltimore) **100**, 436–440 (1968).

Discussion

The results suggest that an intimate relationship exists between intravesical pressure and volume and ureteral dynamics.

The ureteral changes during bladder filling in the absence of reflux closely resemble the obstructive pattern described by DAVIS and ZIMSKIND (1963) in clinical subjects and by BOYARSKY and Martinez (1962) in animals. The sequence of events during bladder filling may be as follows. With low bladder volumes, urine is readily conducted from the ureter into the bladder. As the bladder fills, trigonal stretching serves to attenuate and narrow the intramural ureter, resulting in a physiological ureteral obstruction, as suggested by TANAGHO, et al. (1965). Also, as the intravesical pressure increases, additional resistance to ureteral emptying may develop. The ureter responds to this physiologic obstruction almost instantly. Ureteral activity increases compensatorily as evidenced by increased frequency of peristalsis with or without a rise in ureteral resting pressure. Eventually this compensatory response is overcome and urine pools in the ureter. Distension of the ureteral musculature impairs contractility so that ureteral peristalsis disappears and a high resting pressure remains (ZIMSKIND, et al., 1969).

The urodynamic phenomena can be correlated with radiographic evidence collected by intravenous pyelography in patients with a filled bladder. NOGRADY, *et al.*, (1963) showed that voluntary retention of bladder urine in normal children produced ureteral dilatation by intravenous urograhy, in some cases simulating hydroureteronephrosis. PIERCE and BRAUN (1960) reported comparable findings in adults. The present findings provide a physiologic explanation for this phenomenon in that ureteral peristalsis was probably inhibited by the degree of bladder filling. From these observations it would appear that the presence of dilated ureters by intravenous urography should be interpreted with caution if the bladder is full.

The ureteral response to bladder filling in refluxing patients differed from that in non-refluxing individuals in that refluxing ureters exhibited pressure changes simultaneous with bladder pressure elevation. In non-refluxing ureters, the effect was in the same direction but delayed.

The ability of atonic refluxing ureters to empty at least partially after bladder emptying probably reflects the same mechanisms which apply to the non-refluxing ureter. The degree to which gravity and somatic muscular contractions contribute to urinary conduction in such ureters remains to be determined.

The findings emphasize the importance of maintaining low intravesical volumes and pressures in patients with defective ureteral peristalsis due to reflux, hypotonia of the ureteral musculature and/or chronic ureteritis. Therapeutic measures such as timed voiding or bladder drainage by tube or catheter promote ureteral emptying by reducing the resistance to urine flow into the bladder.

Congenital Physiopathology of the Ureterovesical Junction

W. Gregoir and G. Debled*

It has long been known that reflux does not occur normally in adults, children or new-borns. The protecitve mechanism of the ureter against vesical pressure has not yet been clearly explained and is a constant subject of discussion.

In mammalian and human embryos reflux cannot be established by the end of the first trimester of pregnancy, i.e. before any diuresis is possible and long before the development of ureterotrigonal musculature. Therefore, this musculature cannot be assumed to play a primordial and exclusive part in the anti-reflux mechanism which is the result of the anatomical disposition of the uretero-vesical junction, of the fixity of the ureteral orifice during bladder filling and distension and also of the quality and integrity of the detrusor musculature. In fact, reflux is found frequently in bladder weakness and almost invariably in megalobladder.

However, the development of the uretero-trigonal musculature and elastic tissue apparently maintains the fixity of the ureteral orifice in regard to the bladder neck. Therefore, it can be easily understood that any insufficiency of this musculature eventually causes reflux when the bladder is submitted to repeated pressure increases by regular micturitions, distending the bladder wall and pulling the ureteral orifice away from the bladder neck.

Histology demonstrates muscular insufficiency may be detected by histological examination in primary congenital reflux, and clinical examination suggests the existence of 2 types of reflux:

1. *Ordinary Reflux :* the muscular insufficiency is limited to the uretero-trigonal implantation, and reflux provokes moderate dilatation. The surgical prognosis is good because dilatation regresses and even disappears completely after surgical correction. Functional and anatomical recovery is excellent and usually permanent.
2. *Reflux with Myasthenia :* the muscular insufficiency is not limited to the ureterovesical implantation. Such pronounced dilatation early in life cannot be explained by simple congenital reflux in the absence of any lower urinary tract obstruction or neurogenic disorders. The enormous size of the megalo-ureter suggests the anomaly extends to the entire ureteral musculature. Reconstructive surgery requires multiple procedures and no matter how good the technique, late functional results remain a problem because of primary muscular insufficiency.

---

* Clinique Urologique, Hôpital Universitaire Brugman, Université Libre de Bruxelles, Belgium

We have to recognize that various points are still quite obscure. The part played by the connective tissue and particularly by the elastic tissue in normal physiology is imperfectly understood, and nothing is known about the possible role of the elastic tissue in congenital pathological phenomena such as reflux or functional obstruction. Interest has recently been awakened by the study of ureteral nervous tissue. The part of the nervous system in congenital anomalies is still enigmatic.

Recent systematic histological studies have increased our knowledge of obstructive congenital malformations such as primary megalo-ureter, ureterocele and uretero-hiatus hernia so that we are now in a position to propose some hypotheses.

Primary congenital megalo-ureter is by definition a megalo-ureter without any apparent cause, i.e. without stenosis, reflux, lower obstruction or neurogenic disease. Many hypotheses have been elaborated on its pathology

The histologic study of the ureterovesical junction by serial cross sections on over 50 cases of megalo-ureters has shown that:

1. The normal histological structure is not present.
2. The intravesical or juxtavesical segment of the ureter is always subject to severe tissular malformations, which can be divided into 3 groups:

a) Connective hypertrophy: diffuse hypertrophy of the connective tissue in association with muscular reduction. Hypertrophy of the connective tissue and the muscular reduction induce immobility of the pathologic segment, so that the ureter becomes achalasic (Fig. 1).

b) Muscular hypertrophy: diffuse hyperplasia and variable aspect of the ureteral musculature (Fig. 2).

c) Disembryoplasia: the malformation affects the muscular, connective and epithelial tissues of the terminal ureter; it takes the form of diffuse hypertrophy of the connective tissue with muscular atrophy and epithelial anarchy. The terminal ureter presents a normal lumen and various irregular blind lumens (Fig. 3).

These anomalies are often spectacular, suggesting that the tissular alteration is a functional obstructive disorder responsible for the chronic stasis causing the megalo-ureter.

Ureterocele is another primary obstructive congenital malformation. It has been the subject of several embryonary theories which cannot be confirmed. Its morphologic development and selective emplacement in the submucosal ureteral segment proves the occurrence of an obstructive factor at the level of the ureteral orifice. Histology confirms this fact. In the normal ureter, there are decussations of muscular fibers in the upper and lower lips of the ureteral orifice. These decussations are highly hypertrophied in ureterocele. Thus the growth mechanism of the ureterocele is similar to the primary megalo-ureter mechanism, except that the functional obstruction affects only the decussation of the ureteral orifice instead of the intravesical ureter; as in the case of any obstruction, dilatation develops directly above, i.e. in the submucosal ureter.

The uretero-hiatal hernia is another malformation of the ureterovesical junction and is caused by an abnormally broad detrusor hiatus. This too broad hiatus allows the vesical mucosa to insinutate out of the bladder alongside the ureter. The main difference between uretero-hiatal hernia and congenital diverticulum, in which one ureter has its implantation, is that the uretero-hiatal hernia develops and ascends inside the ureteral sheath, while the vesical diverticulum remains inside the bladder adventitia.

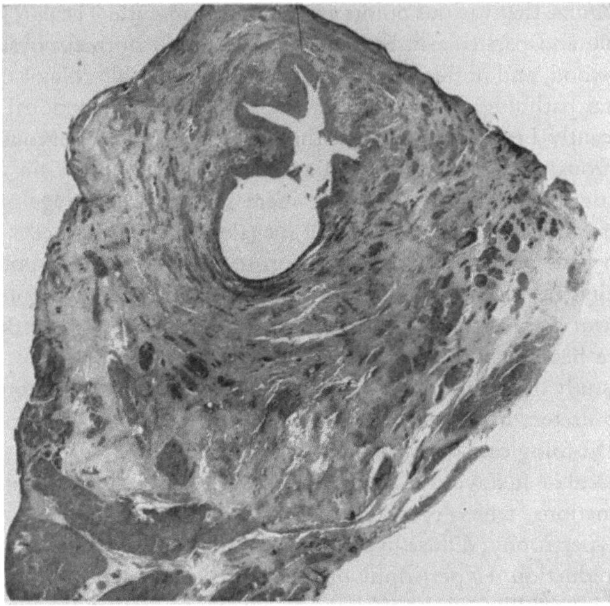

Fig. 1. Primary megalo-ureter. Cross section of the terminal ureter. Note the severe hyperplasy of the collagenous tissue, which contains only a few muscular fibers

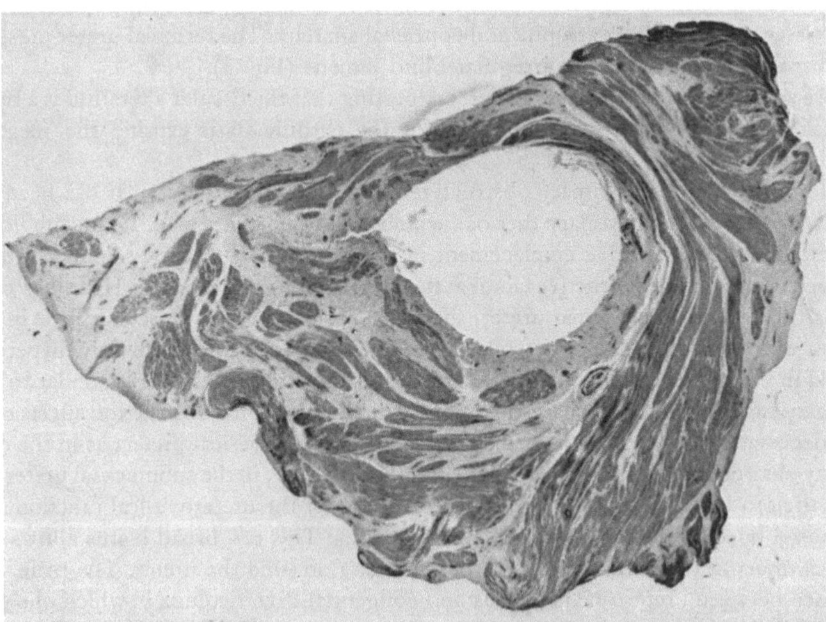

Fig. 2. Primary megalo-ureter. Cross section. Submucosal ureter. Hypertrophy of the musculature forming a pseudo-sphincter

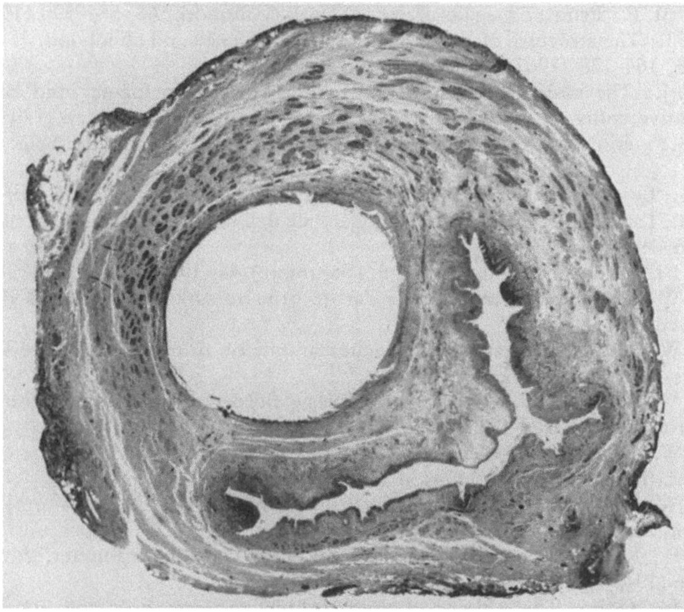

Fig. 3. Primary megalo-ureter. Cross section. Submucosal ureter. Fibro-epithelial dysembryoplasy. Note the ureteral lumen and the fibro-muscular anarchy. The circular lumen has been catheterized; the principal ureteral lumen is continuous with the dilated ureter

Uretero-hiatal hernia is accompanied either by reflux or by chronic stasis without reflux. In the first case, the uretero-trigonal musculature is primarily pathologic and secondly dislocated by the developing hernial saccule; this dislocation causes secondary reflux and dilatation. In the second case, (dilatation without reflux), we observe extensive connective hypertrophy associated with muscular weakness of the terminal ureter, the same process as in primary megalo-ureter.

Until results are obtained from more exhaustive studies on connective, elastic and nervous tissues, it seems justifiable to propose some hypotheses on the growth mechanism of primary malformations of the ureterovesical junction:

- congenital reflux originates in abnormal anatomical malformation of the ureterovesical junction and insufficient uretero-trigonal musculature.
- primary congenital megaloureter originates in tissular anomaly (hypertrophy and disembryoplasia).
- ureterocele originates in congenital hypertrophy of the ureteral orifice decussations.
- uretero-hiatal hernia is caused by a too broad hiatus in association with reflux or chronic obstruction producing parallel tissular alterations in the ureteral wall itself.

References

Bischoff, P.: Betrachtungen zur Genese des Megaureters. Z. Urol. **85**, 268–274 (1961).

Bischoff, P., Strohmenger, P.: Betrachtungen zur Frage der operativen Behandlung des kongenitalen Megaureter beim Erwachsenen. Urologe **5**, 218–221 (1966).

Campbell, M. F.: Primary megalo-ureter. J. Urol. (Baltimore) **68**, 584–590 (1952).
Cussen, L. J.: The structure of the normal ureter in infancy and childhood. J. Urol. (Baltimore) **5**, 164–178 (1967).
Cussen, L. J.: The structure of the normal human ureter in infancy and childhood. A quantitative study of the muscular and elastic tissue. Invest. Urol. **5**, 179–194 (1967).
Debled, G.: L'étiologie du méga-uretère congénital primaire. Acta urol. belg. **36**, 129–136 (1968).
Debled, G.: La structure normale de l'uretère terminal. Acta urol. belg. **36**, 119–128 (1968).
Debled, G.: La pathologie obstructive congénitale de l'uretère terminal. Acta urol. belg. **39**, 371–465 (1971).
Gayet, R.: L'uretère intramural. J. Urol. (Baltimore) **44**, 193–217 (1937).
Gregoir, W.: Les troubles moteurs de l'arbre urinaire supérieur. Congrès Belge d'Urologie. Mai 1951.
Gregoir, W.: L'urokymographie et la radiomanométrie urinaire. Bruxelles: Office International de Librairie 1953.
Gregoir, W., Debled, G.: La dysembryoplasie fibroépithéliale de l'uretère intramural. Acta urol. belg. **34**, 229–232 (1966).
Gregoir, W., Debled, G.: L'étiologie du reflux congénital et du méga-uretère primaire. Urol. int. (Basel) **24**, No 2, 119–134 (1969).
Gregoir, W.: Traitement chirurgical du reflux congénital et du méga-uretère primaire. Urol. int. **24**, No 6, 502–526 (1969).
Irvin, G. E., Kraus, J. E.: Congenital megalo-ureter and hydro-ureter. Arch. Path. **45**, 752–756 (1948).
Murnaghan, G. F.: The physiopathology of mega-ureter. Proc. roy. Soc. Med. **51**, 9, 776–779 (1958).
Nesbit, R. M., Withycombe, J. F.: The problem of primary megalo-ureter. J. Urol. (Baltimore) **72**, 162–171 (1954).
Swenson, O.: A new concept of the pathology of megalo-ureters. Surgery **32**, 367–371 (1952).
Tanagho, E. A., Pugh, R. C.: The anatomy and function of the uretero-vesical junction. Brit. J. Urol. **35**, 151–165 (1963).
Woodburne, R. T.: Anatomy of the uretero-vesical junction. J. Urol. (Baltimore) **92**, 431–435 (1964).

The Growth of Ureterovesical Unity. A Contribution to the Study of Muscular Arrangement in the Terminal Ureter

D. Ruano Gil, G. Domenech Ratto, and A. Tejedo Mateu*

Introduction

The existence of vesicoureteral reflux and the problems raised by its pathologic physiology make examination of the terminal ureter's muscular arrangement a matter of keen interest. Hence the flood of studies published on the connection between the ureter and the bladder. Nevertheless controversy rages over the distribution of muscle fibers in the terminal ureter. Thus all kinds of argument go on about the periureteral sheath, some declaring it a single anatomical structure (Versari, 1911; Lo Cascio, 1914; Ruotolo, 1948; Gil Vernet, 1953, 1968, 1970; Hutch, 1961; Tanagho and Pugh, 1963; Debled, 1968; Gregoir, 1968), while others deny that it exists at all (Woodburne, 1964).

As to the ureter's own musculature, Tanagho and Pugh (1963) maintain that it continues beyond the ureteral meatus, ending in man at the colliculus seminalis and in woman at the neck of the bladder. On the other hand Gil Vernet (1953, 1968, 1970) holds that the ureter's own fibres all enter the labia of the ureteral meatus and the vesical mucosa adjoining the meatus.

As there is so much controversy it seems worthwhile to examine the problem from an embryological point of view, bringing forward data that shed light on the muscular arrangement of this vital anatomical cross-road.

Subject-Matter and Methods

All the data in the present article come from 35 vesicoureteral sections from fetuses measuring between 35 and 180 mm in length, vertex to coccyx, newborn infants, and children up to two years of age; after fixation in 10% formalin the sections were cut to a thickness of 10 to 11 microns and stained with hematoxilyn-eosin or by Azan method. With the latter method we refrained in certain cases from using the aniline mixture of orange-blue acetic acid, so as to stain nothing but muscle fibres. The cuts were made on a horizontal or sagittal plane, but all of them somewhat obliquely so as to section the whole course of the ureter's intramural segment, including the meatus.

---

* Second Chair of Anatomy, Faculty of Medicine, Barcelona, Spain

Observations

In the 73 mm fetus (three months' gestation) the vesical musculature was found to be remarkably well developed. Fig. 1 clearly shows a stout periureteral sheath, emerging from the detrusor, which surrounds the ureter from where it enters the bladder upward. On the other hand the ureter's own musculature has not as yet undergone any differentiation.

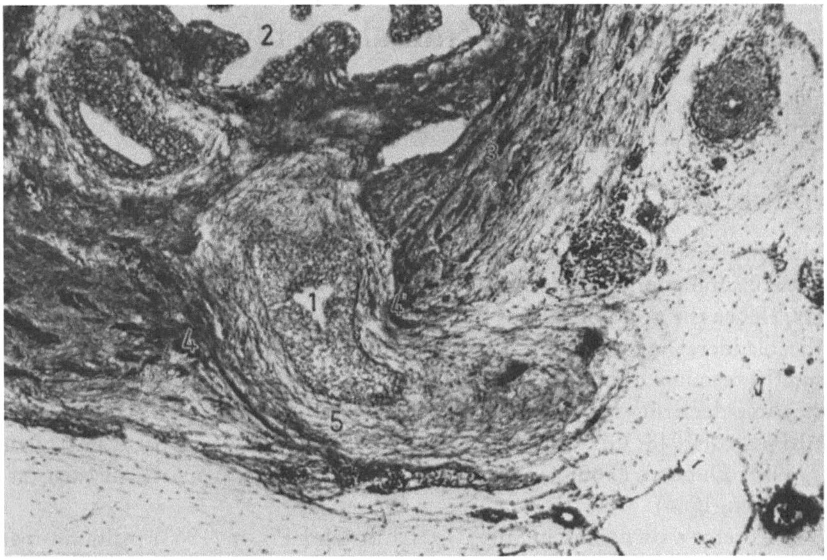

Fig. 1. 73 mm. fetus (3 months pregnancy). 1. Ureteral lumen. 2. Bladder. 3. Detrusor 4. and 4′. Detrusor emanations forming the extramural portion of the periureteral sheath. 5. Embrionic connective tissue, in the recess of which muscular fibers are still not differentiated

This begins when the fetus is about 118 mm long ($3^1/_2$ months' gestation). Fig. 2, a horizontal section of a fetus that length, just shows the presence of the ureter's own muscle fibers in its walls. It is also possible to distinguish the muscle fibers of the periureteral sheath emerging from the detrusor. Between the fibers of the sheath and the ureter's own fibers there is a gap, which we take to be the substantial cavity described by Waldeyer (1892).

When the fetus is about 132 mm long (four months of pregnancy) the intramural segment of the ureter is beginning to be ringed by fibers from the detrusor which enclose it downwards, forming the intramural segment of the periureteral sheath (Fig. 3).

In the four-day-old infant the greater thickness of this musculature brings out these details more strikingly. Fig. 4 shows the course of the ureter's own fibers, ending in the corium of the meatus and of the area adjoining the meatus. Apart from these fibers and clearly separated from them by Waldeyer's cavity we find other muscle fibers of different configuration, which belong to the intramural segment of the periureteral

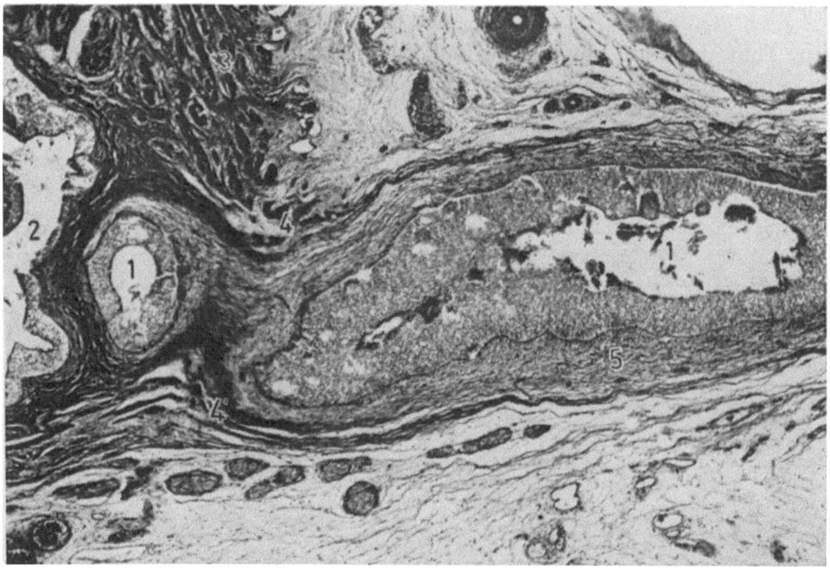

Fig. 2. 118 mm. fetus (3 and half months of gestation). 1, 2, 3, 4 and 4′ the same as for Fig. 1. 5. Embryonic connective tissue, in the recess of which the intrinsic muscular fibers of the ureter are beginning to differentiate themselves

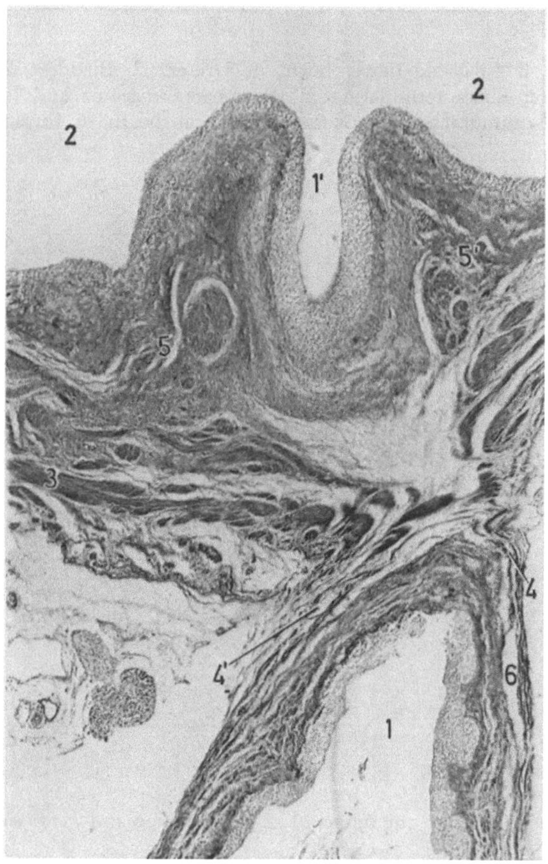

Fig. 3. 132 mm. fetus (4 months pregnancy). 1. Ureter. 1′. Ureteral meatus. 2. Vesical cavity. 3. Detrusor. 4. and 4′. Extramural portion of the periureteral sheath. 5. and 5′, intramural of said sheath. 6. Waldeyer space

sheath. Thus there are obviously significant differences between the ureter's own fibers, which are very thin and few in number, and these others, which form much stouter fasces.

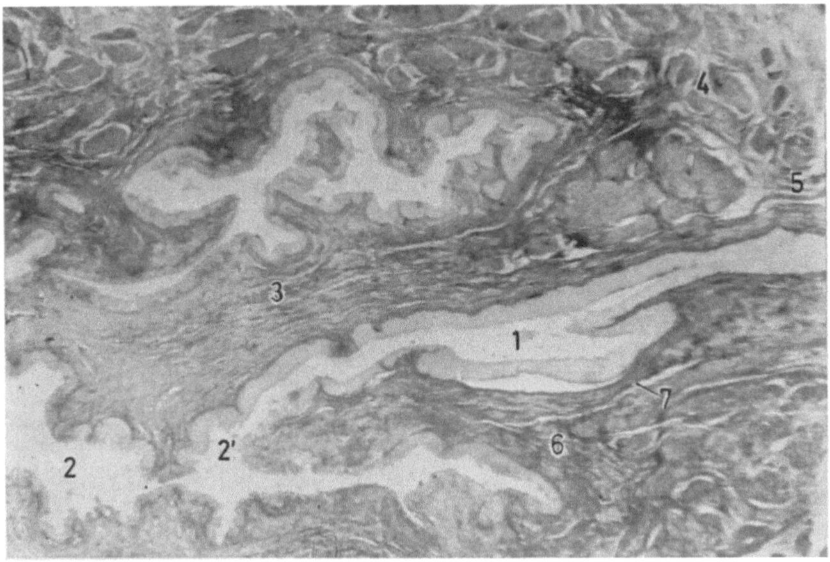

Fig. 4. 4-day-old newly born. 1. Ureter. 2. Bladder. 2'. Ureteral meatus. 3. Intrinsic ureter fibers terminating in the meatus margins and juxtameatic region. 4. Detrusor. 5. Extramural portion of the periureteral sheath. 6. Intramural portion. 7. Waldeyer space

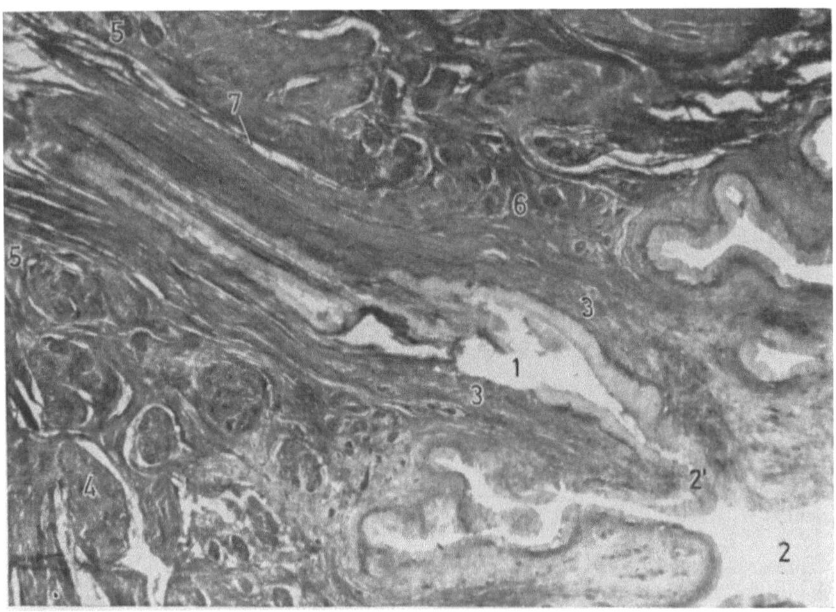

Fig. 5. Child of one month. 1, 2', 2, 3, 4, 5, 6 and 7, the same as in Fig. 4

Fig. 5 shows a section taken from a month-old child; we observe how the fibers making up the ureter's own musculature run along the segment adjoining the bladder and the segment within the vesical wall. We find their course paralleled by the periureteral sheath, which WALDEYER's cavity separates from it. Fig. 5 shows beautifully how the ureter's own fibers run lengthwise in the intramural segment, sometimes not quite straight but with slight undulations, to end in the labia of the ureteral meatus.

Summary

In this paper we have studied the muscular structure of the terminal ureter. The material studied consisted of 35 ureterovesical blocks taken from 35 to 180 mm long fetuses and newborn children of up to 2 years of age, sectioned at a thickness of 10–11 microns. They were stained with hematoxilin-eosin and by the Azan method. In some cases the mixture of blue aniline orange acetic acid was not used with this method so that only the muscular fibers would take up the color. Careful study of all the sections allowed us to observe the complicated development of the ureterovesical region from the point at which it is outlined to more advanced levels of the ontogeny where it acquires a more complex character. In all the cases studied, we observed that the ureter, at its juxta- and intravesical portions, is made up of two types of musculature: one is peripheral, with similar characteristics to the detrusor musculature developing in close connection with this and forming the periureteral sheath. This sheath is made up of two distinct portions: one extravesical, which surrounds the proximal section of the ureter and terminates by insertion into the adventitia. The other portion is intravesical and its fibers run along the distal end of the ureter.
The other musculature appears later and is the intrinsic system made up of a reduced number of strongly inclined spiroid fibers which appear longitudinal in the sections and which, following the ureter distally, terminate in the chorion of the margins of the ureteral meatus and the juxtameatic region (observed by GIL VERNET). This musculature is responsible for the closing and retraction of the ureteral meatus after urinary ejaculation in the bladder. In none of the examples studied we did observe that the intrinsic muscular fibers of the ureter terminated in the female vesical cervix or in the male veru montanum, (TANAGHO and PUGH, 1963).
Between the two muscular layers we consistently found an acellular space, which we identified as the WALDEYER cavity.

References

DEBLED, G.: Structure musculaire de l'uretère terminal. Acta urol. belg. **36**, No 2, Avril (1968).
GIL VERNET, S.: Biología de la Prostata. Cap. V. Madrid: Paz Montalvo 1953.
GIL VERNET, S.: Plan estructural y dinámico de la musculatura del uréter. Conferencia pronunciada en la Real Academia Nacional de Medicina. 25 de Marzo de 1969.
GIL VERNET, S.: Unión uretero-trigonal. Esfinter ureteral prevesical. Arch. esp. Urol. Tomo XXIII, No 3 (1970).
GOMEZ BOSQUE, P.: Anatomía funcional de la musculatura uretérica. Arch. esp. Urol. Tomo XI, No 39, Valencia (1954).
GREGOIR, W.: L'Etiologie du reflux vesicoureterale congenite et du megaureter primaire. Urolog. int. (Basel) **24**, 119–134 (1969).

Hutch, J.: Theory of maturation of the intravesical ureter. J. Urol. (Baltimore) **86**, 354 (1961).

Lo Cascio, G.: La Morfogenesi della tunica muscolare dell'uretere umano, del bacineto renale e dei grandi e piccoli calici. Ricerche del Lab. di Anat. Um. di Roma. XVIII, Fasc. 1–2 (1914).

Ruotolo, A.: Sulla costituzione e sul significato funcionale della guaina e della cosi detta fessura di Waldeyer. Estratto dalle Ricerche di Morfologia. Roma, 1948.

Tanagho, E., Pugh, R. C. B.: The Anatomy and Function of the ureterovesical junction. Brit. J. Urol. **35**, 151 (1963).

Versari, R.: La Morfogenesi della guaina dell'uretere umano. Richerche del Lab. di Anat. Um. di Roma XV, fasc. 3, 4 (1911).

Waldeyer, W.: Über die sog. Ureterscheide. Verh. anat. Ges. Wien, 1892.

Woodburne, R. T.: Named by Debled.

Anatomical Aspects of Vesicoureteral Reflux

S. GIL VERNET*

1. The Prevesical Ureteral Sphincter

Until recently it was supposed that the circular muscle fibers of the ureter disappeared at the point where it entered the bladder wall, leaving only the longitudinal ones, which run slantwise into the ureteral meatus.

A recent detailed investigation made me to think in a different way. No doubt the ureter shows only longitudinal fibers in its parietal segment; however these are not an extension of the fibres bearing the same name in the extravesical ureter, but spring exclusively from the circular fibers of the ureter. In the segment adjoining the bladder,

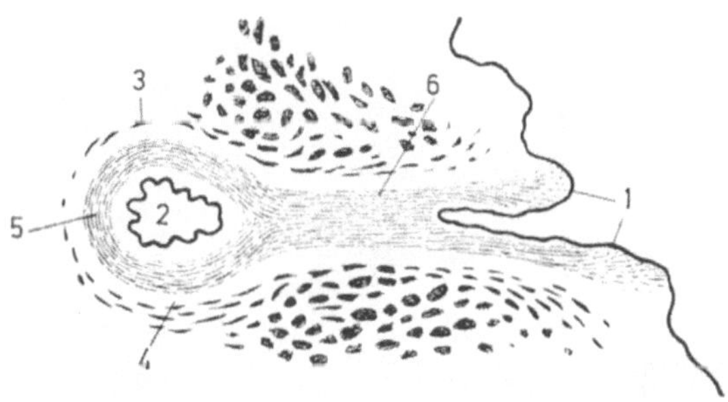

Fig. 1. Scheme of the prevesical ureteral sphincter: 1. Fibers proper of the ureter insert into the ureteral orifice. 2. Ureter upon penetrating into the bladder. 3. Extravesical periureteral sheath. 4. Waldeyer's space. 5. Prevesical ureteral sphincter. 6. Longitudinal fibers the continuation of sphincterial fibers

where it is about to enter the vesical wall, the ureter is composed solely of circular muscle fibers, which fact suggests the idea of a sphincter: the prevesical sphincter (GIL VERNET). On entering the bladder wall these circular fibers change direction, run lengthwise and all end at the labia of the ureteral meatus and the area of the vesical mucosa adjoining the meatus. Fig. 1 sketches this arrangement. In

* Sociedad Internacional de Urologia, Barcelone, Spain

other words, the prevesical sphincter and the longitudinal fibers emerging from it to terminate by insertion into the ureteral meatus are two distinct components of a single anatomical structure. The two components combine to make up an anatomical and functional unit – the sphincteral system of the ureter's terminal segment, designed to control the flow of urine and prevent vesicoureteral reflux.

This sphincter lies in the segment of the ureter adjoining the bladder. It measures about 2 cm in length and is made up of very slender muscle fibers arranged in a regular pattern. Only the lowest ones change direction, run lengthwise, and compose the parietal ureter's own fibers, clustering about it. These fibers cross the detrusor urinae, continue inward, and unravel below the vesical mucosa, thus beginning the submucosan stretch. All these fibers and in the labia of the ureteral meatus and in the vesical mucosa adjoining the meatus.

This basic description is valid for all age-groups. Detailed illustrations appear in an article published by *Archivos Españoles de Urología* (1970).

2. Where the Ureter's own Fibers End

The ureter's own fibers end in the ureteral meatus. None of those fibers extend as far as the trigonum vesicae, the neck of the bladder, or the colliculus seminalis.

Fig. 2 shows a series of sagittal sections from the terminal part of the ureter supporting this statement. The appearance of the ureteral fibers, all longitudinal, very slender

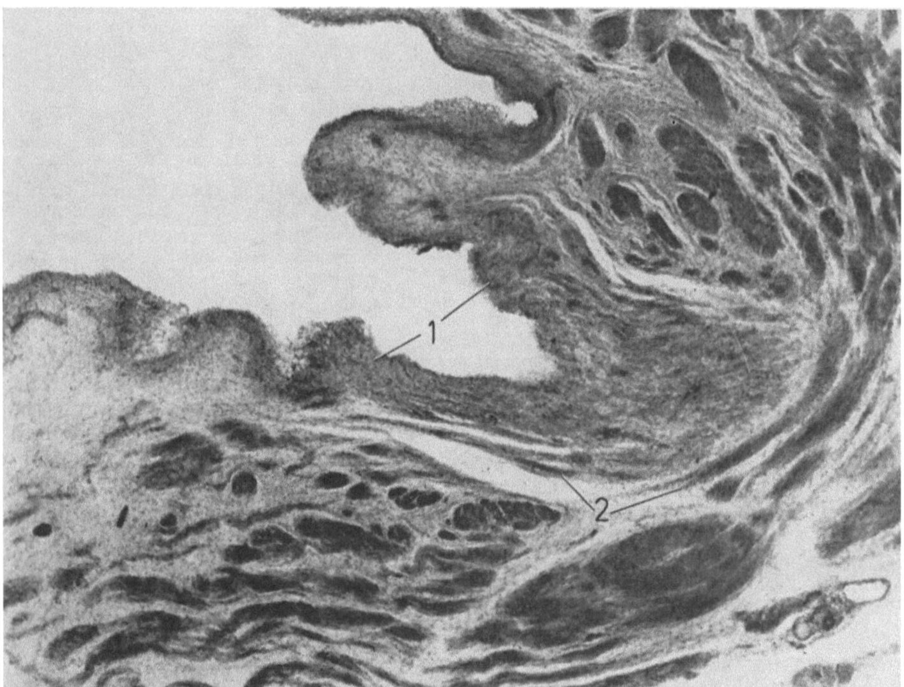

Fig. 2. Parasagittal cut that involves the ureteral orifice. 1. Fibers proper of the ureter inserting into the labia of the ureteral orifice. 2. Fibers of the ureteral sheath

with hardly any interfascicular tissue is in marked contrast to the thick fasces of the detrusor urinae, separated by a large amount of connective tissue. An open space intervenes between the two structures – that is, between the ureter's own fibers and the fasces of the detrusor: this is WALDEYER's cavity. Some fasces of the detrusor, in the shape of slender fascicles (GIL VERNET), approach the ureter. They belong to the periureteral sheath (Fig. 2).
In the sections adjoining the meatus it is possible to observe some very thin fibers which look ureteral and which enter the corium of the vesical mucosa. At the same time muscular elements belonging to the periureteral sheath emerge more noticeably. These will form the slanting segment of the sheath and the interureteral ring.

3. The Periureteral Sheath

In the segment adjoining the bladder and the parietal segment the ureter runs inside a tunnel known as the periureteral sheath. The two structures are separated by an area of lymphatic cells, described by WALDEYER in 1892.
The periureteral sheath comprises two segments, the extravesical and the intravesical, which are simply two different parts of a single structure. It serves to protect the ureter to it, hold in place, and to facilitate its movements in the terminal segment.
The extravesical segment encases about 2 cm of the ureter's length. Its fibers emerge from the detrusor, run upwards, and end at the adventitia of the ureter (Fig. 3).

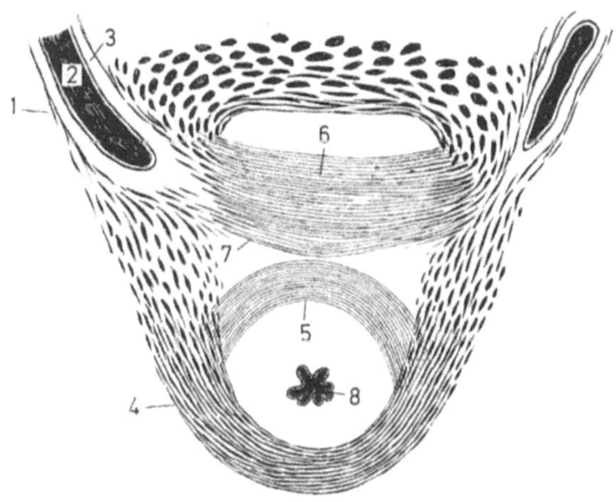

Fig. 3. Scheme periureteral sheath. 1. Extravesical periureteral sheath. 2. Ureter. 3. Waldeyer's space. 4. Loop of the detrusor. 5. Trigonal loop. 6. Interureteral ridge. 7. Oblique portion of the periureteral sheath. 8. Vesico-ureteral orifice

The intravesical segment exhibits more complexity. All its fibers also emerge from the detrusor. Some of them run horizontally and meet those from the other side midway, to form the interureteral ridge. Beyond this ridge other fibers run slantwise to meet those from the opposite side in the middle of the trigonum. All these fibers occupy only the posterior half of the trigonum.

4. Dynamic Interpretation of the Ureteral Musculature

To understand how this musculature works we must admit that the prevesical ureteral sphincter and the longitudinal fibers emerging from it and ending in the ureteral meatus make up an anatomical and functional unit. They relax to allow the passage of urine into the bladder; afterwards they contract, cause closure of the parietal ureter and withdrawal of the meatus, and thus prevent reflux.

If the longitudinal fibers of the parietal ureter were a continuation of the fibers bearing the same name in the extravesical ureter, they would have to contract at the same time, thus causing withdrawal of the meatus at the moment when urine is passed into the bladder and not after. If we assume that these fibers emerge from the sphincter and therefore form an integral part of the sphincteral system, then everything falls into place: closure of the terminal ureter and withdrawal of the meatus are caused by the same muscular structure, the ureteric sphincteral system. Endoscopic examination shows that urine brought along by the ureter empties into the bladder at regular intervals. The column of urine shoots out forcefully as an ejaculation, after which the ureter closes and the meatus withdraws. It seems clear that urinary ejaculation is brought about solely by contraction of the extravesical ureter's musculature: it is the end of the peristaltic wave which begins with a systolic phase in the musculature of the renal pelvis.

Thus the normal prevention of reflux is largely the work of the ureteric sphincteral system.

A subordinate factor preventing reflux is the crosswise segment of the parietal ureter and the segment below the mucosa. At the moment of micturition an increase in bladder pressure closes the ureter by coapting its walls. The technique for preventing reflux consists in the opening of a channel below the mucosa by ureterocystostomy.

The ureter and neighbouring structures must be in a normal state for the walls of the ureter to coapt perfectly. Edema, congestion, inflammation, or sclerosis cause reflux in some cases and ureteral ecstasis in others.

5. Vesicoureteral Reflux

The dynamic interpretation of the sphincteral system adjoining the bladder fits in with the findings of pathologic anatomy; including the hypertrophy of the sphincter and the longitudinal fibers emerging from it in cases of vesicoureteral reflux, where these muscle fibers swell up in the compensatory phase.

Fig. 4 shows a deictic example in a five-year-old child. Congenital and progressive obstruction of the bulbus urethrae caused dilation of the whole of the lower urethra, the vesical neck, and the bladder, and vesicoureteral reflux appeared. This figure shows a horizontal section of the ureter adjoining the bladder. The mucosa is normal and there is extensive hypertrophy of the ureter's own musculature, which forms a solid mass. The muscular fasces are arranged in a circle, looking like a hypertrophied sphincter.

Where the ureter enters the vesical wall it is very small, whether in a normal state or a pathologic one. It corresponds to the lowest part of the ureteral sphincter, at the exact point where the very slender muscle fibers in their circular arrangement carry on as longitudinal fibers (Fig. 5).

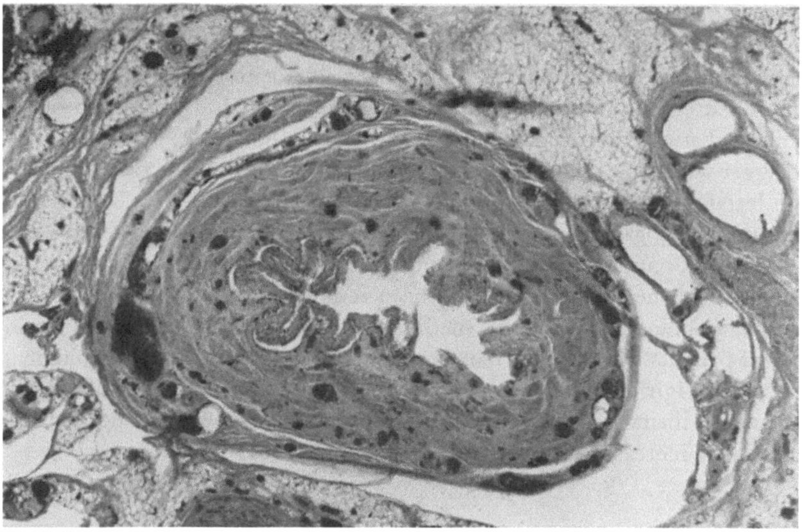

Fig. 4. Prevesical ureteral sphincter hypertrophie

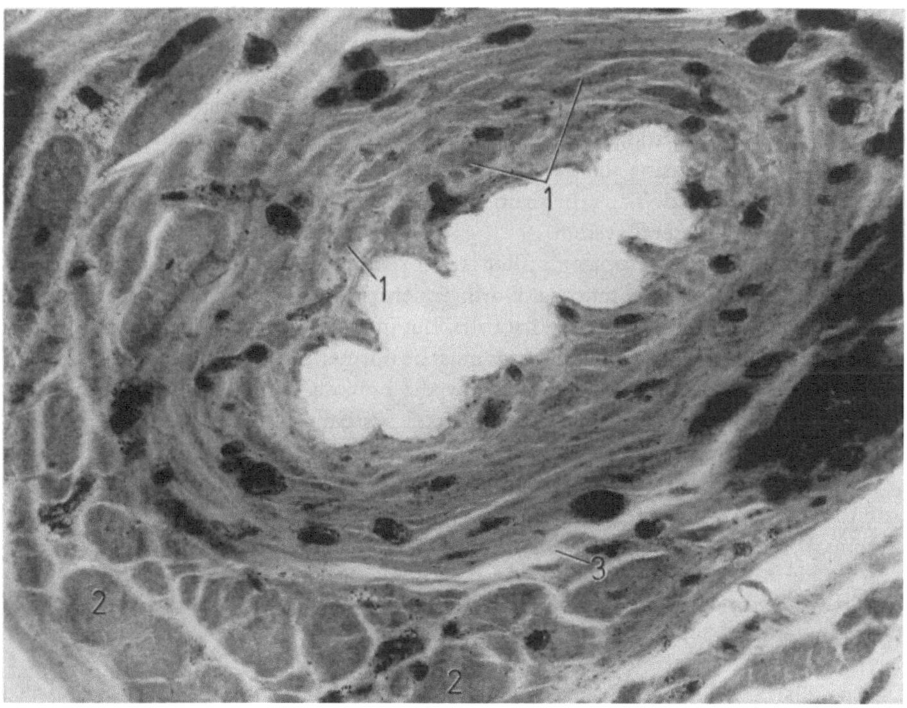

Fig. 5. Ureter where it enters the vesical wall. 1. Circular fibers carrying on as longitudinal fibers. 2. The periureteral sheath. 3. Waldeyer's cavity

Sections taken from the ureter where it passes through the bladder wall appear dilated, as if apersistently stronger vesical pressure had stretched the ureter. This is the first phase of an active reflux brought about during micturition by a sudden increase in vesical pressure. The musculature of the parietal ureter itself, made up of longitudinal fibers that emerge from the sphincter, appears much hypertrophied. It shows the struggle which the sphincteral system has put up to prevent reflux and looks as though intravesical pressure had finally got the better of it.

In short, morphologic examination of the ureter's terminal segment reveals an active ureteral reflux occurring at the moment of micturition. This would certainly have disappeared had the urethral blockage been removed.

Meticulous examination using the method of histotopographic sections during the autopsy of patients who suffered from vesicoureteral reflux, would indubitably give a great deal of insight into such problems. Direct study of the lesions is the best way of understanding them. Knowledge will be increased by exhaustive study of our cases rather than by increased numbers of the cases we study.

Summary

In the segment adjoining the bladder the ureter is made up of circular fibers, which suggest the idea of a sphincter: the prevesical sphincter. These circular fibers enter the bladder wall, change direction to run lengthwise, then all end at the labia of the ureteral meatus and in the area of the vesical mucosa which adjoins the meatus.

This means that the prevesical ureteral sphincter and the longitudinal fibers emerging from it to end in the ureteral meatus are two different parts of a single anatomical structure designed to control the flow of urine and prevent vesicoureteral reflux. When urine shoots into the bladder they relax and allow the urine through; after the ejaculation they contract, causing the parietal ureter to close and the meatus to withdraw so that reflux is prevented. Thus the main factor preventing reflux in the normal person is the ureteric sphincteral system.

A subordinate factor preventing reflux is the slanted segment of the parietal ureter and the segment below the mucosa. During micturition increased vesical pressure closes the terminal ureter by coapting its walls. But in order for this to happen the tissues of the ureter and neighboring structures must be normal. The technique for preventing reflux consists in opening a passage below the mucosa by ureterocystostomy.

The dynamic interpretation of the sphincteral system adjoining the bladder fits in with the findings of pathologic anatomy, hypertrophy of that system in cases of vesicoureteral reflux at the compensating stage. The case we have cited exhibits extensive hypertrophy of the ureter's own musculature, which forms a solid mass as though the sphincter had hypertrophied in its struggle to prevent reflux.

References

Debled, G.: Structure musculaire de l'uretère terminal. Acta urol. belg. **36**, No2 (1968).

Gil Vernet, S.: Biologia y Patologia de la Prostata, Cap. V. Madrid: Paz Montalvo 1953.

Gil Vernet, S.: Morphology and fonction of Vesico-Prostato-Urethral Musculature. Treviso: Libr. Edit. Canova 1968.

GIL VERNET, S.: Plan estructural y dinamico de la musculatura del ureter. Conferencia pronunciada en la Real Academia Nacional de Medicina el dia 25 de Marzo, 1969 Madrid.
GIL VERNET, S.: Union urèterotrigonal. Esfinter ureteral prevesical Archivos Españoles de Urologia. XXIII, Nª 3 (1970).
GOMEZ BOSQUE, P.: Anatomia funcional de la musculatura ureterica Archivo Español de Morfologia. XI, No 39 Valencia (1954).
GREGOIR, W.: L'Etiologie du refluxe vesicureteral congenite et du megaureter primaire. Urol. int. (Basel) **24**, 119–134 (1969).
HUTCH, J.: Theorie of Maturation of the Intravesical ureter. J. Urol. (Baltimore) **86**, 354 (1961).
RUOTOLO, A.: Sulla costituzione e sul significato funzionale della guaina e della cosi detta fessura di Waldeyer. Estratto dalle Ricerche di Morfologia, Roma 1948.
TANAGHO, E., PUGH. R. C. B.: The Anatomy and function of the ureterovesical junction. Brit. J. Urol. **35**, 151 (1963).
VERSARI, R.: La Morfogenesi della guaina dell' uretere umano. Ricerche del Lab. di Anat. Um. di. Roma. XV, fascicoli, 3, 4 (1911).
WALDEYER, W.: Über die sog. Ureterscheide. Verh. anat. Ges. Wien 1892.

Discussion

Vesicoureteral Reflux

1. Anatomy of the Ureterovesical Junction

The reason for the vesicoureteral reflux is insufficiency of the valve mechanism of the ureterovesical junction. In this case BISCHOFF differentiates an anomaly of the ureterovesical junction as a malformation due to a non-matured valve mechanism in connection with an intramural oblique canal and a submucosal part of the terminal ureter is a maturation of the perinatal period which should be finished during the first year (GIL-VERNET).

Histological investigations have shown that in patients with vesicoureteral reflux the musculature of the terminal ureteral segment is damaged (GREGOIR, TANAGHO). Beyond this point more collagenic tissue is found. These morphological changes of the reflux ureter are very similar to those of the primary non-obstructive ureter. A histological differentiation of both malformations is not possible (GREGOIR).

TANAGHO believes that the muscular anomaly of the ureterovesical junction is an embryonal congenital malformation. During embryonary development, the muscles of the ureter are built from the top towards the bottom. Therefore, the primary lesion, of the muscles is in the lowest ureteral segment, the vesicoureteral junction. Such an embryonal congenital malformation can display all stages, from a slight reduction of the muscles to total muscular defect. Thus, the most varied types of vesicoureteral reflux can be encountered.

Besides the microscopic structure of the vesicoureteral junction it is also very important for the sufficiency of the valve mechanism to know where the ureter inserts into the bladder (AUVERT, GREGOIR). It is well known that the intramural course of the ureter is more oblique when the ureter inserts more cranially into the bladder wall. The height of the ureteral insertion into the bladder wall depends on where the embryonal WOLFFian duct inserts into the cloaca: the more caudal the insertion of the

Wolffian duct, the closer to the bladder neck is the insertion of the ureter and the shorter the intramural segment (GREGOIR).
The further development of the trigonum vesicae essentially depends on the height of the origin of the ureteral bud out of the Wolffian duct (MELCHIOR). The length of the trigonal ledges and the symmetry of the trigonum vesicae depend on the location of the ureteral buds.

2. *Reflux and Infection*

As vesicoureteral reflux and urinary tract infection are usually observed together, the following questions have to be asked (AUVERT):

a) Can secondary vesicoureteral reflux be induced by an urinary tract infection?
b) Can vesicoureteral reflux cause urinary tract infection?

AUVERT has often observed vesicoureteral reflux in patients who suffer from trigonitis. The reflux disappeared with the complete cure of the trigonitis. He explains this type of vesicoureteral reflux by edema in the area of the ureteral ostium, followed by secondary changes of the muscle architecture in this region.
One of the main valve mechanisms operating against vescioureteral reflux is the compression of the intramural, or rather the submucosal ureteral segment during the increase of intravesical pressure. When infections induce reactions in the tissue, the compressibility of both the bladder wall itself and the intramural ureter is lost (HINMAN).
AUVERT also thinks that every primary vesicoureteral reflux is complicated sooner or later by an infection. With this type of the vesicoureteral reflux in general, the infection vanishes after a successful surgical treatment.
GRASSET considers inflammatory reactions of the mucosa not important in inducing vesicoureteral reflux. He thinks the decisive sites of action are the submucosal or muscular area. A disturbance of the muscular dynamics in the ureterovesical junction is said to cause insufficiency of the antireflux mechanism. This disturbance of the muscular dynamics is the result of physiological laws, according to which the muscles can be temporary paralysed as well as stimulated by inflammatory irritation.
GREGOIR believes that paralysis of the trigonal muscles is as poor an explanation of temporary vesicoureteral reflux as edema of the mucosa. GREGOIR demonstrated in experiments that iatrogenic myoparalysis of the ureter and the trigonum cannot provoke reflux with novocaine. Even infiltration of the mucosa with a solution of sodium chloride causes a local edema but no reflux. GREGOIR and TANAGHO regard the primary muscular defect as a "conditio sine qua non" for vesicoureteral reflux. Against this, appearance of temporary reflux in connection with urinary tract infection is a result of hypertonia of the bladder caused by inflammation.

Primary Vesico-Ureteral Reflux in Children

J. Auvert and B. Court*

Three anatomical structures contribute to the prevention of reflux: the mucosal valve, the Waldeyer muscle and the strong floor of the trigone (Fig. 1). Destruction of two of these three structures is necessary and sufficient to produce reflux (Fig. 2). This report will deal mainly with primary reflux in children, which has a serious effect on the kidney.

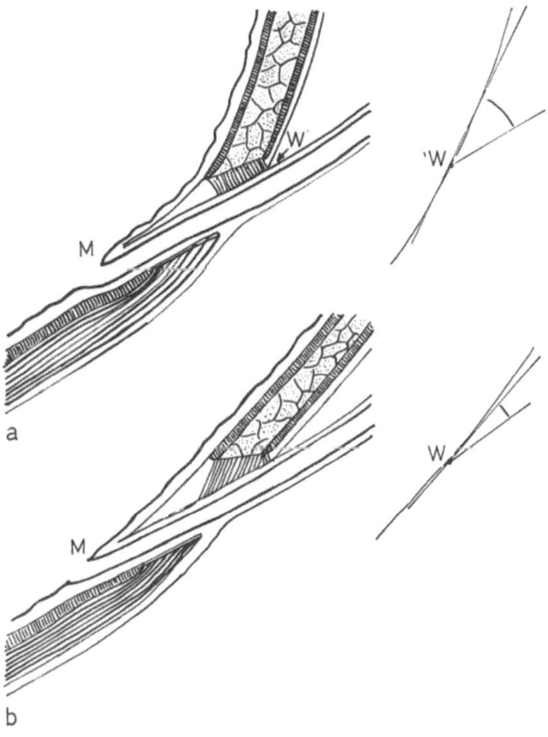

Fig. 1. Antireflux structures. The trigonal floor presents a good resistance against ureteric wall and mucosal valve. (*M* = mucosal valve; *W* = Waldeyer muscle; *a* = empty bladder; *b* = full bladder)

* Service d'Urologie Hôpital Henri Mondor, 94010, Creteil, France

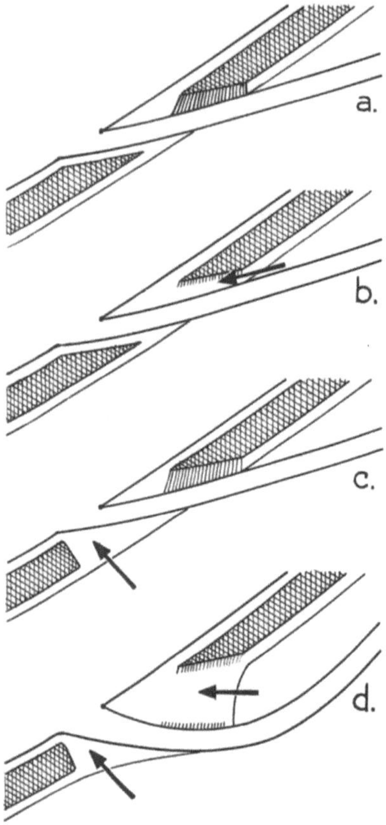

Fig. 2. Experimental production of reflux. Two of the three antireflux structures must be destroyed to obtain reflux. (*a* = normal structure; *b* = section of Waldeyer muscle; *c* = section of the trigonal floor; *d* = destruction of Waldeyer muscle and a lack of trigonal floor produces reflux even if the mucosal valve remains intact)

Primary Reflux in Children

This type of reflux causes no trouble in emptying the bladder. It depends on a congenital abnormality of the ureteral meatus, the incidence of which is twice higher in females than in males.

Symptoms are fever, cystitis, pyuria, proteinuria, sometimes lumbar pain at the time of voiding or enuresis.

Clinical examination is normal. Particular attention must be paid to micturition. Successive micturition may be observed, where the patient needs to micturate two or three times before the bladder is empty. This usually indicates a dilated upper urinary tract from which urine enters the bladder after micturition with reflux.

Intravenous pyelogram (I.V.P.) is the most important examination. The example in Fig. 3 shows: distorted calyces (chronic pyelonephritis or papillary necrosis), renal parenchyma are thin and irregular, pelvis and ureter enlarged. Following I.V.P. one can observe reflux at micturition directly but reflux is shown more accurately by *retrograde cystography* (Fig. 4).

The delayed technique is the best one: after retrograde injection of 100 ml of contrast medium through the urethra a few hours are allowed to elapse until diuresis has filled the bladder (the child should drink to shorten this time). When the child wants

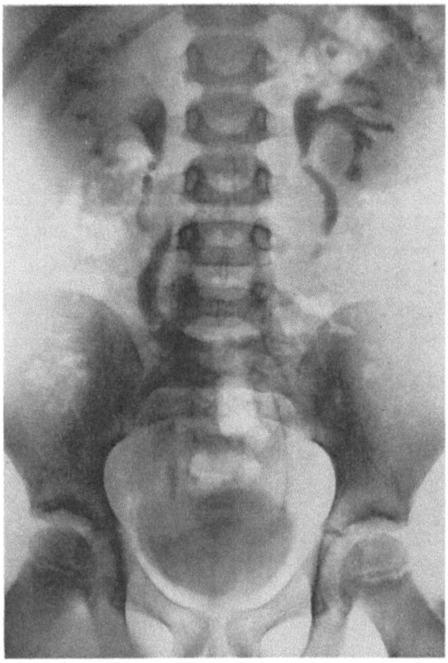

Fig. 3

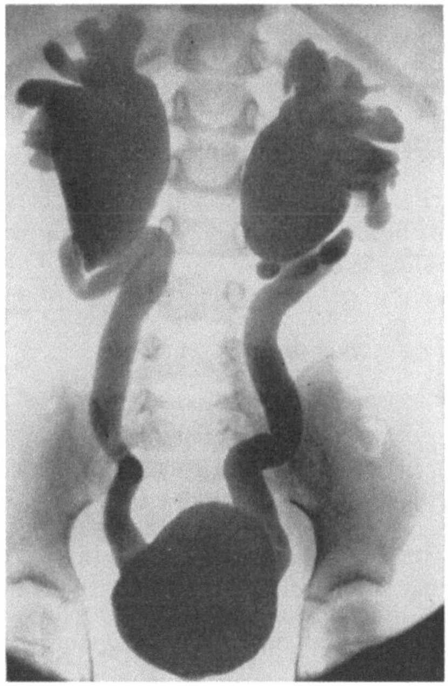

Fig. 4

Fig. 3. I.V.P. in a 5-year-old boy with fever and repeated attacke of cystitis

Fig. 4. Retrograde cystography during micturition (same case)

to void, pictures are taken before, during and exactly at the end of micturition. It is helpful to take another picture 30 sec later, showing the false residual urine. Radiocinematography is useful. Cystitis must be carefully excluded before a diagnosis of organic reflux is made. Reflux may be unilateral on one occasion and bilateral on another. We normally perform 2 cystographies 6 months apart. Bilateral reflux is twice as common as unilateral.

Associated abnormalities were observed in 63 patients subjected to operation: 4 duplicities with 2 ureters on each side, 3 mega-ureters with renal failure, 2 cases with bifidity and 1 bladder diverticulum.

Lateromeatic diverticulum as described by Hutch in paraplegic neurogenic bladder is uncommon. It must be differentiated from a parietal hernia of the bladder. The bladder is usually very large and its wall is thin, but often a mega-bladder empties well without residual urine. The trigone is very large and the orifices are for a part and a long way from the neck, which is normal and non-sclerosed. The diameter of the meatus may be as much as 6 to 7 mm. Ureteral peristalsis is reasonably powerful. Megalo-ureter is seen in small children.

Classification

It is necessary to distinguish between low and high pressure reflux; in low pressure reflux micturition causes very high pressure in the ureter, renal pelvis and calyces. The most dangerous type of reflux appears when pressure in the bladder is low and persists at micturition. The two forms are called an active reflux. Passive reflux which stops on micturition is usually well tolerated.

Renal Consequences

Reflux causes renal damage by inducing pyelonephritis with an interstitial inflammatory process which later becomes scarred tissue and leads to renal atrophy. The process is slow and takes about 20 years.

We observed in children:

between 0 and 5 years	no renal lesion
between 5 and 10 years	slight lesions
between 10 and 15 years	30 percent of cases with renal insufficiency
between 15 and 20 years	50 percent of cases with renal insufficiency

After 20 years, every patient with bilateral congenital reflux is suffering from severe renal failure, and almost all patients die before they reach 30 years of age.

The best methods of detecting renal damage are:

— Reduction in the size of both kidneys (seen by tomography), usually asymetric.
— Creatinine and urea clearance, which are best estimated on a 24 hour-study because of the false residual urine in patients with reflux and dilated upper urinary tract.
— Radioactive scintigraphy.

Treatment

We had 2 failures in 6 cases treated by the first Hutch technique.
The technique of Bischoff is less safe: the mucosal covering of the low ureter frequently disappears.
A combination of the above two techniques resulted in 5 failures in 31 ureters treated (84 per cent resulted in success).
All failures were treated by the Politano-Leadbetter technique, which is the best one. In 31 patients (50 refluxing ureters), there were only 8 failures, which are explained in 6 cases by associated megalo-ureter. We tried to avoid failures by reducing the diameter by plastic reconstruction.
The Politano-Leadbetter technique is also the best method of treating reflux in a thin ureter: 96 per cent success.

Ureteral Duplicity

Reflux in the upper ureter often disappeared after successful treatment of a ureterocele in the lower ureter.
Reflux in the lower ureter can be treated by pyelo-ureteral anastomosis or by reimplantation into the bladder after careful separation of the two canals.

Evolution of Kidney Function after Cure of the Reflux

Urinary infection disappeared in 40 patients except in 2 cases without antibiotics. Febrile attacks did not recur and growth started again.
Later results on kidney function are less favorable. Only 16 of 40 cases improved: all were younger than 10 years of age. In 24 cases damage remained unchanged. Pyelonephritic lesions seemed to progress slowly despite correction of the reflux.
It is important to correct reflux surgically before pyelonephritis is detected.
Children between 0 and 2 years of age are a special group. Very often they present reflux with abnormally long and broad ureters and renal dysplasia. We perform bilateral cutaneous ureterostomy in such cases.

Summary

Primitive reflux appears as a benign disease before 5 years of age. In young children, if reflux persists in spite of antibiotic therapy for two or three months and cure of associated cystitis, and if it is clearly seen in two retrograde cystographies more than three months, apart, the reflux must be corrected surgically before pyelonephritis develops and becomes irreversible.
It is too late to correct reflux surgically when there is renal failure (blood urea nitrogen over 80 mg percent and creatinine level over 2 mg percent); pyelonephritis will progress inexorably. In our series, 3 patients over 20 years of age underwent kidney transplantation.

References

AMAR, A. D.: Cystoscopic demonstration of vesico-ureteral reflux: évaluation in 250 patients. J. Urol. (Baltimore) **95**, 777, (1966).
AMAR, A. D., SCHEER, C. W.: Ureterocele with associated reflux. J. Urol. (Baltimore) **92**, 197 (1964).
AMBROSE, S. S., NICOLSON, W. P., The causes of vesico-ureteral reflux in children. J. Urol. (Baltimore) **87**, 688 (1962).
AMBROSE, S. S., NICOLSON, W. P.: Ureteral reflux in duplicated ureters: J. Urol. (Baltimore) **92**, 439 (1964).
AUVERT, J.: Calcul du rein et reflux vésico-urétéral chez l'enfant. Disparition du reflux après ablation du calcul. J. Urol. Néphrol. **71**, 482 (1965).
AUVERT, J., SAVARY, M.: Le reflux vésico-urétéral. J. Radiol. Électrol. **44**, 704 (1963).
AUVERT, J., XERRI, A., FEKETE, C., OTT, R.: Le reflux vésico-urétéro-rénal. Urol. int. (Basel) **22**, 141 (1967).
AUVERT, J., ZMERLI, S., CASTETS, J. B., MAGDER, E.: Reflux vésico-urétéral congénital pur (12 cas opérés et contrôlés). Congrès Européen d'Urologie pédiatrique. Bruxelles, 27 mai 1976. Acta urol. belg. **30**, 548 (1962).
BETTEX, M.: Über den vesiko-ureteralen Reflux beim Säugling und Kind. Bern/Stuttgart: H. Hubern 1965.
BETTEX, M., GENTON, N., SCHÄRLI, A.: Results of uretero-cystoneostomy in vesico-ureteral reflux in infants. Arch. Dis. Childh. **41**, 160 (1966).
BITKER, M. P.: Le reflux vésico-urétéral chez le paraplégique. Classification des reflux dans les vessies neurologiques. J. Urol. Néphrol. **64**, 208 (1958).
BITKER, M. P., PAYNOT, J.: L'exploration radio-manométrique de la vessie dans l'étude du reflux vésico-urétéral. Classification des reflux et indications thérapeutiques. J. Urol. Néphrol. **70**, 485 (1964).

BRUEZIERE, J.: Le reflux vésico-urétéral chez l'enfant et le nourrisson. Concours méd. **85**, 3343 (1963).

BRUEZIERE, J.: Opération de Leadbetter-Politano dans le traitement du reflux vésico-urétéral chez l'enfant. Ann. Chir. infant. **7**, 67 (1966).

CENDRON, J.: Reflux vésico-urétéro-pyélique chez l'enfant. J. Urol. Néphrol. **72**, 153 (1966).

CHATELAIN, C., BOURGEAU, M., MATHIEU, H., PROT, P., GUEDENEY, J., LABURTHE, TOLRA, Y., MARTIN, P.: A propos du reflux vésico-urétéral chez l'enfant. Une revue de 140 cas. 15ème Congrès de la Soc. Int. Urol. (Tokyo) 12–18 juillet 1970.

CHAUVIN, H. F.: Le reflux vésico-rénal. Rapport du 58ème Congrès de l'A.F.U. Paris 1964.

GRASSET, D.: Réflexions à propos d'une série de 42 reflux vésico-urétéraux opérés chez l'enfant. 15ème Congrès de la Soc. Int. Urol. (Tokyo) 12–18 juillet 1970.

GREGOIR, W.: Le reflux vésico-urétéral congénital. Acta urol. belg. **30**, 286 (1962).

GREGOIR, W., DEBLED, G.: L'étiologie du reflux congénital et du méga-urétère primaire. Urol. int. (Basel) **24**, 119 (1969).

GREGOIR, W.: Traitement chirurgical du reflux congénital et du méga-urétère primaire. Urol. int. (Basel) **24**, 502 (1969).

HIMMAN, F., MILLER, E. R.: Mural tension in vesical disorders and ureteral reflux. J. Urol. (Baltimore) **91**, 33 (1964).

HUTCH, J. A.: Vesico-ureteral reflux in the paraplegie: cause and correction. J. Urol. (Baltimore) **68**, 457 (1952).

HUTCH, J. A., BUNGE, R. S., FLOCKS, R. H.: Vesico-ureteral reflux in children. J. Urol. (Baltimore) **74**, 607 (1955).

HUTCH, J. A., MILLER, E. R., HINMAN, F.: Perpetuation of infection in urolstraded urinary tracts by vesico-ureteral reflux. J. Urol. (Baltimore) **90**, 88 (1963).

HUTCH, J. A., TANAGHO, E. A.: Primary reflux. J. Urol. (Baltimore) **93**, 158 (1965).

JOHNSTON, J. H.: Reflux treated by ureteric reimplantation. J. pediat. Surg. **1**, 145 (1966).

LATTIMER, J. K., LEUTZINGER, D., JUSTICE, M., KNAPPENBERGER, S.: When should the child with reflux be operated upon? J. Urol. (Baltimore) **85**, 275 (1961).

LEADBETTER, W. F.: Vesico-ureteral reflux in children. Rapport du 15ème Congrès de la Société Internationale d'Urologie. Tokyo 12–18 juillet 1970.

MARSHALL, F. C.: Excretory urography changes in children which suggest occurence of reflux. J. Urol. (Baltimore) **87**, 681 (1962).

MATHISEN, W.: Technik bei vesiko-ureteraler Anastomose und zur Korrektion von vesiko-ureteralem Reflux. Verh. dtsch. Ges. Urol. **19**, 150 (1961).

MELICK, W. F., BRODEUR, A. E., NARYKA, J. J.: Use of ureteral pacemaker in treatment of ureteral reflux. J. Urol. (Baltimore) **95**, 184 (1966).

POLITANO, V. A., DURHAM, N. C.: Vesico-ureteral reflux in children. J. Amer. med. Ass. **171**, 1252 (1960).

POLITANO, V. A., HARPER, J. M.: Experiences and results with conservative management of vesico-ureteral reflux. J. Urol. (Baltimore) **92**, 445 (1964).

POLITANO, V. A., LEADBETTER, W. F.: An operative technique for the correction of vesico-ureteral reflux. J. Urol. (Baltimore) **79**, 932 (1958).

SCOTT, J. E. S.: Results of operations for ureteric-reflux. Arch. Dis. Childh. **41**, 165 (1966).

TANAGHO, E. A., HUTCH, J. A.: Primary reflux. J. Urol. (Baltimore) **93**, 158 (1965).

TANAGHO, E. A., HUTCH, J. A., MEYERS, F. H., RAMBO, O. N.: Primary vesico-ureteral reflux: experimental studies of its etiology. J. Urol. (Baltimore) **93**, 187 (1965).

VIVILLE, CH.: La mise en évidence du reflux vésico-urétéral. J. Urol. Néphrol. **71**, 921 (1965).

WILLIAMS, D. I., ECKSTEIN, H. B.: Surgical treatment of reflux in children. Brit. J. Urol. **37**, 13 (1965).

WINTER, C. C.: A new test for vesico-ureteral reflux: an external technique using radio-isotopes. Arch. Dis. Childh. **36**, 130 (1961).

Ureteral Dynamics in Obstructive Uropathy

D. GRASSET*

Urodynamic exploration represents an essential stage in the diagnosis of obstructive uropathy. It provides information about the functional capability of the urinary tract and, consequently, about the chances of recuperation after removal of an obstacle. It is useful in formulating therapeutic indications and it generally leads up to the prognosis.

Three of the methods used are:

1. *Cineradiography:* this provides information about the peristaltic activity of the upper urinary tract, without any risk of infection or interference with pyelo-ureteral dynamics due to intraluminal instruments. The filmed or televised pictures can be recorded on magnetic tape for convenience of interpretation.
Cineradiography, however, has its limits:
a) Pyelo-ureteral dynamics does not depend solely on peristaltic activity but on other parameters which are not recorded by cineradiography.
b) Cineradiography is not always feasible: patients with obstructive uropathy frequently lack adequate kidney function, so that opacification of the urinary tract cannot be effected by the intravenous route.
2. *Electro-ureterography* is an interesting method but we have no personal experience of it.
3. *Electromanometry* is the technique most frequently employed for the investigation of the dynamics of obstructive uropathy. Following other investigators (AUVERT, LUTZEYER, RUTISHAUSER), we introduced this method at the Urological University Clinic of Montpellier in 1964. For the past few years, and following initial work with adults, we have concentrated our study of pyelo-ureteral dynamics on infants and children.

Pyelo-Ureteral Electromanometry or Uretero-Pyelo Manometry (U. P. M.)

Methods and Techniques

U.P.M. is effected by means of a "pressure chain" composed of three elements

1. a pressure-sensitive cell (pressure transducer) upon which the pressure to be measured is applied;

* Clinique Urologique, Hôpital St-Charles, C.H.U. Montpellier, France

2. an electromanometer which transforms variations of pressure into variations of electric current[1];
3. an oscilloscope which displays the electric current variations in graphic form as a mathematical function.

In order to ensure correct transmission of the endo-ureteral pressure to the transducer, the following important points should be observed:

1. There should be no air bubbles in the system and
2. the catheter must be of the correct size. Its rigidity is essential. Its caliber does not directly influence transmission of pressure, according to the hydrostatic law of Pascal. Contact of the liquid with the catheter wall, however, can produce a resonance phenomenon which will interfere with the transmission of peristaltic waves. This phenomenon is negligible in the large-caliber catheters used in adults, but it can result in significant errors with the narrow catheters used in infants and children. This problem is solved by using silicone catheters.
3. The introduced catheters must not be used for retrograde pyelography until after the urodynamic study has been made, as this might create artificially elevated pressure.

Results

1. Pyelo-Ureteral Dynamics in Normal Conditions

The dynamics in the adult are practically identical to that in the child. They can be represented as a function of two parameters:

a) The *diastolic pressure*, or base pressure, is equal to or lower than 10 cm H_2O. This is a permanent pressure maintained by the pyelo-ureteral musculature. The manometric expression of this tension may be called "pyelo-ureteral tonus" in analogy to the

Fig. 1

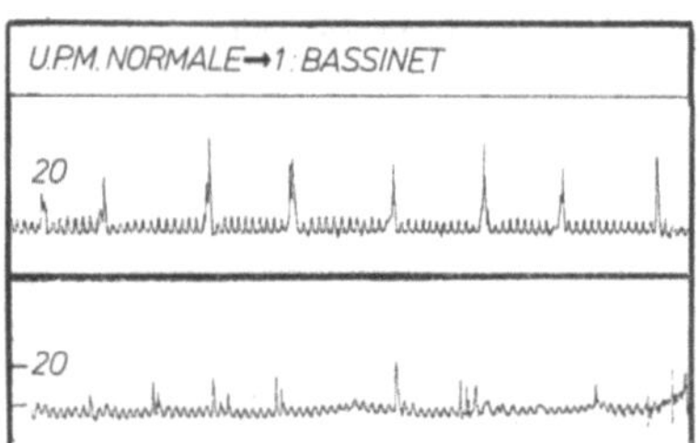

Fig. 2

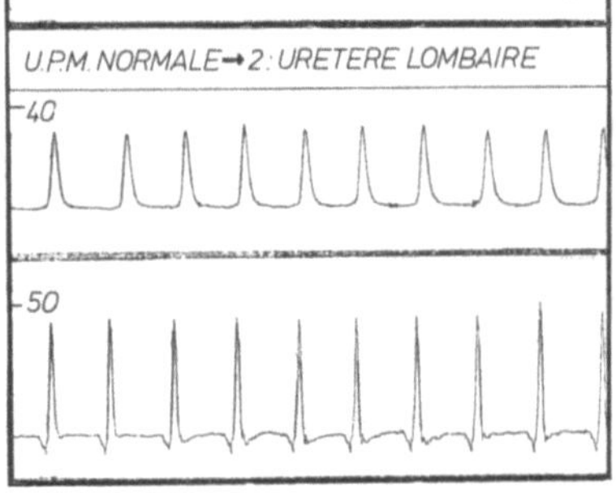

Fig. 3

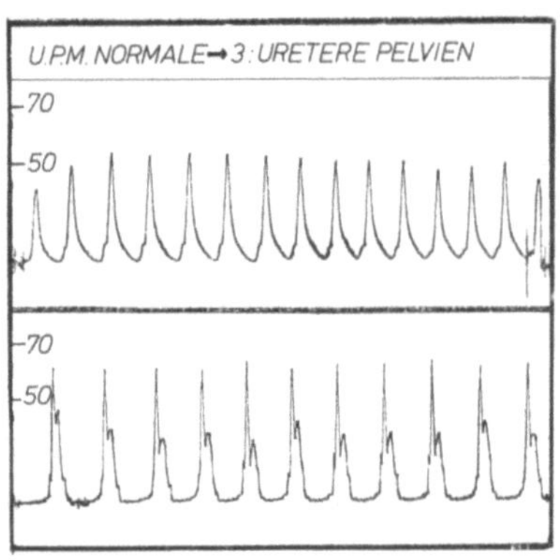

[1] Schwarzer (Polyphysiograph) or Electronique Marcel Dassault (Roche Bio-electronique)

term "tonus" used by physiologists in describing the residual contraction characteristics of a striated muscle at rest.

b) The *systolic activity* is represented by sudden variations in pressure, induced by peristaltic waves. The frequency, amplitude and morphology of the peristaltic waves varies from the top to the bottom of the pyelo-ureteral tree. The variations observed in the child are similar to those seen in the adult: "Pyelitic fibrillations" are recorded in the renal pelvis at regular intervals. They represent systoles with amplitudes not exceeding 20 cm H_2O (Fig. 1). In the lumbar ureter the systoles reach an amplitude of 40 to 50 cm H_2O with an average frequency of 3 systoles per minute (Fig. 2). In the pelvic ureter, however, the systoles are more frequent (4 to 6 per min) with an amplitude of 50 to 60 cm H_2O. The only difference noted between infants and adults is that in infants one rarely sees the polyphasic waves which characterize the normal activity of the pelvic ureter in adults (Fig. 3).

2. Pyelo-Ureteral Dynamics in Obstructive Uropathy

One may distinguish three stages of development which correspond to particular therapeutic indications.

a) The first stage, or stage of *compensation*, is characterized by peristaltic activity of exaggerated frequency and amplitude (*hyperkinesia*) (Fig. 4).

Fig. 4

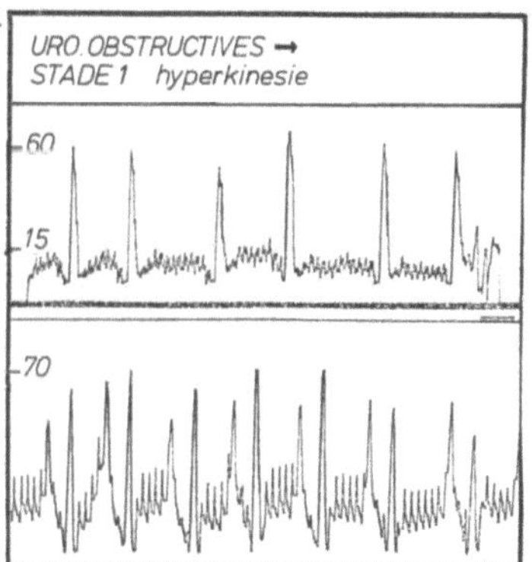

Fig. 5

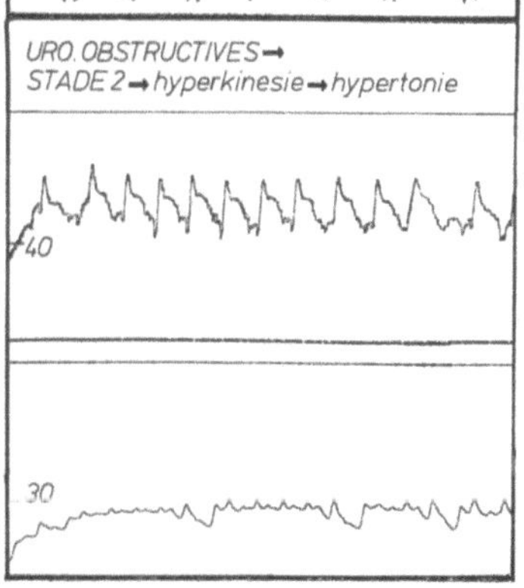

Fig. 6

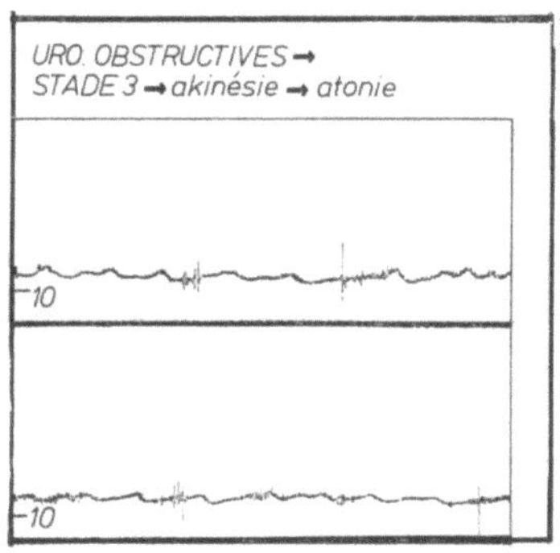

Fig. 7

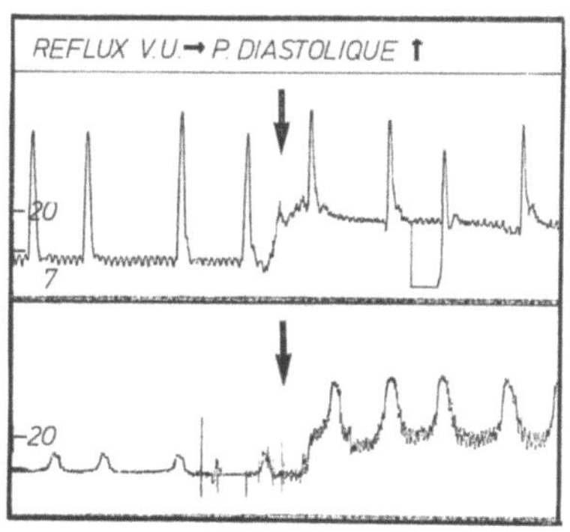

This manometric phenomenon cannot be demonstrated absolutely by cineradiography because it is not proportional to the caliber of the ureter. Electromanometry can confirm that the ureters, although dilated, are still functional. It is evident that at this stage conservative surgery will give the most promising results.

b) In the second or intermediate stage, the stage of *muscular hyposthenia*, one sees a decline in peristaltic activity (*hypokinesia*) although the diastolic pressure is still elevated (hypertonia) (Fig. 5). A certain tonus of the ureteral musculature still exists and this justifies conservative surgery, possibly preceded by temporary urinary diversion in order to demonstrate the recuperative potential of the urinary tract.

c) The third stage, the *decompensated* or *asystolic stage*, is characterized by complete absence of peristaltic activity (akinesia) and a collapse of diastolic pressure (atonia) (Fig. 6). At this stage, there is no hope of recovery and a palliative urinary diversion must be undertaken.

We have also studied ureteral dynamics during *vesico-ureteral reflux*. These studies are possible only when there is passive reflux with a gaping orifice, not obliterated by an ureteral catheter. The ureteral pressure is measured with the bladder empty. Then, as the bladder is filled with radio-opaque liquid, the variations in ureteral pressure are recorded, and the vesico-ureteral reflux is observed by amplified radioscopy. Therefore it is easy to see the reflux and to record the pressure simultaneously. We have always observed an elevation of the diastolic pressure during reflux (Fig. 7). But the systolic activity can vary: it can either stay the same or show hyperkinesia. The latter is evidence of good adaptation by the upper tract to the functional obstacle (the reflux) and antireflux surgery may be undertaken with a reasonable chance of success.

Conclusion

Pyelo-ureteral electromanometry is considered an important step in the urodynamic investigation of obstructive uropathy in infants and children. The information it supplies is complementary to that given by I.V.P. and cineradiography. It justifies the risk of instrumental exploration. The risk can be minimized by observing technical precautions.

Dynamics of Urinary Obstructions

P. F. BISCHOFF, H. BOEHNCKE, and A. LASSRICH*

An evaluation and demonstration of late results of plastic corrections in infantile urinary obstructions is given cinematographically up to 8 years postoperatively. Only corrections of severe and severest urinary obstructions in babies and infants, their operative corrections and the late results are presented.

The results of anatomical and embryological investigations in different stages of urinary obstructions of OESTLING concerning urinary disturbances of varying forms and extent are especially interesting as well as the ideas of MARBERGERS's hypothesis for investigation. Already 3 years ago MARBERGER at the Urodynamic Symposion in Iowa City demonstrated his ideas by film. His new film and parts of the elder one have been demonstrated here.

Referring to these ideas I may call your attention to the so called urethral-valves of the posterior or postprostatic urethra, positively and urinary disturbance only defined to boys.

TURNER-WARWICK demonstrated the great variety of bladder-neck-obstructions, especially in regard to the different circumstances of the anatomic basis in male and female.

Development of these obstructions and valves of the postprostatic urethra can be explained only by the embryonal formation. Perforating the urogenital membrane of the trigonum the posterior urethra mechanically may be kinked by the part of the fascia perinei superficialis. Also here we have a mobile muscular tube the motility of which being restrained by the tough vicinity.

Consequently these are the same conceptions as they have been pointed out by the research of OESTLING concerning the conflict-situation between on one side the uretero-pelvic junction, upper and middle part of ureter and on the other side the tough, unelastic adventitia.

From this point of view also for the distal ureter there is a similar situation facing the tough and independent muscular layer of the bladder (BISCHOFF).

Obviously the turbulences occuring in these obstructed, kinked and restrained muscular tubes support development of the wellknown, often giant dilatations and hypertrophics of the ureter walls, appropriately named "energy killer" by MARBERGER.

---

* Abteilung Urologie, Elisabeth-Hospital (Director: Prof. Dr. P. F. BISCHOFF), Hamburg, Kinderkrankenhaus Hochallee (Director: Dr. H. BOEHNCKE), Hamburg, and Radiologisches Institut, Abteilung Pädiatrie, Universitätskrankenhaus (Director: Dr. A. LASSRICH), Hamburg-Eppendorf, Germany

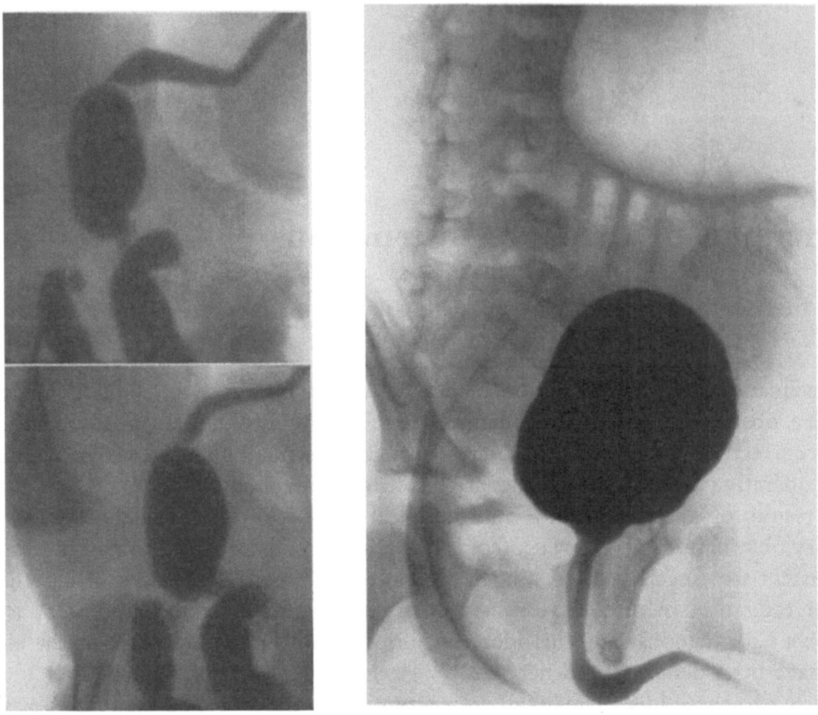

a b c

Fig. 1a and b. M. W., boy of 3 months. Micturition-cystography: balloon-like distention of the distal ureters on account of kinks of the postprostatic urethra by fascia perinei superficialis.
The main part of the contrast has been pressed into the upper urinary tract by the way of insufficient ureter-ostia

Fig. 1c. Micturition-cystography three years after splitting of fascia perinei superficialis (bilateral shortening and modellage of megaureters and bilateral antireflux-plasty after Bischoff-method)

I wondered why GRABER in his excellent report and imposing model-explaining the opening of the trigonum urogenitale especially in females being responsible for urinary incontinence did not make the further step also to refer to the possibility of urinary obstruction in males caused by the tough pelvic fascia.
Basing on this perhaps mechanical seeming conception, we have developed our surgical procedures during the last 20 years, initially more or less unconscious and by intuition, then with more knowledge and experience directed, modified, again abandoned and once more corrected (BISCHOFF and LASSRICH).
Following this conception, we think to have found out the explanation of the successful indwelling catheter-drainage propagated by us for treatment of corresponding obstructions of the postprostatic urethra in babies and infants (BISCHOFF).
We observed indwelling catheter-drainage for some weeks obviously in many cases being sufficient not only to overcome uremia but also definitively to extend the kinked urethra as well as the correction of teeth in anomalies of position is carried out by a jaw-orthopedist.

In severe cases and failures of indwelling catheter-drainage we changed over to splitting of the fascia superficialis perinei by perineal approach without opening of the urethra, technically in babies and infants a very simple procedure.
Exposition of the pars prostatica of urethra is so surprisingly easy that in advance a catheter should be inserted in order to avoid injury of the urethra.
The situation is much more clear and simple compared to adults. Splitting of the fascia perinei superficialis from the distal part of the pars prostatica until passage into the urethra is important.
The micturition-cystography (MCG) of a boy of 3 months shows a typical valve of the posterior urethra, total reflux and excessive megaureter (Fig. 1a, b).
Picture 5 shows the same case three years later after splitting of the fascia perinei superficialis by perineal approach (besides an additional bilateral shortening of megaureters, modellage and anti-reflux-plasty after our own procedure), the MCG reveals a normal pattern (Fig. 1c).
In severe urinary obstructions the early surgical intervention is of greatest importance. This means intervention before termination of the first year of life (BISCHOFF *et al.*). This because we know that there is no technical limit of surgical possibilities, but being induced by extent and severity of renal damage and infection (BISCHOFF and BOEHNCKE). Sclerotisation of the muscular tubes is not to be expected in the first year of life.
In many cases concerning evaluation of the results we are mistaken in putting the spotlights on technical questions obviously important. Fate und future of children early operated on by restauration of free urinary passage only depends on remaining efficiency of renal parenchyma and thoroughly healing of the infection.
The 6 cases demonstrated in the film, without exception operated on 5 to 10 years ago, all have been surgically corrected after methods wellknown to all of us and frequently discussed. Disturbances of the vesical outlet have been treated either by indwelling catheter-drainage, by electro-resection or in the case of one girl by VY-plasty of the bladder-neck. All reflux-cases have been corrected by our own method.
The two megaureter-cases without reflux in single kidneys have been operated on by shortening and modellage obtaining the intramural distal ureter-stump.
Perhaps it may be mentioned that modellage should be less extensive depending on the quality of the ureter-wall, in other words in infants and babies mostly an harmonic oblique adaptation without kinks is sufficient.
I would like to point out that we frequently applied the much more simple procedure of LICH-GREGOIR in simple infantile reflux cases, really indicated and not demonstrated in the film, especially in regard of the position of ureter-ostium and bladder-capacity. This means muffing of the ureter from the outside. In cases of extensively elongated and dilated ureters, especially in proven sufficient bladder-capacity, we prefer our modification of the Leadbetter-Politano and Hutch-procedure, which means extraperitoneal lysis of the megaureter, open pullthrough of the ureter, intravesical modellage, transposition of the ureter-ostium near to the vesical outlet and suture of the mucous-muscular membrane layer (BISCHOFF).
Obviously every statistic comparing the results of only one surgical method without exact explanation of special circumstances of the cases and comparing the quality of different surgical procedures, does not meet the problems, as well as uncritical

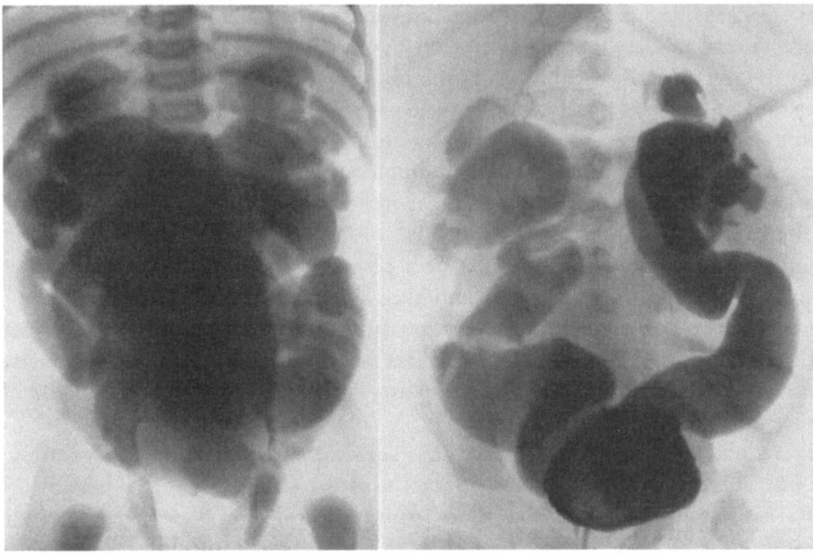

Fig. 2a and b. a Boy of 13 days. Megaureter-megacystis-syndrom and bilateral reflux (intravenous urogramm 24 hours after injection). b Five months after indwelling catheter-drainage operative correction is possible (bilateral dissection of megaureters, modellage, bilateral antireflux-plasty)

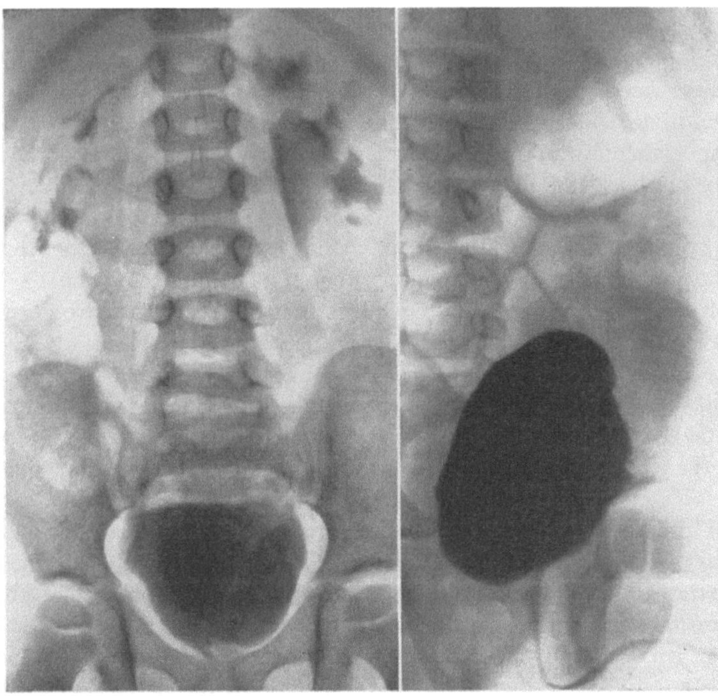

Fig. 3c and d. c Intravenous urogramm six years after termination of treatment. d Micturition-cystography: no reflux. After micturition no residual urine

statitistics with quoted surgical successes of 95 per cent clearly prove that indication for surgical intervention is objectionable (BISCHOFF). Therefore I am very glad understanding from the report of AUVERT as well as from GREGOIR that they have obtained the same results as we did and by the way every experienced surgeon in this most complicated materia.

Only by a keyboard of procedures and modifications harmonized to every case healing and improving in infants can be achieved.

Otherwise in children over 12 years and adolescents we feel glad having obtained the status quo and prevented any further deterioration.

Evaluation of a success of treatment depends on the following criteria:

1. Normalizing of blood-chemistry, no acidosis after Astrup-procedure, normal results of creatinine and blood-urea,
2. thoroughly healing of infection, this means sterile urine without leucocytes and bacteria,
3. normal rate of growth and development (BOEHNCKE *et al.*), further in urography:
4. normal function in time,
5. free urinary passage without reflux,
6. restauration of dilated renal pelvis and calices.

An example: The intravenous urogramm of a 13-day-old newborn shows a severest bladder-neck obstruction, megaureter – megacystis and bilateral reflux (Fig. 2a). Without immediate indwelling catheter-drainage as we know from similar cases the child would have been lost in the first weeks of its life.

After 5 months indwelling catheter-drainage, the infection is under control, the acidosis overcome and a normalisation of creatinine and blood urea have been achieved (Fig. 2b). Surgical intervention may be now without risk.

In several sessions bilateral dissection of the megaureters, modellage and bilateral antireflux-plasty following our own procedure under safeguarding continuation of indwelling catheter-drainage. No surgical correction at the bladder-neck has been carried out.

Six years after conclusion of the treatment the dilated renal pelves and calices and megaureters regenerated and are normalized (Fig. 2c, d). The film demonstrates physiological peristaltic waves, proving functional healing; this also being evident by thoroughly emptying of the bladder without residual urine. There is no reflux as well in stout filling of the bladder as after micturition.

These three days of excellent lectures on basic research-work and clinic with most interesting and original experiments on account of reproduction of the physiology of urodynamics by a scientifically exact model or explaining of pathological deviations by satisfying anatomical bases at least in a measurable and recordable form, proved that still today there is a great gap between clinic and research to be overcome. It is a special merit having brought together teams and research groups over the world here for the purpose of a meeting on urodynamics in Aachen. Only by this way we have a possibility of productive evaluation of more or less open questions, observations and experiences. This only results in combination to a scientific entireness.

Surgically and clinically we have the precarious situation of being restricted in our methods and surgical measures on sick people to step by step development which in failures does not immediately admit diagnosis of a mistake.

Therefore we have to take in mind that our view behind and forward in every direction accepts a scientifically based support and perhaps a new impulse for our measures from any direction. We, as we might say, in the first line of the happening could perhaps be estimated unscientifically in comparison to reckoning, measuring and electronic-microscoping inquirers. But accomplishment of our great goals and problems requires a good function of our cerebellum, generously granted to the surgeons, instead of only an excellent function of the cerebrum.

References

Bischoff, P. F.: Betrachtungen zur Genese des Megaureters. Urol. int. (Basel) **2**, 257–286 (1961).

Bischoff, P. F.: Möglichkeiten, Erfolge und Mißerfolge plastischer Operationen im Kindesalter. Z. Urol. **62**, 12, 883–895 (1969).

Bischoff, P. F.: Eine neue Modifikation zur Behandlung des Megaureters mit vesicoureteralem Reflux. Z. Kinderchir. **8**, 1/2, 93–106 (1970).

Bischoff, P. F.: Problems in treatment of vesicorenal reflux. J. Urol. (Baltimore) **107**, 133 (1972).

Bischoff, P. F., H. Boehncke: Dringliche Urologie im Säuglings- und Kindesalter. Chirurg **39**, 12, 544–548 (1968).

Bischoff, P. F., H. Boehncke, A. Lassrich: Argumente für die Frühoperation der Harnwegsmißbildungen im Kindesalter. Mschr. Kinderheilk. **114**, 4. 252–254 (1966).

Bischoff, P. F., Lassrich, A.: Pyeloskopische Beobachtungen am primären Megaureter vor und nach plastischen Operationen (Film). Langenbecks Arch. klin. Chir. **259**, 168–172 (1960).

Boehncke, H. Bruns, H. A.: Renaler Minderwuchs bei Kindern mit operativ behandelten, kongenitalen Harnwegsmißbildungen. Urologe B **10**, 4, 161–178 (1970).

Marberger, H.: Movie: Flow Characteristics. Hydrodynamics of Micturition. F. Hinman (Ed.), Charles C. Thomas. Publ. Springfield Illinois, 1971.

Oestling, K.: The genesis of hydronephrosis. Acta chir. scand. **86**, Suppl. 72 (1942).

Discussion

The Etiology of the Primary Megalo-Ureter

The non-obstructive primary megalo-ureter always has an abnormal structure in the terminal ureteral segment (Gregoir). While Tanagho sees the reason for this in the circularly arranged muscle fibres of the juxtavesical ureter, Debled refers especially to the pathological changes of the intravesical ureter. Bischoff shows that the primary megalo-ureter is the most frequent special form of the functional ureterostenosis. The embryo has an ureter which is folded up like an accordion and stretches later in its adventitia. If the adventitia adheres at one point, then the functional disturbance develops into a real megalo-ureter.

Clinical Experiences with Antireflux Procedures

J. G. MOORMANN and P. MAY*

The etiology of vesico-renal reflux and megaureter and the problems of diagnosis of reflux will be excluded from this report of our own experience with anti-reflux procedures. These problems are dealt with in other contributions in this monograph. We now use only three of the numerous surgical methods for the correction of reflux:

1. the anti-reflux operation after LICH/GREGOIR,
2. ureteroneocystostomy (UCN) with a submucous tunnel modified after POLITANO/LEADBETTER and PAQUIN,
3. the Boari-flap operation (BOARI/KÜSS) with submucous tunnel.

In our experience there are specific indications for each technique and good results may be achieved with consistent use of any of these techniques.

As will be shown later in the analysis of our cases and of the surgical techniques we usually prefer the operation after LICH/GREGOIR for treatment of vesico-ureteral reflux. The indication for this operation is given by the morphology of the upper urinary tract on IVP and on the reflux cystogram. If the ureters are only moderately dilated and curved our operation of choice is the LICH/GREGOIR procedure.

In cases where the ureters are more ectatic and more curved on IVP, where we must suspect fixation by adhesions and in megaureters with a narrow prevesical segment we prefer UCN with tunnel in the region of the trigone.

Our indication for surgical treatment of vesico-ureteral reflux is determined by the morphological picture of the upper urinary tract and by refractory or recurrent secondary urinary tract infection despite sufficiently extensive chemotherapy according to antibiograms. We would like to specify our indications as follows:

1. pathological alterations of the ureters and the collecting system,
2. changes at the border between parenchyma and calyces that are visible on x-ray,
3. persistent urinary tract infection,
4. persistent vesico-ureteral reflux in spite of non-infected urine after chemotherapy.

All surgical cases were handled according to these indications. All children had been treated with antibiotics for at least 5 months but usually 2–3 years prior to surgery without success.

Retrospective analysis of the two largest collectives shows that in most patients the

---

* Urologische Universitätsklinik (Director: Professor Dr. C. E. ALKEN), Homburg/Saar, Germany

symptoms had been present for several years, and that the urinary tract infection had been treated without success.
Considering the protracted attempts at chemotherapeutic correction, the number of patients with manifest pyelonephritic changes and the number of check-ups and x-ray examinations it must be said that surgical treatment of reflux is usually too late. The effects of back-pressure on the renal parenchyma must also be taken into consideration. In several cases we were able to record pressures of up to 60 cm water in the renal pelvis during voiding.
All children operated upon were followed up in our own pediatric urology outpatient clinic, where we are able to administer precise chemotherapy. The follow-up examinations follow a special schedule.
Since 1967 we have performed 413 vesico-ureteroplasties in 284 children. The series includes 186 girls and 98 boys in the age range newborn to 14 years.
The Lich/Gregior technique was used in 265 cases and UCN with tunnel in 134. Seven ureters with tunnel were implanted into the fundus of the bladder, in 7 other cases we used the Boari/Küss flap technique.
Of 176 children subjected to the Lich/Gregoir technique 89 had bilateral and 87 unilateral vesicoureteral reflux. In 70 cases bilateral operations were carried out in one session. In the remaining 19 children two operations were done.
Of 265 ureters treated by the Lich/Gregoir technique reflux recurred in 2 cases and stenosis was observed in 5 cases. In 4 of these cases the reason was hemorrhage into the area of operation. The failure was due to the surgeon's technique and not to the Lich/Gregoir operation.
Secondary ureterovesicoplasty was carried out successfully on 5 ureters.
Of 136 children being followed up, 130 are free of infection after one year without chemotherapy. In the remaining six children the surgical result shown by x-ray is good.
31 of 96 children treated by UCN with tunnel had advanced ureteral dilation and 65 had a mega-ureter. Sixteen of the 31 children with bilateral disease had both ureters treated in one operation.
In this group the initial situation is worse and naturally the results are not as good. Of 134 ureters treated recurrent reflux was seen in 5, stenosis requiring surgical treatment in 13 instances. In 6 children a temporary stenosis led to dilatation of the ureter and to fever but did not require surgery. 72 of 82 children followed are free of urinary tract infection 6 months after surgery. Chemotherapy has been discontinued in these patients.
In 5 children we reimplanted 7 ureters in the fundus of the bladder according to the principles of the Leadbetter operation. Recurrent reflux was found in 4 children and 6 ureters. We discontinued use of this technique.
In 7 children 7 ureters were reconnected to the bladder using the Boari/Küss bladder-flap technique with tunnel. There was no postoperative reflux. In one case the dilatation of the ureter has remained unchanged after surgery. The child continues to have urinary tract infections.

Summary

413 anti-reflux operations were performed in 284 children. The surgical techniques and indications are explained. The postoperative results are discussed and analyzed.

Table 1. Clinical material

	Total No. cases	Ureter total	female	male	VUR	bi-lateral	uni lateral	Mega-ureter	bilat. op. 1 Session	bilat. op. 2 Sessions
	284	413	186	98	211					
Lich/Gregior	176	265	134	42	176	89	87		70	19
UCN with tunnel	96	134	47	49	31	31	65	65	16	15
UCN fundus	5	7	5		4	1				2
Boari/Küss with tunnel	7	7		7						

Table 2. Results

	Number of Ureters	VUR post op. No. of Ureters	Stenosis post op. (No. of Ureters)	temporary stenosis with fever	successful secondary op. (No. of Ureters)	free of infection 6 months after op.
Lich/Gregoir	265	2	5		5	130/136
UCN with tunnel	134	5	13	6	13	72/82
UCN fundus	7	6				1/5
Boari/Küss with tunnel	7		1			6/7

Discussion

Indications for Operation in Cases of Vesicoureteral Reflux

When vesicoureteral reflux is associated with recurrent urinary tract infections which cannot be controlled by a conservative method, surgical correction is necessary. There are no differences of opinion about that. More controversial is the therapeutic method in case of vesicoureteral reflux without an infection.

Auvert has demonstrated that all patients with vesicoureteral reflux show more or less serious renal insufficiencies after 25 years. That is why Auvert believes that, besides bacterial infection, hydrodynamic factors are involved in the development of interstitial nephritis. The acute rise in intravesical pressure during micturition (more than 50 mmHg) causes the urine in the bladder to be pressed back into the renal tissue. Interstitial nephritis caused by hydrodynamics usually develops very slowly and can take more than 20 years.

In children, hydrodynamic factors alone can lead to destruction of the kidney. This statement is based on the observation that, in children with bilateral megaloureters and with one-sided vesicoureteral reflux, the kidney on the reflux side is generally much more seriously damaged than the contralateral kidney (Hinman).

Severe albuminuria (GREGOIR), arterial hypertonias (GREGOIR) and serious disturbances of growth in children (HINMAN) have also been observed, besides various types of progressive renal insufficiency following nephritis in cases of vesicoureteral reflux without infection. Here the symptoms of vesicoureteral reflux, which cannot be demonstrated after surgical treatment, are reversible.

BISCHOFF pleads for surgical treatment of vesicoureteral reflux in children before the age of 4 or 5 to avoid serious renal insufficiencies.

In adults, however, vesicoureteral reflux without infection (for instance, after splitting of the ostium or stone removal) rarely induces destruction of the kidney (BOYARSKY). That is why BOYARSKY advises to look for other associated illnesses, total bladder, ureteral und renal status or congenital anomalies in children, for instance, functional ureteral stenosis (ZIMSKIND).

Normal Micturition

G. RUTISHAUSER and P. GRABER*

Understanding the act of micturition necessitates consideration of the structure of the bladder muscle and its nervous regulation as well as the hydrodynamic principles concerned with storage and emptying of the bladder contents. The important facts about bladder function are now generally accepted. Knowledge of the physiology of smooth muscle makes it apparent that the best starting point for forceful micturition is the mediumfull bladder, when the muscle fibers are stretched to about twice their normal length in the resting position (ANDERSON *et al.*, CARPENTER). Urgency is no longer thought to indicate that intravesical pressure has increased and it is now accepted that even massive pressure peaks do not lead to involuntary loss of urine in a healthy individual (BAUMANN). The law of *Laplace*, which correlates volume and wall tension, is applicable to urine storage. This explains why quite different quantities of urine can be stored under similar, generally low, pressure conditions (LAWSON and TOMLINSON). Bladder voiding through the urethra is known to conform to the laws of turbulent currents, so that even slight variations in diameter may be important (SMITH).

Although the complicated mechanism of micturition has by no means been fully explored yet and many aspects are still controversial, we will try to give a short review of the theories which are of general interest.

Physiological micturition can be described as the voluntary and complete emptying of the bladder whenever the individual decides to do so. The details of micturition are, to a large extent, under the individual's own control.

While, today, bladder function is analyzed by the most up-to-date biochemical and electronic techniques – morphological studies having rather decreased in popularity – former generations of investigators were forced to base their studies on anatomy and embryology. This accounts for the fact that some functional ideas were connected with the myo- and neuroarchitecture, and that it is difficult to eliminate these from modern urological thinking and up-to-date urological nomenclature.

The classic myoarchitectural concept of the bladder as a hollow organ consisting of three layers of smooth muscle can no longer be maintained.

We know now that the same muscle fiber can be followed up in all layers of the detrusor (HUNTER). The muscle bundles cross each other vertically and horizontally with no regular pattern over wide areas of the bladder wall and continue down to the

* Urological Clinic (Chief: Prof. Dr. G. RUTISHAUSER) of the Department of Surgery (Director: Prof. Dr. M. ALLGÖWER), University of Basle, Switzerland

urethra; the proximal part of the urethra extends as far as the urogenital trigone in the male and towards the external meatus in the female and forms an anatomic unit with the bladder. In the muscle fibers themselves, the cells form a functional, though not a true anatomical syncytium (PROSSER *et al.*).
Even though GIL VERNET and others (HEISS, LAPIDES, UHLENHUT *et al.*) have took the trouble for the identification and functional classification of muscle bundles at the bladder outlet, it has never been possible to demonstrate an actual circular sphincter muscle at the vesico-urethral junction. However, an abundance of elastic fibres is evident there, as well as a definite increase in thickness of the circular muscle layers, some of them are arranged in sling-form.
The so-called inner sphincter system and the trigone, where mesodermal and entodermal structures are in close contact, are key structures for the understanding of the act of micturition and their functional importance has been disputed for over 100 years.
Some scientific investigators are of the opinion that the muscle bundles at the bladder neck form a true sphincter system. Others refuse to accept any sphincteric function of these muscle fibres at all. A third group admit that there is no fiber arrangement specifically suitable for constricting the bladder outlet, but postulate that the muscle slings and bundles are functionally coordinated in some way to exert a closing and opening effect. HUTCH, one of the representatives of the last-mentioned theory, whose examinations were based on radiological and clinical observations, recognized that the deep trigone muscle and the semicircle of UHLENHUT radiating from it and enclosing the urethra ventrally at rest forms a flat "base-plate" which guarantees continence. With the onset of micturition, however, this flat structure "ruptures" and becomes funnel-shaped, mainly due to traction of the dorsal and ventral longitudinal fiber bundles of the detrusor, but also to contraction of the trigonal muscles themselves (HUTCH).
The corollary to this theory is that the neck of the bladder is not only *not* tightened by the constant tonus of a "sphincter apparatus", but its musculature is to a great extent actively engaged in the opening of the bladder outlet.
The striated muscle fibers of the pelvic floor, which extend a long way up the urethra, seem to have nothing to do with the actual maintainance of continence (LAPIDES *et al.*). This is proved by the experiments of LAPIDES, who shortened the urethra proximal to the urogenital trigone step by step, close to the bladder neck, without jeopardizing continence. This theory is also supported by the fact that micturition can still be interrupted when the striated muscle part of the outer sphincter system is paralyzed, when the process of interruption takes considerably longer.
In the child, micturition is a purely spinal reflex whose afferent signals from the receptors in the bladder muscle and the mucosa are being switched to the efferent nerves in synapses of the sacral spinal cord. In later life, this reflex is influenced and controlled by higher centres which act as initiators or inhibitors.
Piloting the act of micturition takes place in three stages: Topmost cerebral regions act as sensory and motor integration centres. Their subordinate is the spinal reflex centre in the sacral part of the cord. Finally, the lowest stage is formed by the peri- and intravesical chain of ganglions.
We can, for instance, observe the influence of the highest coordination centres in clinical practice after neurosurgical intervention in the region of the hypothalamus or in

patients suffering from disease of the extrapyramidal system. In both cases, inhibitory impulses are predominant. The cause of the well-known postoperative retention of urine, too, is probably a disturbance in the equilibrium of the cerebral conditioning mechanism of micturition.

As to the peripheral distribution of nerves, the separation of two antagonistic autonomic systems – whether from a neuroanatomical or a functional point of view – can no longer be maintained. It may be that the parasympathetic fibres predominate in the motor afferents of the bladder. Yet, EL BADAWI and SCHENK have shown the close anatomical connection between the sympathetic and the parasympathetic fibres and the synapses in the uretero-vesico-urethral region and their mutual influence by histochemical methods. The results of their work explain the ambivalent findings in stimulation experiments as well as the difficulties encountered in the classification of sympathetic influences on tonus and contraction of the bladder (EVARDSON).

In the course of the last 15 years, micturition has been analyzed by a great variety of methods. The best insight into this process is obtained from methods of examination combining X-ray cinematography with the simultaneous measurement of intravesical, intra-urethral and intra-abdominal pressure, the flow of urine and an electromyogram of the pelvic floor (HOLM).

This type of functional analysis of micturition reveals that individual patterns of micturition often vary considerably. It may be that micturition follows a specific "micturition type" with every individual, a fact that has even been compared with individual handwriting.

When trying to interprete the various micturition patterns, we should be very careful in considering the pictures, graphs and figures registered as being absolutely "physiological". As BORS and COMARR correctly state, "The lower urinary tract is not simply an array of communicating rigid spherical or cylindrical vessels, but consists of living substance, to which the laws of hydrodynamic physics apply in a general way."

Bearing the facts above in mind micturition can be described in five phases.

In the *resting phase*, the shadow of the bladder is seen to lie transversely and is oval and indented. The bladder neck cannot be recognized. The intravesical pressure is low and constant. The EMG of the pelvic floor shows very slight activity or none at all.

In the *premicturitional phase*, pressure increases slightly. The horizontal oval shadow on the X-ray becomes round (HINMANN *et al.*). Rapid electromyographic activity can now be recorded from the top of the detrusor, introducing the isometric *initial phase*. At the bladder neck, electromyographic activity begins a few seconds later, while at the pelvic floor there is complete inactivity, according to the electromyograph.

During this phase, the base of the bladder becomes more and more funnel-shaped due to the action of the anterior and posterior longitudinal parts of the detrusor, the "Fundusring" as well as the longitudinal muscle fibers continuing on to the urethra. The contrast medium can now be recognized in the posterior part of the urethra. In a strictly lateral X-ray view, the enlargement and finally the flattening-out of the posterior cystourethral angle is observed. This development takes place simultaneously with the formation of the funnel whilst, due to contraction of the trigone, the distance between the interureteral muscle pad and the colliculus seminalis diminishes.

Contraction of the trigone and increase of pressure in the bladder cause tightening of the ureteral ostiums. Whilst the transport of urine in the ureter continues during the whole process of micturition, the transported spindles of urine themselves cannot be emptied into the bladder (RUTISHAUSER).

As urine begins to leave the bladder, the isotonic *emptying phase* starts. The bladder base moves slightly downwards, especially in the female, the round shadow of the bladder decreasing constantly due to concentric contraction. The flow of urine reaches its average maximum of 20-30 ml/sec and then rapidly diminishes, paralleling the intravesical pressure drop. The urine flow stops a few seconds before the intravesical pressure has reached zero.

In the *emptying phase*, no action potentials can be registered from the pelvic floor, as is the case with *interruption of micturition*, where marked potential differences arise. Urine flow is interrupted momentarily at the level of the urogenital diaphragm. In the subsequent slower phase, which lasts several seconds, the contents of the proximal urethra are retrogradely transferred into the bladder. In the male, the distal end of the urethra is emptied at this point by contraction of the bulbo-cavernous muscle, whereupon the cycle starts anew with the *period of rest*.

By no means does this review pretend to give full explanation of the act of micturition and its principal factors. Its purpose was to give an idea of the complexity of the process which, as outlined at the beginning, can only be understood as a function of the whole organism: Apart from bladder, urethra, connecting tissue, and muscle structures of small pelvis and urogenital diaphragm, all sections of the central nervous system and the peripheral innervation play an equal part in the process of micturition. There are also important psychological influences, extremely difficult to reproduce in figures and graphs.

References

ANDERSEN, G. F., PIERCE, J. M., BLAIR, L. L.: Tension changes in rabbit bladder muscle: Effect of stretch. Invest. Urol. **6**, 267–272 (1968).

BAUMANN, W.: Prüfung der Blasenphysiologie mit Cystometrie und Sphincterometrie. Urol. int. (Basel) **1**, 427–439 (1955).

BORS, E.: Neurogenic bladder. Urol. Surv. **7**, 177–250 (1957).

BORS, E., COMARR, A. E.: Neurological Urology, p. 44. Basel: S. Karger 1971.

CARDUS, D., QUESEDA, E. M., SCOTT, F. D.: Studies on the dynamics of the bladder. J. Urol. (Baltimore) **90**, 425–433 (1963).

CARPENTER, F. G.: Motor responses of bladder smooth muscle in relation to elasticity and fibre length. Invest. Urol. **6**, 273–283 (1968).

DENNIG, H.: Die Innervation der Harnblase, Physiologie und Klinik. Heft 45, Mschr.Ges. Geb. Neurol. und Psych. Berlin: Foerster und Wittmanns 1926.

EL BADAWI, A., SCHENK, E. A.: Dual innervation of the mammalian urinary bladder. A histochemical study of the distribution of cholinergic and adrenergic nerves. Amer. J. Anat. **119**, 405–411 (1966).

EL BADAWI, A., SCHENK, E. A.: A new theory of the innervation of bladder musculature. Part 1. Morphology of the intrinsic vesical innervation apparatus. J. Urol. (Baltimore) **99**, 585–587 (1968).

EL BADAWI, A., SCHENK, E. A.: A new theory of the innervation of bladder musculature. Part 2. The innervation apparatus of the ureterovesical junction. J. Urol. (Baltimore) **105**, 368–371 (1971).

EL BADAWI, A., SCHENK, E. A.: A new theory of the innervation of bladder musculature. Part 3. Postganglionic synopses in the ureterovesical urethral autonomic pathways. J. Urol. (Baltimore) **105**, 372–374 (1971).
EVARDSEN, P.: Nervous control of urinary bladder in cats. A survey of recent experimental results and their relation to clinical problems. Acta neurol. scand. **43**, 543–550 (1967).
GIL VERNET, S.: Morphology and function of vesico-prostato-urethral musculature. Treviso: Edizioni Canova 1968.
HATAM, A., LICH, R., HOWERTON, L. W., PIRKEY, E. L.: Voiding, its mechanism. J. Amer. med. Ass. **178**, 847–848 (1961).
HEISS, R.: Über den Sphincter vesicalis internus. Virchows Arch. path. Anat., 367–383, Jhrg. 1915.
HINMAN, F. JR., MILLER, G. M., NICKEL, E., MILLER, E. R.: Vesical physiology demonstrated by cine-radiography and serial roentgenography. Radiology **62**, 713 (1954).
HOLM, H. H.: The hydrodynamics of micturition, Suppl. Acta radiol. **231** (Stockh.) (1964).
HUNTER, D. W.: A new concept of urinary bladder musculature. J. Urol. (Baltimore) **71**, 695–704 (1954).
HUTCH, J. A.: A new theory of the anatomy of the internal urinary sphincter and the physiology of micturition. Invest. Urol. **3**, 36–45 (1965).
HUTCH, J. A.: The internal urinary sphincter: A double loop system. J. Urol. (Baltimore) **105**, 375–383 (1971).
LANGWORTHY, O. R., KOLB, L. C., LEWIS, L. G.: Physiology of Micturition. Baltimore: Williams & Wilkins Co. 1940.
LAPIDES, J.: Structure and function of internal vesical sphincter. J. Urol. (Baltimore) **80**, 341–353 (1958).
LAPIDES, J., GRAY, H. O., RAWLING, J. C.: Function of striated muscle in control of urination. Surg. Forum **6**, 611–615 (1955).
LAPIDES, J., SWEET, R. B., LEWIS, L. W.: The role of striated muscle in urination. J. Urol. (Baltimore) **77**, 247–250 (1957).
LAWSON, J. D., TOMLINSON, W. B.: Consideration of some biophysical factors in the distension of thin-walled hollow viscera. J. Urol. (Baltimore) **71**, 446–449 (1954).
LEWIN, M. L., CULP, D. A., FLOCKS, R. H., SPIER, P. F.: Dynamics of the lower urinary tract: A method of study. J. Urol. (Baltimore) **96**, 756–762 (1966).
MUELLNER, S. R.: The physiology of the micturition. J. Urol. (Baltimore) **65**, 805–810 (1951).
PROSSER, C. L., SMITH, C. G., MELTON, C. E.: Conduction of action potentials in the ureter of the rat. Amer. J. Physiol. **181**, 651–660 (1955).
RUTISHAUSER, G.: Druck und Dynamik in den oberen Harnwegen, p. 81. Fortschritte der Urologie, Bd. 2. Darmstadt: D. Steinkopff 1970.
SCHULTHEIS, TH., RUTISHAUSER, G.: Kontinenz und Miktion. Z. Urol. **56**, 191—206 (1963).
SMITH, J. C.: Some theoretical aspects of urethral resistance. Invest. Urol. **1**, 477–481 (1964).
TANG, P. C.: Levels of brain stem and diencephalon controlling micturition reflex. J. Neurophysiol. **18**, 583–689 (1955).
TANG, P. C., RUCH, T. C.: Localisation of brain stem and diencephalic areas controlling the micturition reflex. J. comp. Neurol. **106**, 213–246 (1956).
UHLENHUT, E., HUNTER, D. T., DE WITT, T., LOECHEL, W. E.: Problems in the anatomy of the pelvis. An Atlas. Philadelphia: J. B. Lippincott & Co. 1953.

Studies of Mechanisms of Continence, Incontinence and Voiding

E. R. MILLER*

To understand continence and voiding, a number of problems of anatomy, forces, pressures and flows must be studied. These may be considered as the "static" and the "dynamic" states during continence and the events leading to initiation and cessation of voiding.

What are the forces and pressures in the bladder and in the urethra during the resting continent phase as the empty bladder fills?

What changes occur in these forces in what might be called the dynamic continent state produced by increased intra-abdominal pressure associated with coughing and straining in various patients? e.g. In coughing the intra-abdominal pressure may be raised over 100 cm of water. This pressure is transmitted passively to the bladder. Why does the normal person not leak urine?

What changes take place during the voluntary effort to "hold urine" when the bladder is either partially filled or filled to the level of voiding?

What mechanisms operate during preparation for voiding and during micturition?

How are these changes related to a patient's clinical diagnosis?

What are the levels of conductance for patients with various diagnoses?

What things affect conductance?

How does straining affect voiding?

Where does the physiologic "valve" of the urethra begin?

How do the forces opposing the head of pressure in the bladder affect voiding and how do these opposing forces come into being?

What stops voiding?

Our present approach to the study of the lower urinary tract has resulted from eighteen years of experience and study. At the beginning, radiologic studies of the lower urinary tract during voiding were carried out by means of rapid film changers. This led soon to the use of x-ray movies. These studies were carried out primarily with Dr. FRANK HINMAN, Jr. It was he who first brought to my attention the physiologic studies of BODO VON GARRELTS in the late fifties. It was the acquaintance with his work which led inexorably to the conclusion that radiologic studies by themselves would never allow us to understand the function of the lower urinary tract.

---

* Radiological Research Laboratory, Department of Radiology, University of California School of Medicine. San Francisco, Calif., USA

In 1962 and in 1964 GORAN ENHORNING joined the Radiological Research Laboratory for two periods of work and study and taught us the basic techniques for using additional physiologic data to help interpret the roentgen findings.
At the present time we use ten channels of information during the study of the lower urinary tract, only one of which is the x-ray image.
This is the minimum number of channels that we require for the analysis of continence and voiding.

Information from the Radiologic Image

The bladder and urethra during filling and emptying as well as the position of the measuring points on catheters is observed fluoroscopically at intervals throughout the entire procedure. The times of observation are guided by the amount of material put into the bladder, by the need to know the position of the catheters and from the data being observed on the physiologic recording charts as the data is written out.
From the radiologic image we observe the size and shape of the bladder as a function of filling, the presence or absence of reflux, the catheter position, the bladder and urethra during emptying, and the presence or absence of residual opaque material in the bladder after voiding.
Anatomical data so obtained does give some information about the lower urinary tract, particularly with respect to such things as bladder shape, bladder trabeculation, reflux, urethral shape, relation of the urethra to the bladder, and the presence of residual urine.
This information, however well obtained, is unfortunately so incomplete, that I find myself unable to interprete a voiding cystogram without addition of adequate physiologic data.

Data Required for the Combined Radiological and Physiological Study of the Lower Urinary Tract

Continuous recording of:

A. Bladder pressure
B. Urethral pressure
C. Rectal pressure (recording of intra-abdominal pressure)
D. Anal pressure (recording of intrinsic anal pressures as well as perineal activity)
E. Difference between the bladder and rectal pressure (detrusor pressure)
F. Difference between the urethral and bladder pressures (closure pressure)
G. Amount of opaque material injected and the amount and rate at which it is voided
H. Time.

Intermittent recording of:

I. Sound
J. X-ray channel of information displaying the image of the bladder and urethral regions as well as the position of the devices measuring the above various pressures.

Apparatus (Fig. 1)

Simultaneous presentation of the anatomic and physiologic information is accomplished by mixing the televised x-ray signals from an image intensifier with those from a second television camera observing the output from a physiologic data recorder. These data are presented on a group of monitors and are recorded with sound, both kinescopically and by video-tape recording.

The x-ray channel consists of a fluoroscopic unit used in the erect position, an image intensifier capable of being operated in the 6 and 9 inch modes, and an image Orthocon television camera.

During the procedure the patient sits on a toilet which consists of a chair with a toilet seat. (Fig. 2). The toilet and patient can be rotated by remote control. Beneath the toilet seat is a large plastic funnel through which the voided fluid flows into a receptacle. Immediately above the funnel is a double grid for electronically recording the initiation of the flow of fluid from the urethra. A catheter holder is mounted in front of the patient and above the funnel. In the collection receptacle, a scale measures the amount of voided fluid; and a pressure transducer is used for recording the amount and the rate of flow of the fluid.

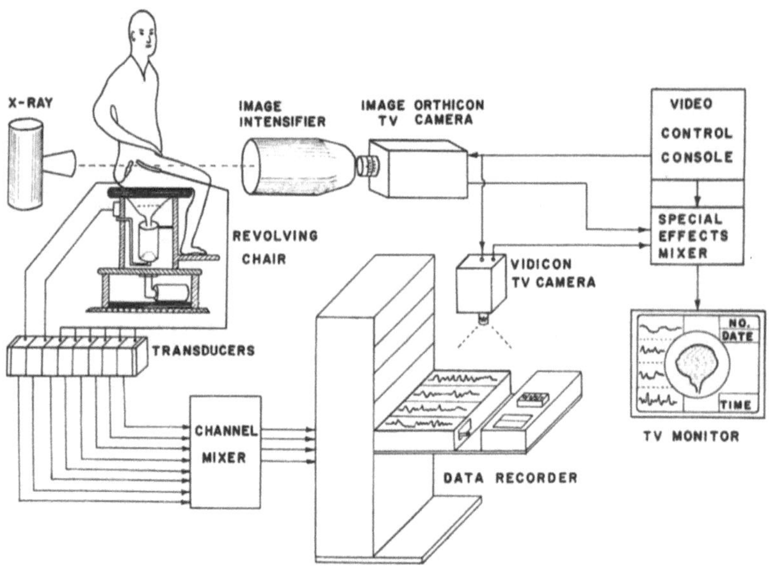

Fig. 1. Block diagram of apparatus used for combined monitoring of anatomical and physiologic data. The output of the image intensifier showing x-ray data is looked at by an image Orthocon TV camera. This information is presented in the middle of the monitor, shown in the right lower corner.

The patient is seated on a toilet chair. This can be rotated by remote control. A funnel catches any voided fluid which is collected in a chamber below. Both the amount and the rate of voiding are determined from a transducer in the chamber.

The bladder-urethral catheter and the rectal-anal channels are connected to transducers and the information is presented to a Sanborn Data Recorder. Above the recorder paper is a Vidicon TV camera which observes the data as it is being written out. The signals from the two television cameras are mixed in a special effects generator and are presented on all monitors, as shown in the right lower corner of the figure.

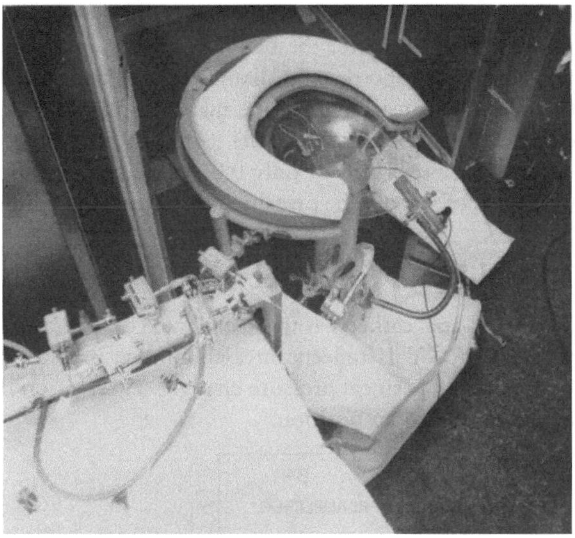

Fig. 2. The toilet chair is seen in the middle of the picture, the image intensifier is seen in the right upper corner. In front of the toilet chair is a catheter holder; to the left are some of the transducers for transforming pressure data to electrical signals

The Urinary Tract Catheter (Fig. 3)

The catheter is four-channelled; its outside diameter is 3 mm. Two channels at its end are open, one for filling the bladder, and the other for determining bladder pressure. A radio-opaque marker one cm long is placed at the end of this catheter. Five cm distal to the bladder opening is an opening for urethral pressure (U-1), also with a radio-opaque marker one cm in length. This opening is covered by a thin rubber balloon 0.5 cm long, and approximately 4 mm in diameter in its distended state.

A similar opening, (U-2) two cm distal to U-1, also has a one cm marker and is covered by a balloon similar to that of U-1. All catheters and balloons are filled with fluid

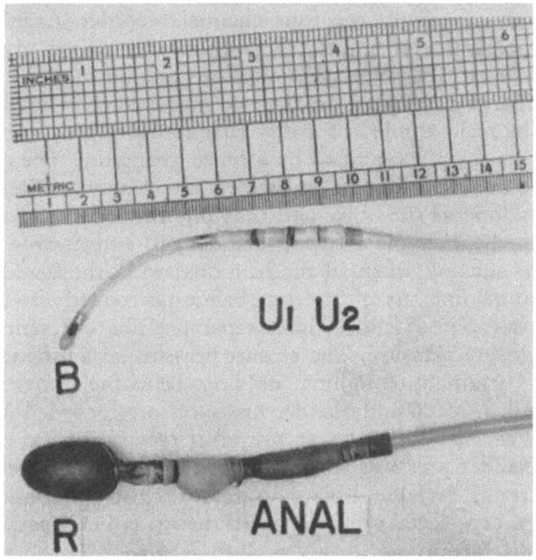

Fig. 3. Catheters used during study. The upper catheter is 4-channeled, used for monitoring bladder and urethral pressures. The lower catheter is 3-channeled, used for monitoring rectal and anal pressures

and all gases are carefully excluded at the beginning. At the time of the insertion of the urethral catheter the balloons are collapsed and at zero pressure and thus do not significantly increase the diameter of the catheter. The filling catheter is connected to a pump for instilling opaque material into the bladder. The bladder and urethral catheters are connected to the Sanborn transducers. These record linearly and require the movement of only 0.04 cubic mm of fluid to indicate a pressure change of 100 cm of water. The pressure transducers are connected to the Sanborn recorders.

Rectal-Anal Catheter

The rectal-anal catheter is three-channelled. One balloon is used for measuring rectal pressure (*R*). This reflects intra-abdominal pressure. The anal balloon (*A*) reflects both anal and perineal pressure changes. A retention balloon between *R* and *A* holds a catheter in proper position.

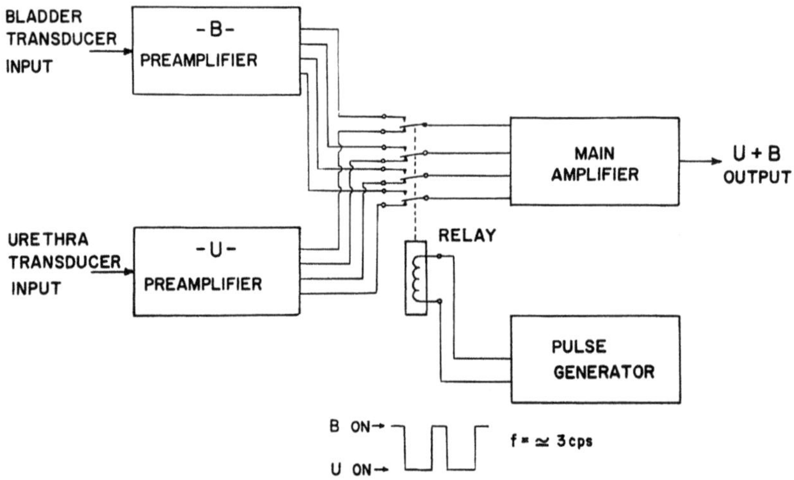

Fig. 4. Block diagram of circuit used for displaying two lines of data on each of the channels of a Sanborne four channel recorder. Each line of data gives the absolute value of the parameter and the separation of the lines gives a constant recording of the difference between the parameters all as a function of time.
The preamplifiers for the two parameters to be displayed on one channel are connected to the main amplifier for that channel through a four channel double-pole double-throw relay. The relay is operated by a pulse generator. The generator is set to connect the transducers to the recording galvanometers alternately at a rate of approximately 3c/sec. One is on twice as long as the other and thus one line on the chart is twice as dark as the other.
In the example showing bladder (B) and urethra (U) pressures, the pressures are adjusted to equality when all the balloons are in the bladder and the B and U recordings appear as single line. As the urethral balloon is moved into the urethra, the urethral pressure normally rises. The B and U lines separate. The difference between the B and U pressures is the closure pressure. The closure pressure as a function of time and position, and as a function of straining, coughing, holding, urinating, moving, etc., are all recorded continuously.
When rectal and bladder pressures are recorded on the same channel after being equalized when the patient is at rest, a) if rectal pressure changes without concommitant change in bladder pressure, it means that there is an independent rectal contraction; b) if the patient strains, rectal and bladder pressures rise simultaneously and equally; c) if bladder pressure exceeds rectal pressure this is due to either a passive stretch of the bladder or to a detrusor contraction

The Recorders

Two Sanborn recorders are used to record the seven channels of physiologic data. They are lighted by a fluorescent light. A clock which records in seconds is mounted beside one of these. Above one of the recorders plotting rectal pressure, flow rate, bladder pressure, urethral and anal pressures is mounted a Vidicon television camera which looks at the data as it is being recorded. A switching circuit (Fig. 4) allows the presentation of two lines of data on each section of the four channels of the recorder. The video signals from the image Orthocon and Vidicon television cameras are mixed, processed through a special effect generator and presented to several monitors. One of the monitors is employed for kinescopic recording by a 16 mm Auricon sound camera for permanent record. The mixed television signals are fed also to an Ampex Videotape recorder for use in immediate interpretation. The appearance of the monitor during one part of one study is shown in Fig. 5.

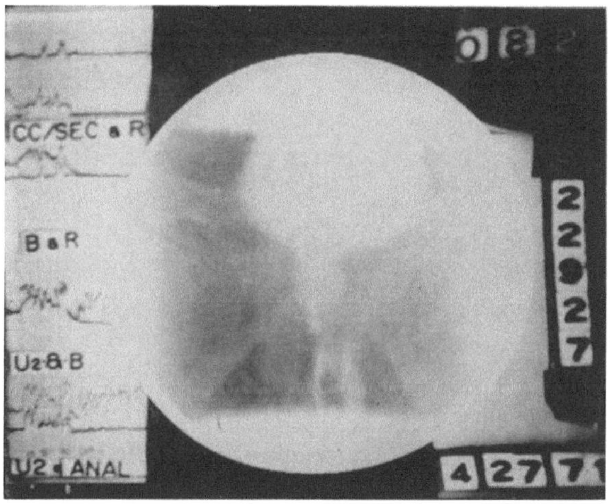

Fig. 5. Appearance of the monitor during a physiologic study. In the middle is a circle showing the x-ray image of the base of the bladder and the two markers on a bladder-urethral catheter; to the left the physiologic data is shown as it is being written out. The top line is rectal pressure. Going along the zero line of this channel is a line of data indicating voiding rate in cc per second. The second channel shows bladder and rectal pressures simultaneously. The next channel shows urethral and anal pressures. In the right upper corner is a digital clock showing time in seconds

Procedure

The catheters are calibrated by subjecting them to pressures of 0,50, and 100 cm of water in a calibration chamber.

The patient empties the bladder. Under sterile conditions the four-channelled urinary tract catheter is introduced to the bladder when the balloons are under 0 cm of water pressure. The three-channel rectal-anal catheter is then introduced into the rectum, and is adjusted so that the anal balloon is in the anus. The patient is then moved to the toilet chair. At the start of the examination 100 to 150 cc of 15% hypaque are instilled into the bladder. All urinary tract catheter measuring points are at first in the bladder and the patient is commanded to bear down for testing the operation of the apparatus. The urethral balloons are then pulled slowly and continuously through the proximal 4–5 cm of the urethra for recording the urethral pressures under "static" conditions (this maneuver is called the "straight pull-through"). The catheter is replaced in the bladder and then moved through the urethra at 0.5 cm intervals throughout the urethra. At each stopping point the patient is requested to cough once, to bear down and to hold urine (this is called the "graduated pull-through"). The catheter is then positioned so that the distal urethral balloon (U-2) is in the previously determined high pressure zone and the bladder is filled until the patient is ready to void. After the patient voids, the examination is terminated.

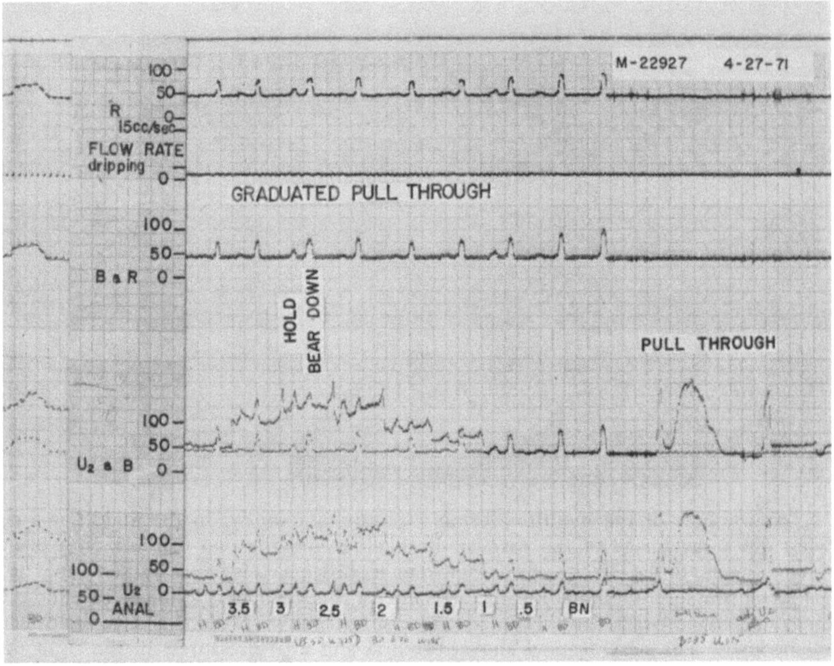

Fig. 6. Data obtained from a patient, who complained of incontinence, in the continence phase. 100 cc of opaque material was in the bladder. Time goes from right to left. Each major division is five seconds. During the pull-through, toward the right of the figure, one can see that thc urethral pressure rose above the bladder pressure throughout the entire length of the urethra. Each major division is equivalent to 25 cm of water, thus the closure pressure reached a maximum of about 140 cm of water.
On the left is shown the results of the graduated pull-through. We observe that the physiologic urethra starts 1 cm below the bladder neck and extends to 3 and one half cm below the bladder neck. The resting closure pressure in the high pressure zone is about 90 cm of water and drops to about 75 cm of water on bearing down. The closure pressure rises to 100 cm of water on holding urine. This closure pressure is well within normal limits, does drop a little on bearing down and does rise on holding urine. This is inconsistent with the diagnosis of stress and continence

Data Extracted from the Records

1. "Static" closure pressure during the straight pull-through and graduated pull-through throughout the length of the urethra.
2. Change of closure pressure with bearing down, coughing and "holding urine", also throughout the length of the urethra.
3. Differences in closure pressure with these maneuver as a function of bladder filling.

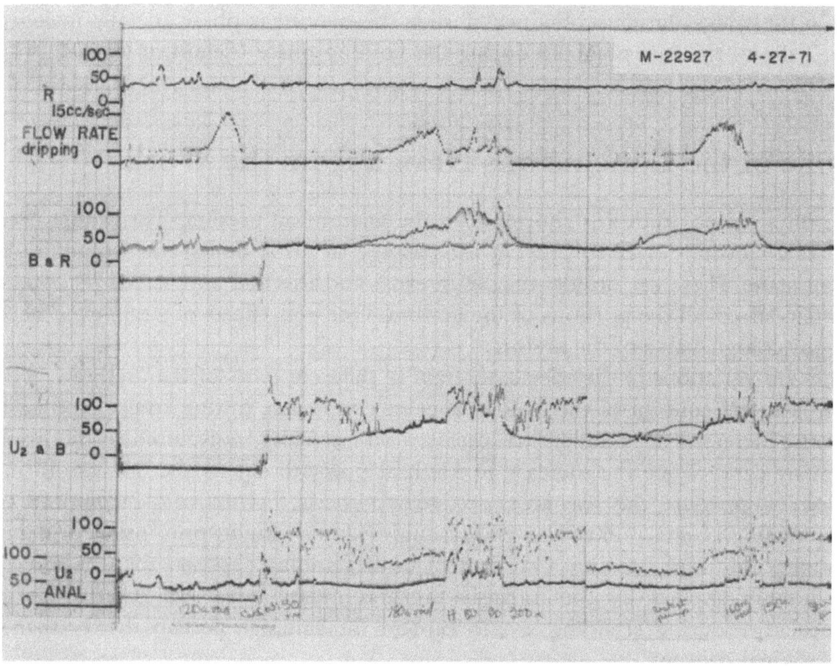

Fig. 7. Data obtained during the voiding phase. Three voids are shown. The earlier void to the right is accomplished with the urethral catheter in place. The second void in the middle has been interrupted by commands to bear down and to hold urine. The third void to the left is accomplished after the urethral pressure catheter has been removed. The voiding rate in the first and the third void with the catheter in and out, differ by only 2 cc per second and thus the catheter is not producing significant obstruction.
In the first void to the right, note that this void is accomplished by first a drop in urethral pressure followed by a detrusor contraction and is accomplished without strain. There is a so-called after-contraction which is apparently simply the detrusor operating on an empty bladder since the patient had no residual on this void.
The second void is initiated in exactly the same way as is the first void. On straining, the patient could increase the flow rate on straining and could stop urination on command to hold urine.
Thus, because the patient's closure is adequate, we saw no evidence of incontinence during the study, the bladder capacity was only 200 cc and the patient voided without strain, this data is inconsistent with the diagnosis of stress urinary incontinence. It is consistent with an urge incontinence. Because of this data, the patient did not undergo a Marshall-Marchetti operation and immediate follow-up showed that she has no difficulty with her bladder or voiding and does not have incontinence

4. Changes in pressures and closure pressure during the patient's preparation for voiding.
5. Voiding rate and pressures during voiding.
6. Mechanisms used to stop voiding.

Example of Data Obtained from a Patient Referred with a Complaint of Stress Urinary Incontinence

Fig. 6 illustrates the data obtained during the continent phase (details in legend). Fig. 7 shows the data obtained during the voiding phase (details in legend).

Some of the Conclusions Drawn from the Work

The physiologic urethra is shorter than the anatomical urethra. The whole urethra acts as a closure device to prevent the passage of urine during the voluntary continent state. There are no identifiable internal and external sphincters.
A strain affects all the portions of the urethra. One gets effects of strain even beyond the physiologic valve of the urethra.
There are variations in the effects of strain in different parts of the urethra.
ENHORNING's concept of the maintenance of continence during strain hypothesized that the increased intra-abdominal pressure was actually transmitted to the urethra. In many cases, since the increase in urethral pressure equals the increase in intra-abdominal pressure, this was an attractive hypothesis. ENHORNING recognized early that certain voluntary muscular effects might play a part in preventing urinary incontinence during straining. Our present understanding suggests that the forces on the urethra applied by the intrinsic urethral musculature and external perineal musculature during straining would explain the findings perhaps better than did the original concept.
Resting closure pressure among incontinent patients is lower than among continent patients.
In incontinent patients, closure pressure decreases with strain. If the closure pressure rises with strain, the patient cannot void by straining.
If closure pressure during strain is less than the resting closure pressure, the patient may void with or without strain.
Leaking or voiding occurs only when closure pressure drops to zero or becomes negative.
Closure pressure can go to zero and voiding occur as a result of a detrusor contraction urethral pressure drop, strain, stretching of the bladder or any combination of these.
A single patient may use more than one mechanism within a void or in different voids, so that all voids are unique.
In general, strain decreases the conductance through the urethra, that is, increases the resistance. The effects, however, are variable. A given patient may even increase the voiding rate and still have a decreased conductance (conductance equals flow squared over bladder pressure).

Strain also produces variable effects as a function of volume of the fluid in the bladder. This is perhaps best seen in patients who are incontinent at high bladder volumes but are not incontinent when the bladder is nearly empty.

Patients who are incontinent tend to use strain more frequently than others to achieve voiding. This is consistent with the fact that patients who are incontinent develop a lower closure pressure as a function of strain and have less static closure pressure to overcome.

There is a marked variation in the amount of time required for a patient to prepare to void. In the case of strain voids, the preparation time may be very short, in others it may exceed a minute.

The voiding rate versus the bladder pressure may be meaningful in patients who void without strain. The calculation of conductance or urethral resistance is not meaningful in patients who void with strain because the patients has voluntary control over conductance during strain.

A drop in urethral pressure is a significant happening in almost all voids. Patients may void without a detrusor contraction, for example, by a sharp drop in urethral pressure or a strain or a combination of these or by pure stretch of the bladder as filling proceeds.

We have not found a correlation between patients' entering diagnosis or complaint with voiding rates, conductance, bladder pressure, or the amount of residual.

As expected, the entering diagnosis of incontinence is much more frequent in the elderly patients than the younger patients.

In our experience most of the patients who have incontinence have been operated on previously or have had multiple pregnancies.

The definition of neurogenic disorders of micturition have been broadened to include, in addition to spinal cord destructions, those which show lack of smooth coordination between the driving and resisting forces normally employed in voiding.

Acknowledgements

I would like to pay tribute to Dr. Frank Hinman, Jr., Dr. John Hutch, Dr. Emil Tanagho, Dr. Goran Enhorning, Dr. Alexander S. Cass and Dr. Guy Drouin. Dr. Frank Himman awakened my interest in the study of the lower urinary tract, and Dr. Goran Enhorning worked in the Radiological Research Lab and first made me aware of the tremendous potential of the simultaneous recording of bladder and urethral pressures as a means of investigating the physiology of the bladder-urethral complex. The others brought in cases for study and spent untold hours in helping with the interpretation of the complex data that came from the studies.

The staff in the Radiologic Research Laboratory, Mssrs. Edward McCurry, Alan Cline and Bernard Hruska, with the technical help of Mrs. Ann Temple and Miss Karen Baner, brought the apparatus, procedure and techniques to their present state. Mr. McCurry's contributions, both in the development and fabrication of the apparatus and catheters, has been of the highest value.

This work was supported by the Department of Radiology, Department of Urology, and the Research Committee of the Medical School of the University of California; Dr. Hinman's private funds; and USPHS grants HD-01843 and AM-10045. The clinical work was carried out in the Cancer Research Institute of the University of California, Dr. DAVID WOOD, Director. Without his support and encouragement, this work would not have been possible.

References

DROUIN, G., MCCURRY, E. M.: Catheters for Study of Urinary Tract Pressure. Invest. Urol. Vol. **8**, No. 2, pp. 195–202 (1970).

ENHORNING, G.: Simultaneous Recording of Intra-vesical and Intra-urethral Pressure. Study on Urethral Closure in Normal and Stress Incontinent Women. Acta chir. scand. **276**, Suppl. I (1961).

ENHORNING, G., MILLER, E. R., HINMAN, F. JR.: Urethral Closure Studied with Cineradiography and Simultaneous Bladder-Urethral Pressure Recording. SG and O. **18**, 507 (1964).

GARRELTS, BODO VON: Analysis of Micturition. A New Method of Recording the Voiding of the Bladder. Acta chir. scand. **112**, 326–340 (1957).

GARRELTS, BODO VON: Micturition and Disorders of the Prostate and Posterior Urethra. Acta chir. scand. **115**, 227–241 (1958).

GARRELTS, BODO VON: Micturition and Urethral Stricture. Acta chir. scand. **14**, 466–489 (1958).

HINMAN, F. JR., G. M. NICKEL, E. MILLER, E. R.: Vesical physiology demonstrated by cineradiography and serial roentgenography. Radiology **62**, 713–719 (1954).

HINMAN, F., JR., MILLER, G. M., NICKEL, E., STEINBACH, H. L., MILLER, E. R.: Normal micturition: Certain details as shown by serial cystograms. Calif. Med. **82**, 6–7 (1955).

HINMAN, F., JR., MILLER, E. R., HUTCH, J. A., GAINEY, M. D., COX, C. E., GOODFRIEND, R. B., MARSHALL, S.: Low pressure reflux: Relation of vesico-ureteral reflux to intravesical pressure. J. Urol. (Baltimore) **88**, 758–765 (1962).

HINMAN, F., JR., MILLER, E. R.: Mural tension in vesical disorders and ureteral reflux. Trans. Amer. Ass. gen.-urin. Surg. **55**, 13–20 (1963).

HINMAN, F., JR., HUTCH, J. A., MILLER, E. R.: Role of reflux in pathogenesis of chronic pyelonephritis. Excerpta Med. Inter. Congr. Series **78**, 305–306 (1963).

HUTCH, J. A., HINMAN, F. JR., MILLER, E. R.: Reflux as a cause of hydronephrosis and chronic pyelonephritis. J. Urol. (Baltimore) **88**, 169–175 (1962).

HUTCH, J. A., MILLER, E. R., HINMAN, F. JR.: Vesico-ureteral reflux. Amer. J. Med. **34**, 338–349 (1963).

HUTCH, J. A., MILLER, E. R., HINMAN, F. JR.: Perpetuation of infection in unobstructed urinary tracts by vesico-ureteral reflux. J. Urol. (Baltimore) **90**, 88–91 (1963).

HUTCH, J. A., MILLER, E. R.: Vesicoureteral reflux: role in Pyelonephritis, pp. 613–630. In: Pyelonephritis, ed. KASS, E. H. (Ed.). Philadelphia: F. A. Davis 1965.

MILLER, E. R.: Techniques for simultaneous display of x-ray and physiologic data. In: Neurogenic Bladder, pp. 79–85. (Ed.): BOYARSKY, S., Baltimore: Williams and Wilkins Co. 1967.

MILLER, E. R.: Combined monitoring for the study of continence and voiding. In: HINMAN, F., JR.: Hydrodynamics of Micturition. Springfield Ill.: C. C. Thomas 1970.

TANAGHO, E. A., HUTCH, J. A., MILLER, E. R.: Diagnostic procedures and cinefluoroscopy in vesicoureteral reflux. Brit. J. Urol. **38**, 435–444 (1966).

TANAGHO, E. A., MILLER, E. R.: Initiation of voiding. Brit. J. Urol. **42**, 175–183 (1970).

TANAGHO, E. A., MILLER, E. R., MEYERS, F. H., CORBETT, R. K.: Observations on the dynamics of the bladder neck. Brit. J. Radiol. **38**, 72–84 (1966).

Vesicourethral Dynamics

E. A. TANAGHO*

The bladder and the urethra have a common embryologic origin-the urogenital sinus. Sacularization of the cranial part and tubularization of the caudal part of the urogenital sinus result in the differentiation of a bladder and a urethra. The former acts as a reservoir and the latter as its sphincter. The two are directly continuous anatomically and are intimately related functionally.

Anatomical dissection and histological studies of the human vesical outlet have demonstrated that there is no complete ring of muscle fibers encircling the vesical neck that can be described as a sphincter. A sphincteric mechanism, however, is offered by the intrinsic musculature of the whole urethra in the female (Fig. 1) and the prostatic urethra in the male, which is the principal urinary sphincteric mechanism. A striated-muscle sphincter surrounds the urethra and functions as a unit with the pelvic floor.

The discussion of vesicourethral dynamics presented herein is based on detailed anatomic and histologic studies and elaborate functional studies of normal and pathological vesicourethral activity in the human, supplemented by physiologic and pharmacologic studies on experimental animals.

Anatomic Considerations

The bladder musculature consists of a mesh of muscle bundles with no definite layering pattern, but which can be traced from one axis or plane to another. These muscle bundles ramify and connect with each other They pursure an irregular course except at the bladder outlet, where they form 3 distinct layers: inner longitudinal, middle circular, and outer longitudinal.

The inner longitudinal layer with the superficial trigone sweeps over the internal meatus to form the inner longitudinal layer of the urethra. The fibers of the middle circular layer condense ventrally and fan out dorsocranially to fuse with and adhere to the deep trigone. They end at the internal meatus and do not extend into the urethra. At the urethrovesical junction, the fibers of the outer longitudinal layer loop almost transversely around the proximal urethra, constituting its outer circular layer.

---

* Division of Urology, University of California School of Medicine. San Francisco, CA, USA

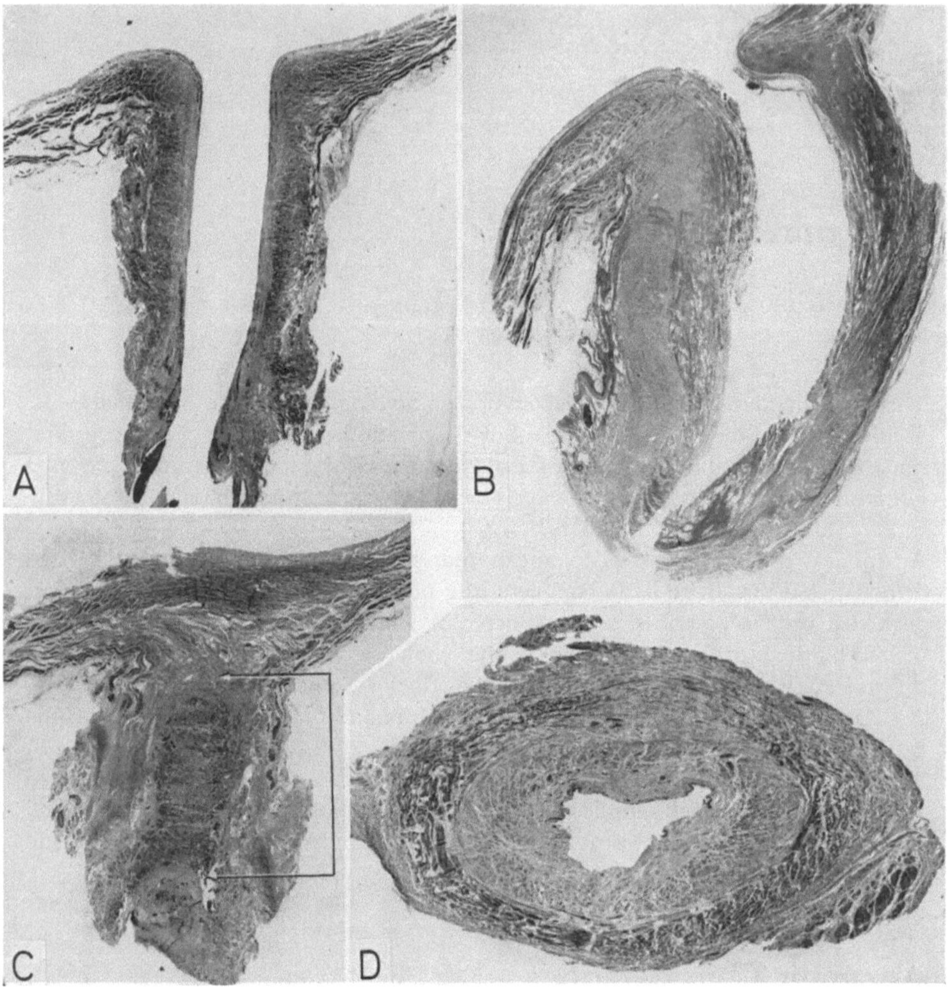

Fig. 1 A–D. A Sagittal section of female urethra shows 2 muscle layers: 1. inner longitudinal layer embedded in heavy collagen and 2. outer semicircular fibers, which offer sphincteric mechanism. (Courtesy of British Journal of Urology, c 1970.) B and C Parasagittal section of female urethra shows circularly oriented striated external sphincter maximally condensed around midurethra. D Cross-section of female urethra shows intrinsic smooth musculature arranged in inner longitudinal and outer semicircular sphincteric layers

Anatomic and histologic studies of the proximal urethra show only 2 muscle layers: 1. an inner longitudinal layer and 2. an outer circular layer consisting of "sling fibers", which never form a complete ring around the urethra and are oblique rather than circular (Fig. 1). Most recent investigators agree that there are no muscle fibers arranged in a truly circular fashion arround the bladder neck to function as an internal sphincter.

The only fibers able to exert sphincteric action are the sling fibers of the outer circular layer. These fibers surround almost the entire length of the female urethra. In the male, they are confined to the prostatic urethra. The prostate as it develops from and around the urethra carries along the outermost sling fibers, which then become the muscular prostatic stroma. About half the thickness of the outer circular layer remains wrapped around the prostatic urethra. These fibers and the muscular prostatic stroma are the direct continuation of the outer longitudinal layer of the bladder and form the urethral sphincteric mechanism in the male.

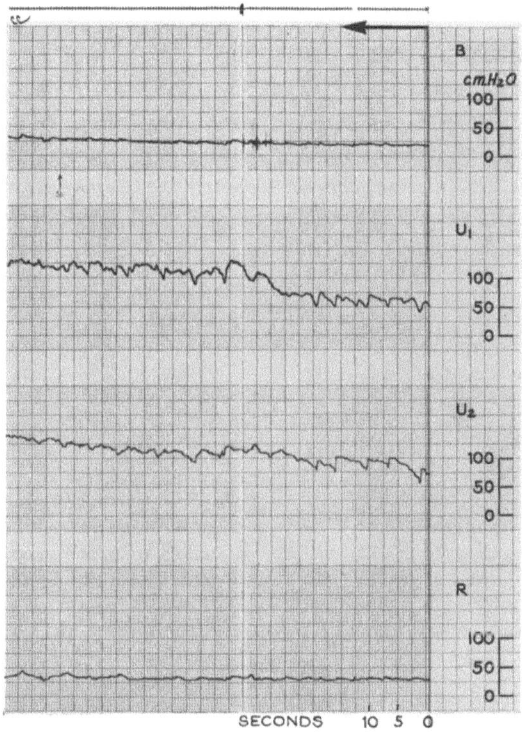

Fig. 2. Simultaneous recording of resting bladder (*B*), proximal urethral (U_1), midurethral (U_2), and rectal (*R*) pressures in adult female at end of bladder filling. As bladder fills, bladder pressure shows minimal rise, with progressive rise in proximal and midurethral pressure. (Courtesy of British Journal of Urology, c 1970)

The almost circular orientation of the sling fibers around the urethra and their direct continuity with the outer longitudinal layer of the bladder enable them to perform a sphincteric function and to maintain sufficient closure pressure. When the bladder distends, its fibers stretch and their increased tension is directly transmitted to the urethral muscle fibers, thereby increasing the urethral closure pressure (Fig. 2).

The urethra passes through the urogenital diaphragm, which is part of the pelvic floor. Part of the striated musculature of the pelvic floor is directly related to the urethra and forms the striated external sphincter. The rest of the pelvic floor is not directly related to the urethra anatomically, but plays a major functional role in voiding dynamics. The pelvic floor, including the urogenital diaphragm, curves anteroposteriorly and laterally. Its concavity diminishes when it contracts and the pelvic contents are pushed upwards. When the pelvic floor relaxes, the pelvic contents sag downward, usually resulting in a slight funneling of the bladder outlet. In the resting state, the pelvic floor is midway between the 2 positions and there is no funneling of the bladder outlet. The urethra is well fixed to the urogenital diaphragm and therefore reflects changes in the position of the pelvic floor.

Physiologic Considerations

Although there is no anatomic sphincter, there is a physiologic sphincteric mechanism throughout the entire urethra in the female and in the prostatic and membranous urethra in the male. Two distinct elements contribute to this sphincteric mechanism: 1. the intrinsic urethral smooth musculature and 2. well developed circularly oriented muscle fibers surrounding the urethra and maximally condensed around its midportion.

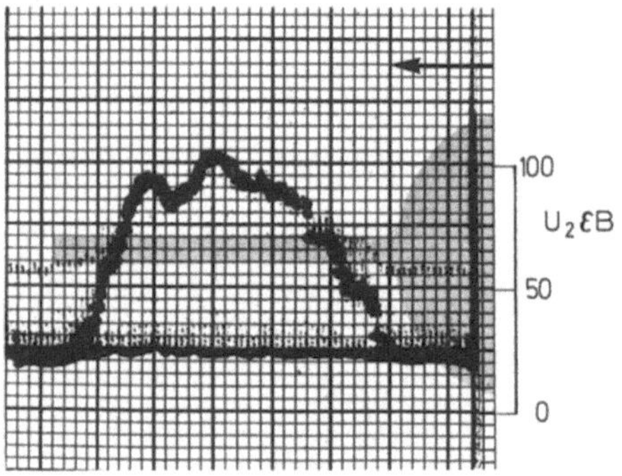

Fig. 3. Profile of female urethral pressure. Note minimal rise at internal meatus. Pressure progressively increases and reaches maximum at midurethra, then progressively diminishes to reach zero at external meatus. Proximal urethral pressure is almost 50 cm H_2O, while maximum midurethral pressure is about 100 cm H_2O

A profile of urethral pressure in the resting state obtained by withdrawing a pressure-recording catheter from the bladder through the urethra to the external meatus clearly reflects this sphincteric mechanism (Fig. 3). Such a profile has always the same pattern. As the pressure recording balloon enters the urethral canal, it registers a minimal rise in pressure at the level of the internal meatus. Proceeding downward, pressure increases progressively, reaching a peak at the midurethra in the female and the membranous urethra in the male, and then progressively drops. Such a pressure profile superimposed on a sagittal section of the female urethra shows that the closure pressure of the proximal urethra is primarily due to the smooth musculature. The rise in pressure in the midurethra is due to both the smooth musculature and the surrounding striated muscle fibers and can be confirmed by functional testing.

The smooth sphincteric element of the urethra is of detrusor origin and receives its motor supply through the pelvic parasympathetic. When bladder and urethral smooth musculature is stimulated by parasympathomimetic drugs such as ethacholine (Mecholyl), pressure recordings show a rise in bladder pressure, as well as in proximal and midurethral pressures (Fig. 4A). The same changes can be obtained from direct pelvic nerve stimulation (Fig. 4B).

Sympathomimetic stimulation (epinephrine) occasionally produces an irregular rise in urethral pressure (Fig. 5). The same response can be generated by norepinephrine (Fig. 6A) or more critically by angiotensin (Fig. 6B), which is a pure vasoconstrictor that has no effect on visceral smooth musculature. Blocking urethral smooth musculature by atropine did not interfere with urethral response to sympathomimetics.

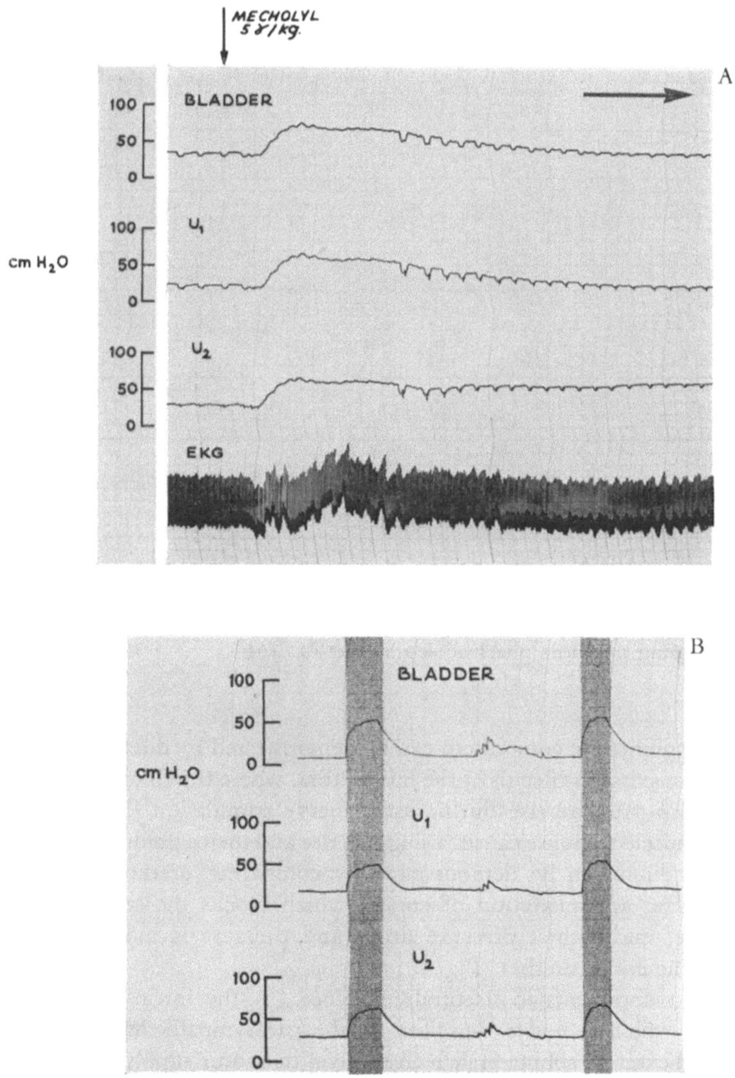

Fig. 4 A and B. A Simultaneous recording of effect of i.v. injection of 5 γ/kg of mecholyl on bladder, proximal and midurethral pressures and E.K.G. Note simultaneous rise in bladder and urethral pressures. Start and duration of effect of drug is monitored by E.K.G. B Pelvic nerve stimulation induced equal rise in bladder and urethral pressures. Shaded area indicates duration of electrical stimulation. Note gradual rise and gradual drop in pressure characteristic of smooth-muscle activity

However, after the sympathetic effect on blood vessels was blocked by dibenzyline, epinephrine did not affect urethral pressure (Fig. 7). These experiments indicate that the sphincteric mechanism is not under sympathetic control. The changes within the sphincteric segment in response to sympathomimetics recorded by previous investigators are due to the effect of the drugs on the blood vessels, not on the smooth musculature.

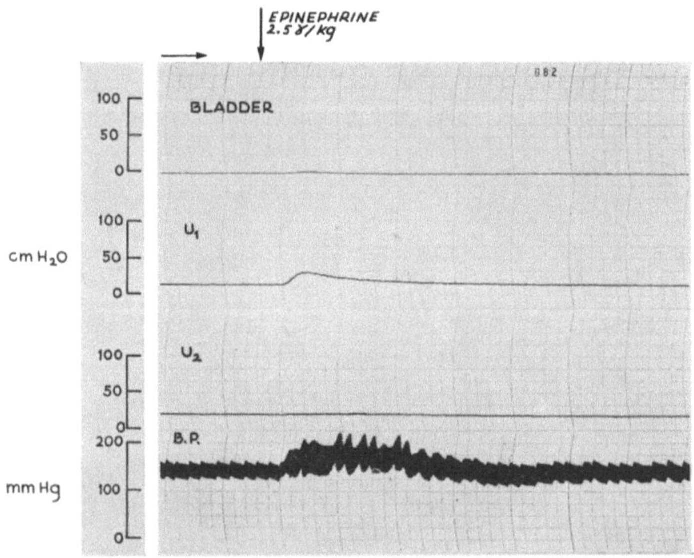

Fig. 5. Effect of epinephrine on bladder and urethral pressures. Proximal urethral pressure rose, but bladder and midurethral pressures showed no change. Simultaneous blood-pressure recording monitors pharmacologic effect of drug

The striated sphincteric component can be demonstrated by direct nerve stimulation (Fig. 8). The maximum effect is at the midurethra, where the striated muscle response causes a sharp pressure rise during sacral nerve stimulation (Fig. 9). In contrast, the smooth muscle response causes a sluggish rise and then a gradual drop in pressure. The same findings can be demonstrated by comparing urethral pressure profiles before and after administration of curare, which blocks the striated musculature. Under curare, midurethral pressure drops and pressure is more or less uniform throughout the entire urethra (Fig. 10).

There are 2 components in urethral resistance: 1. the intrinsic smooth urethral musculature, which is under the effect of the parasympathetic motor supply and 2. the striated external sphincter, which receives its motor supply from the pudendal nerve. Both can be demonstrated by direct nerve stimulation and each can be blocked separately, the smooth muscles by atropine and the striated muscles by curare (Fig. 11). The contribution of each element to the total urethral resistance can be quantitated. Under normal conditions, the critical opening pressure needed to overcome urethral resistance is about 100 mm Hg; it is slightly higher in the male than in the female. The sustained urethral resistance once the urethra is forced open is about 50 mm Hg

(Fig. 12A). When the striated component is blocked by curare, the critical opening pressure drops to about 55 mm Hg and the sustained to 30–35 mm Hg (Fig. 12B). When additionally the smooth muscle component is blocked, the critical opening pressure and the sustained resistance drop to about 25–30 mm Hg (Fig. 12C). Each component contributes about 50 per cent of the total urethral resistance, the striated element slightly more.

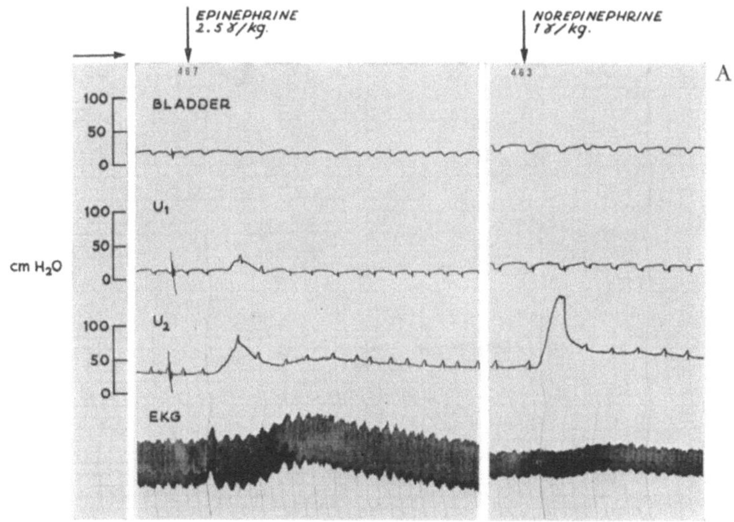

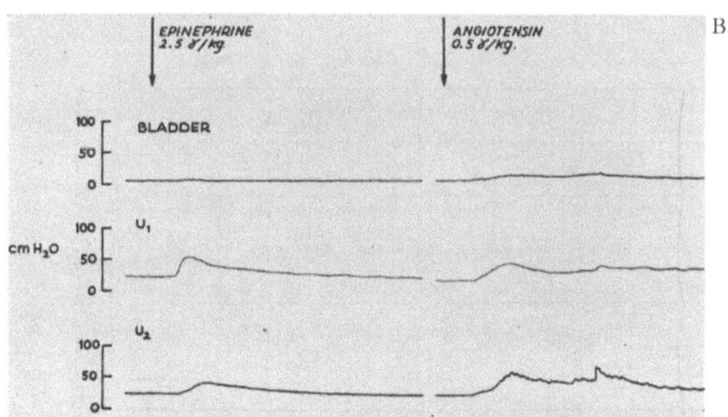

Fig. 6 A and B. A Effect of epinephrine and norepinephrine (which acts more specifically on blood vessels), Epinephrine caused rise in midurethral pressure and norepinephrine produced even greater rise at that level. E.K.G. monitors pharmacologic effect of drug. B Comparison of effect of epinephrine and angiotensin (a pure vasoconstrictor) on bladder and urethral pressures. Both drugs gave same effect, indicating that rise in urethral pressure with epinephrine is primarily due to its vasoconstrictor effect

The critical opening pressure is almost equal to the normal highest resting midurethral pressure, while the sustained resistance is almost equal to the proximal urethral resistance, further demonstrating that the highest urethral resistance is offered by the striated component. To force the urethra open, this resistance must be exceeded. Once the urethra is open, the remaining resistance is that of the smooth muscle sphincter.

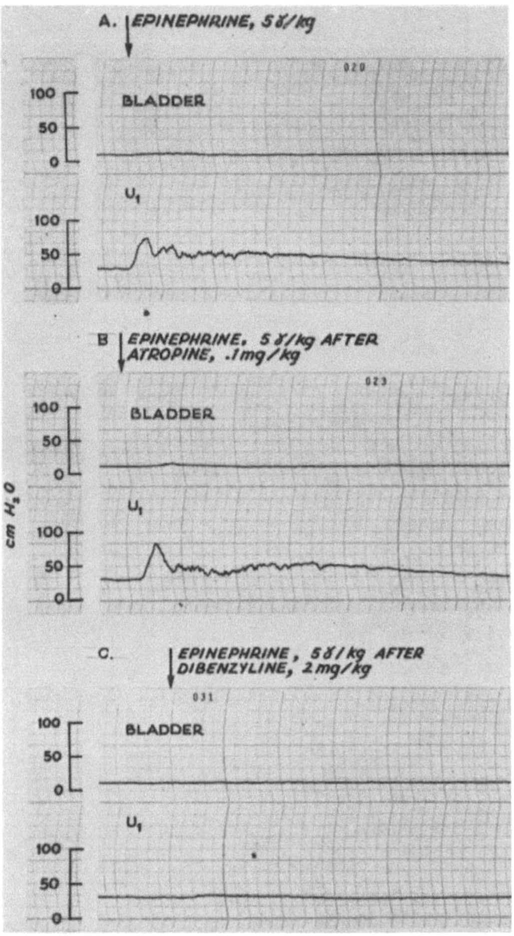

Fig. 7 A – C. Effect of epinephrine on urethral pressure after blocking of smooth muscles by atropine and blocking of vascular sympathomimetic effect by dibenzyline. A Epinephrine produced moderate rise in proximal urethral pressure. B Epinephrine in the same dose, but after complete blocking of smooth muscle activity by atropine, gave same rise in pressure, demonstrating that this pressure rise is not due to urethral smooth musculature. C Epinephrine given after blocking vascular sympathomimetic effect by dibenzyline produced no response in urethral pressure, confirming that epinephrine effect on urethra is primarily on its vascular element

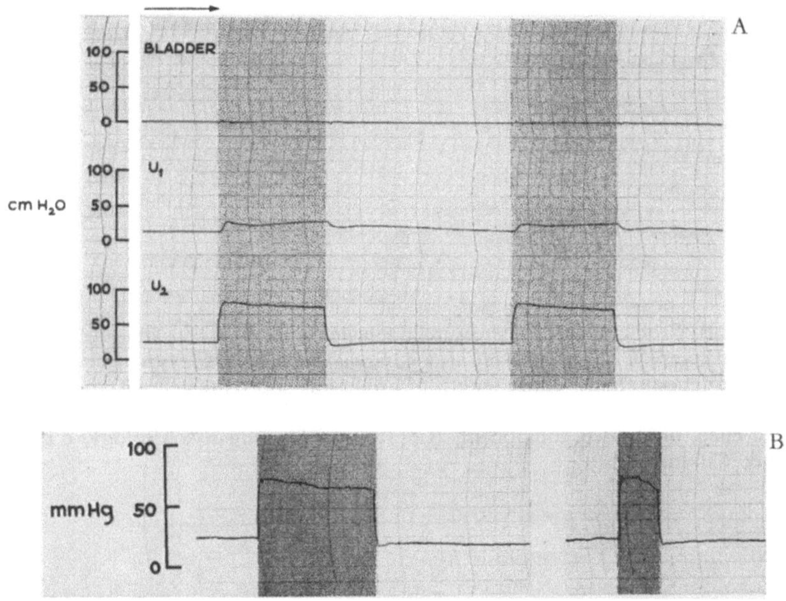

Fig. 8 A and B. A Effect of unilateral electric stimulation of S_2 and S_3 on resting bladder and proximal and midurethral pressures. There is sharp and appreciable rise in midurethral pressure, minimal rise in proximal urethral pressure, and no change in bladder pressure. Shaded area represents duration of electric stimulus. (Courtesy of Investigate Urology, c 1969, The Williams & Wilkins Co.) B Effect of unilateral electric nerve stimulation of S_2 and S_3 on urethral resistance to constant flow. There is sharp and sustained appreciable pressure rise during electric stimulation

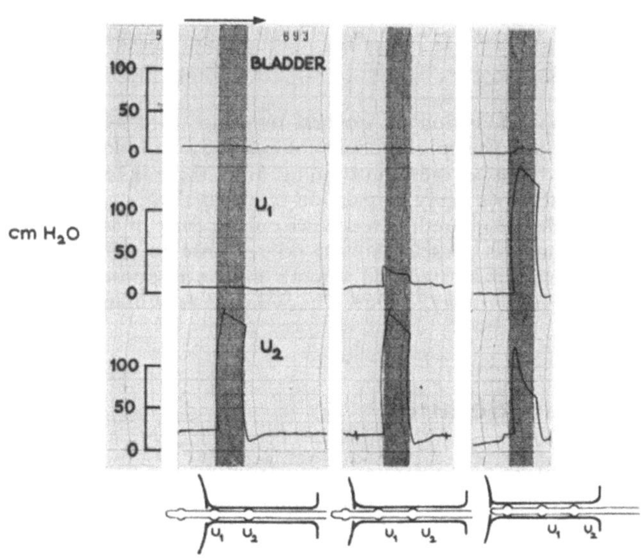

Fig. 9. Effect of unilateral electric stimulation of S_2 and S_3 on urethral pressure during progressive movement of recording balloon along entire length of urethra. Inserts below tracings correspond with level of recording balloons while tracings above are recorded. Activity of striated muscle is maximal in midurethra

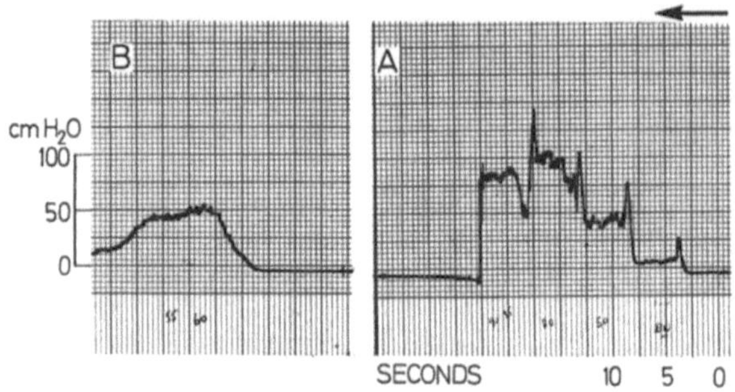

Fig. 10 A and B. Profile of urethral pressure in female child. A Awake and supine. B Under anesthesia and curare. Urethral pressure after curare is due only to smooth musculature and is sustained throughout almost entire length of urethra at approximately 30 cm H_2O, which is equivalent to proximal urethral pressure when child is awake. Pressure in midurethra lost added effect of striated component. (Courtesy of Investigative Urology, c 1969, The Williams & Wilkins Co.)

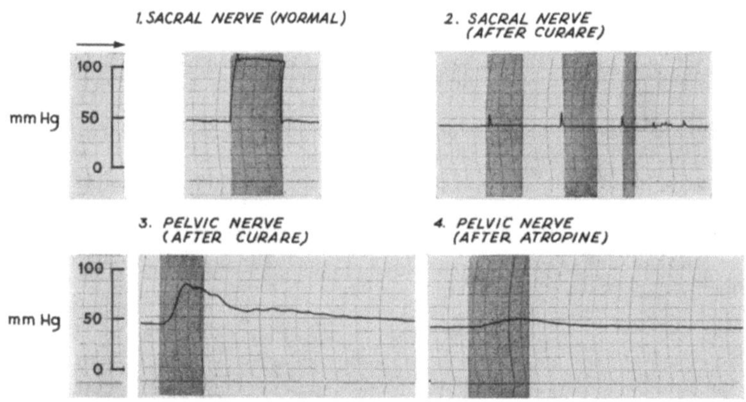

Fig. 11. Effect of nerve stimulation on urethral resistance to constant flow of 20 ml per sec. 1 and 2 show effect of unilateral stimulation of S_2 and S_3. In 1, note sharp rise in urethral resistance when striated muscles were functioning. In 2, there is no response after curare. In 3, effect of unilateral pelvic nerve stimulation on urethral resistance while curare is still active shows appreciable rise in urethral resistance only because of smooth muscle response. In 4, after atropine, smooth muscle showed no response to pelvic nerve stimulation. Compare difference between striated and smooth muscle response to nerve stimulation. (Courtesy of Investigative Urology, c 1969, The Williams & Wilkins Co.)

Functional Consideration

Voluntary activity can be demonstrated by instructing the patient to hold his urine and to contract the perineal muscles. Involuntary activity is seen when the patient bears down or coughs. If we monitor these activities with a pressure-recording catheter, we can further analyze the activity and the magnitude of the urethral closure element at any particular level. This technique is used to identify the anatomical length of the urethra as well as the functionally sphincteric length. Normally both are equal (Fig. 13).

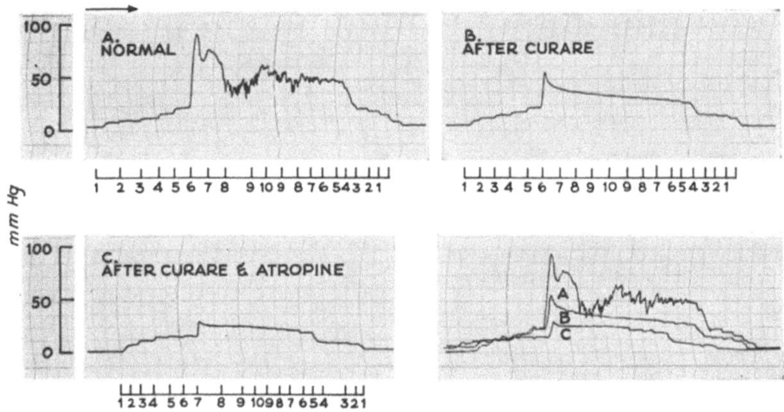

Fig. 12A–C. Urethral resistance to progressively increasing flow rates increasing from zero (1 on bottom scale) to 40 cc/sec (10 on scale) and then progressively decreasing to zero. Peak of pressure curve occurs when urethra is forced to open (critical opening pressure). In A this pressure is 85 mm Hg; it drops sharply once urethra is opened, although flow rate continues to increase. Pressure level is maintained as sustained urethral resistance (50 mm Hg in A) before dropping gradually with progressively decreasing flow rate. B is urethral resistance to same progressively changing flow rate after administration of curare. Difference between A and B is caused by blocking striated muscle component. After curare and atropine (C), there is further drop in critical opening pressure and sustained urethral pressure. Differences between B and C are caused by blocking smooth muscle component. Bottom right tracing is composite of A, B, and C and shows contribution of each component to urethral resistance. (Courtesy of Investigative Urology, c 1969, The Williams & Wilkins Co.)

Changes in the Proximal Urethra. These represent changes in the smooth sphincteric element. Normally with cough or bearing down there is an increase in intravesical pressure. The "hold" maneuver is an active contraction of the entire pelvic floor musculature, which causes an elevation of the bladder base and indirectly augments the closure pressure in the proximal urethra.

Changes in the Midurethra. With cough there is a sharp rise in the midurethral pressure that exceeds the rise in proximal urethral and bladder pressure. With bearing down there is a slow, sustained rise in midurethral pressure equal to or greater than the associated rise in the bladder and proximal urethral pressure. The "hold" maneuver causes a sustained rise in midurethral pressure, with no change in bladder pressure and a minimal rise in proximal urethral pressure. This rise in midurethral pressure is a direct response to contraction of the striated external sphincter and can be used to gauge its efficiency.

Urethral pressure profiles aid in analyzing the elements of urethral resistance, their relative values, and the sites of their weakness or hyperactivity. Usually the basic pathology involves primarily the striated component, with the smooth components affected only indirectly. Less commonly there is partial or complete loss of the smooth component, with the striated component remaining normal or becoming hyperactive in compensation.

Striated Muscle Weakness. Weakness of the striated muscular sphincter occurs in 2 instances: 1. commonly as a part of a generalized weakness of the pelvic floor and 2. in

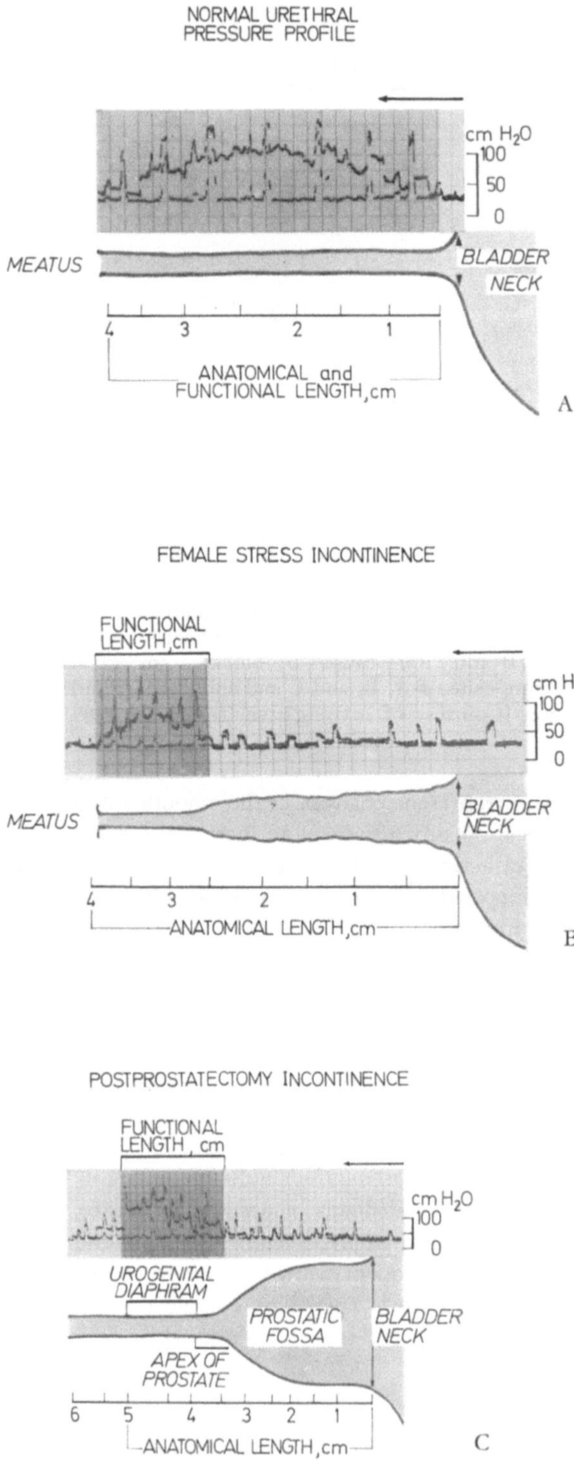

Fig. 13 A–C. A Profile of urethral pressure obtained by gradual pull-through of recording balloon. Distance from internal meatus is noted underneath. Effects of commands of "bear down" and "cough" on bladder and urethral pressures is recorded. Urethral and bladder pressures are overlapped so that differences in pressure under stress and at various levels can easily be seen. Note that anatomical urethral length is equal to its functional (sphincteric) length. B Postprostatectomy incontinence with wide prostatic fossa that offers no closure effect. Sphincteric functional length is limited mainly to striated sphincteric segment (membranous urethra) with adequate closure efficiency. There is minimal closure efficiency at apex of prostate. This smooth muscle deficiency is the basis of incontinence in spite of adequate striated sphincteric role. C Urethral pressure profile with com- in a case of marked stress mands incontinence. Though anatomic length is normal (4 cm), functional length is only $1^1/_2$ cm with loss of any sphincteric role to the proximal $2^1/_2$ cm

specific limited injury to the striated external sphincter, with the rest of the pelvic floor remaining intact.

1. Weakness of the striated external sphincter occurring as part of a generalized weakness of the pelvic floor is most commonly encountered in women with stress incontinence. A urethral pressure profile will reflect the weakness of the striated component around the midurethra. Secondary to generalized pelvic floor weakness, the smooth muscular element is functionally impaired, as is reflected in the pressure profile (Fig. 14).

The weakness of the striated sphincter becomes more apparent when the urethra is subjected to the "hold" maneuver, cough, or bearing down. The urethral musculature contracts minimally or not at all during the "hold" maneuver because of the primary weakness of the striated component (Fig. 15 A). When the urethra is subjected to the stresses of coughing or bearing down, it fails to maintain its closure efficiency and low or negative closure pressure and urine leakage occur (Fig. 15 B). This is due to the primary weakness of the pelvic floor and the secondary weakness of the smooth sphincteric element.

Weakness of the musculature of the pelvic floor may not be apparent with minimal bladder filling, but as the bladder distends, this weakness is reflected in less and less closure pressure as well as further abnormal deviation when the bladder is subjected to stress (Fig. 16).

2. Weakness limited to the striated external sphincter may be encountered in the absence of weakness of the pelvic floor musculature, or a slight degree of the latter

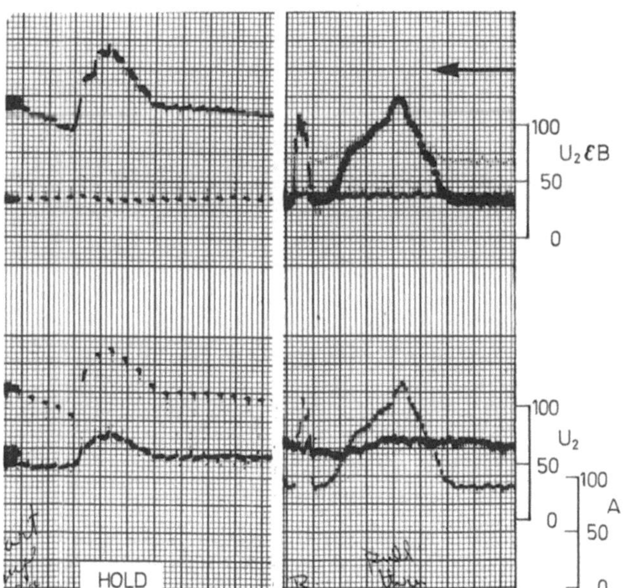

Fig. 14. Pressure profile of 55-year-old female with stress incontinence. Smooth pull-through pressure profile on right is relatively low. Profile with commands on left shows low closure pressure in proximal 1 cm of urethra. Pressure is higher in midurethra, but still below normal. Note lack of rise in urethral pressure with "hold" maneuver. Anal sphincter shows adequate activity

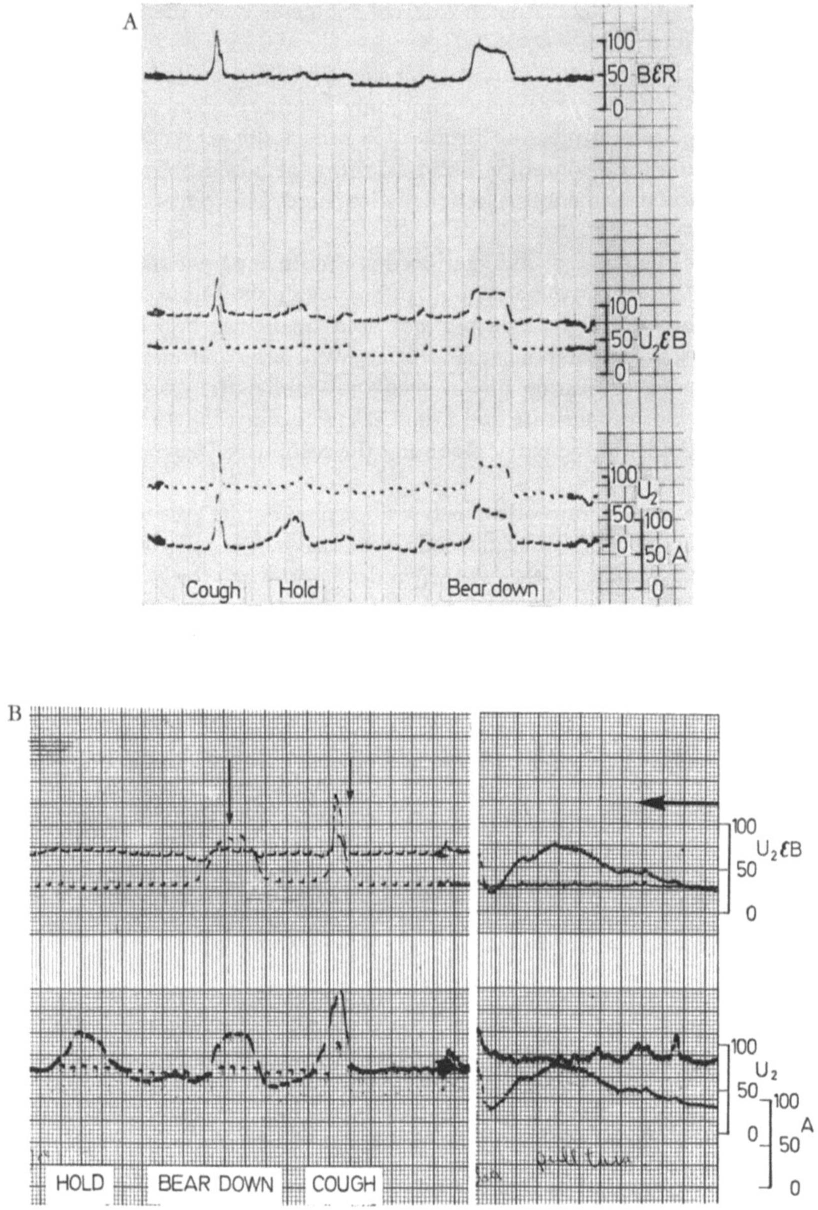

Fig. 15 A and B. A Rapid speed recording of effect of commands on patient shown in Fig. 14. With "cough" and "bear down", urethral pressure rise is slightly less than rise in intra-abdominal and intravesical pressures. With "hold" maneuver, there is no rise in midurethral pressure (vertical arrow), but there is adequate rise in anal sphincter pressure. B 52-year-old female with moderate stress incontinence. Urethral pressure profile at right is below normal range. Effect of commands is shown at left. With "cough" and "bear down", rise in mid-urethral resistance is less than rise in intra-abdominal and bladder pressure, leading to leakage (vertical arrows). With "hold" maneuver note lack of striated muscle response around urethra, though rise in anal pressure is adequate

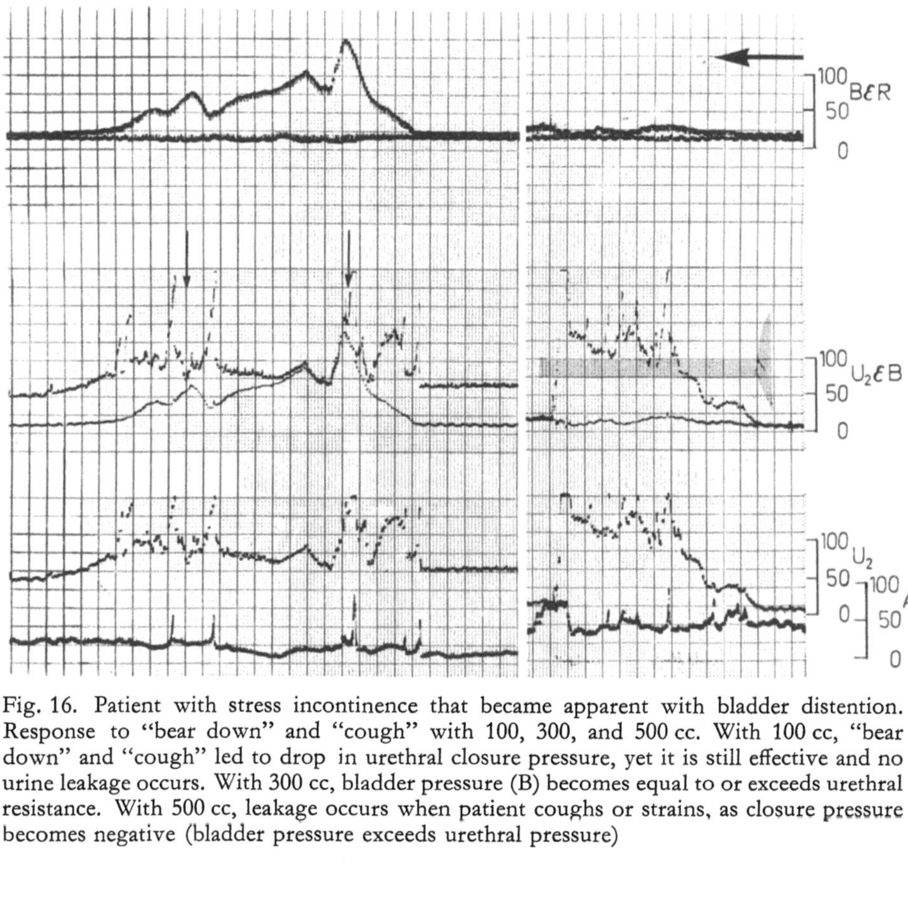

Fig. 16. Patient with stress incontinence that became apparent with bladder distention. Response to "bear down" and "cough" with 100, 300, and 500 cc. With 100 cc, "bear down" and "cough" led to drop in urethral closure pressure, yet it is still effective and no urine leakage occurs. With 300 cc, bladder pressure (B) becomes equal to or exceeds urethral resistance. With 500 cc, leakage occurs when patient coughs or strains, as closure pressure becomes negative (bladder pressure exceeds urethral pressure)

may be present along with a disproportionate and serious degree of external sphincter weakness. In vaginal urethral dissection, surgical damage to the striated external sphincter musculature inevitably occurs. An extreme example of such damage was seen in a case of a wide urethrovaginal fistula involving the distal part of the middle third of the urethra, which occurred as a complication to excision of a urethral diverticulum. Fortunately an intact proximal half (2 cm) of the urethra with adequate, healthy intrinsic musculature permitted a successful repair (Fig. 17). Although the efficiency of the smooth sphincteric element may be nearly normal, continence nevertheless may not be sustained, especially under stress, because of concomitant weakness of the striated external sphincter (Fig. 18).

Striated Muscle Overactivity. Overactivity of the striated musculature can be encountered in 2 instances: 1. functional overactivity and 2. neurogenic overactivity.

1. Functional overactivity is commonly seen secondary to local irritating factors such as recent infection, chemical or mechanical irritation, meatal or distal urethral stenosis, and the "chronic urethral syndrome" in adult females (Fig. 19). Psychosomatic functional overactivity is encountered frequently and should not be ignored. Besides being the commonest basis of recurrent urinary tract infection, especially

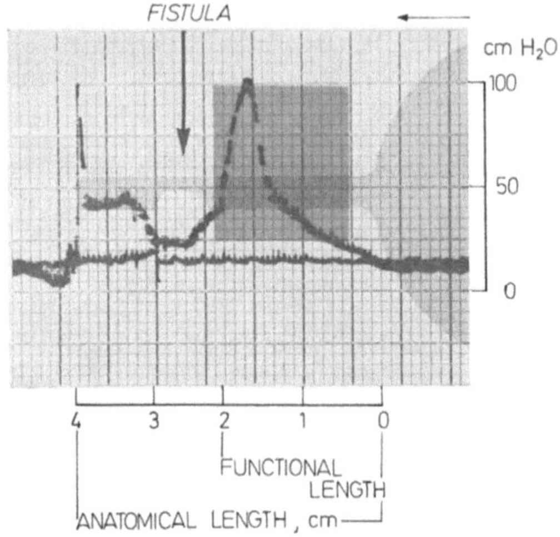

Fig. 17. Urethral pressure profile in a case of a urethrovaginal fistula involving distal third centimeter of urethra, thus reducing sphincteric functional length to proximal 2 cm. Fortunately this proximal segment had adequate closure efficiency that permitted successful repair

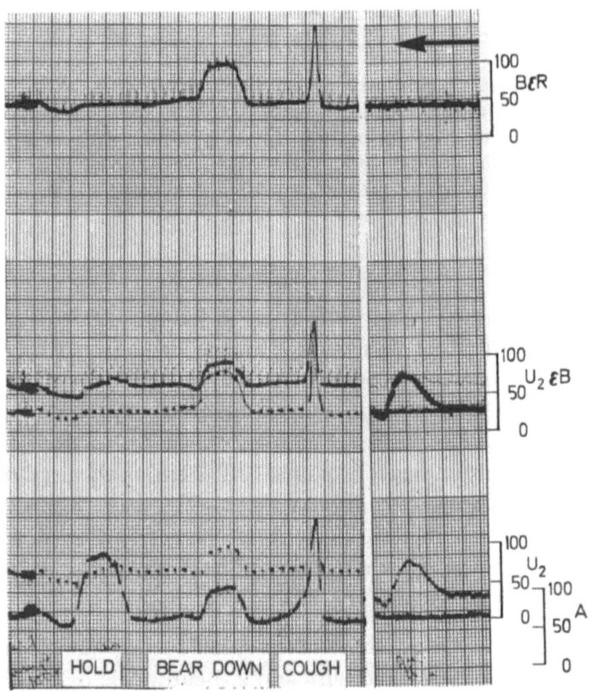

Fig. 18. 58-year-old female with stress incontinence following failure of two anterior vaginal repairs. Urethral pressure profile is low, with "bear down" and "cough" there is drop in urethral closure pressure. Rise in intravesical pressure is greater than rise in urethral pressure. More important is that with "hold" maneuver, in spite of powerful anal response, midurethral pressure dropped instead of increasing

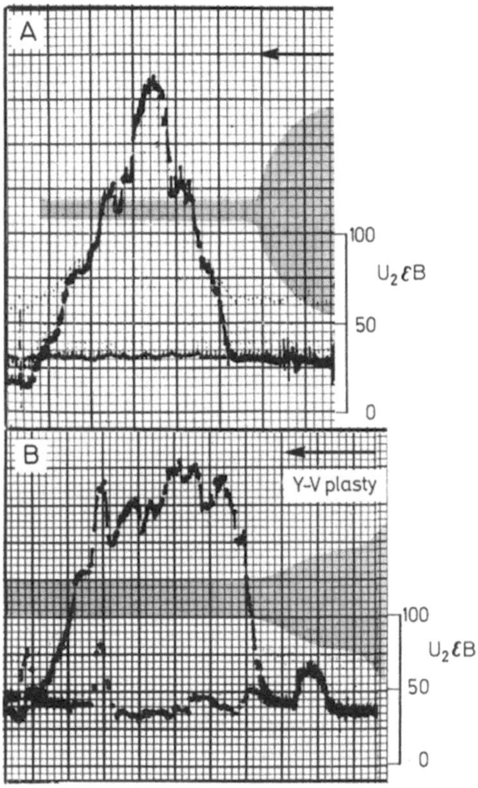

Fig. 19 A and B. A 8-year old girl with recurrent urinary tract infection. Urethral pressure profile reveals abnormally high midurethral pressure. B 18-year-old female with persistent recurring urinary tract infection. Y-V plasty done at age 5 for assumed bladder neck obstruction. Urethral pressure profile reveals very low closure pressures in proximal urethra with abnormally high midurethral pressure, which is the primary functional obstruction

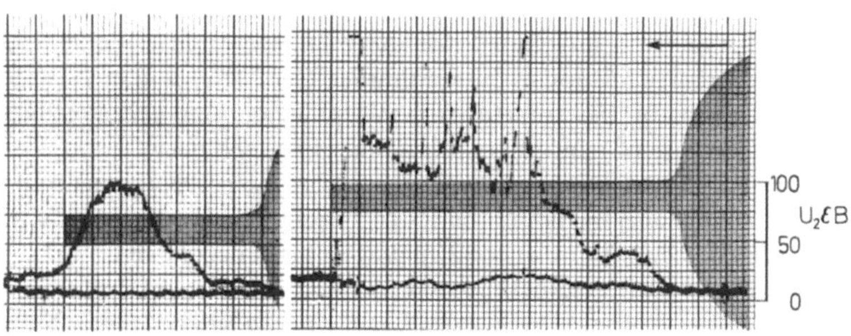

Fig. 20. 45-year-old male patient with a partial upper motor neuron lesion. Urethral pressure profile at right reveals normal closure pressure in prostatic urethra and marked rise at apex of prostate and in region of spastic external sphincter. Its hyperirritability is clearly seen. After pudendal block (left), urethral pressure profile of the striated sphincter segment reverted to normal. Proximal urethral pressure diminished

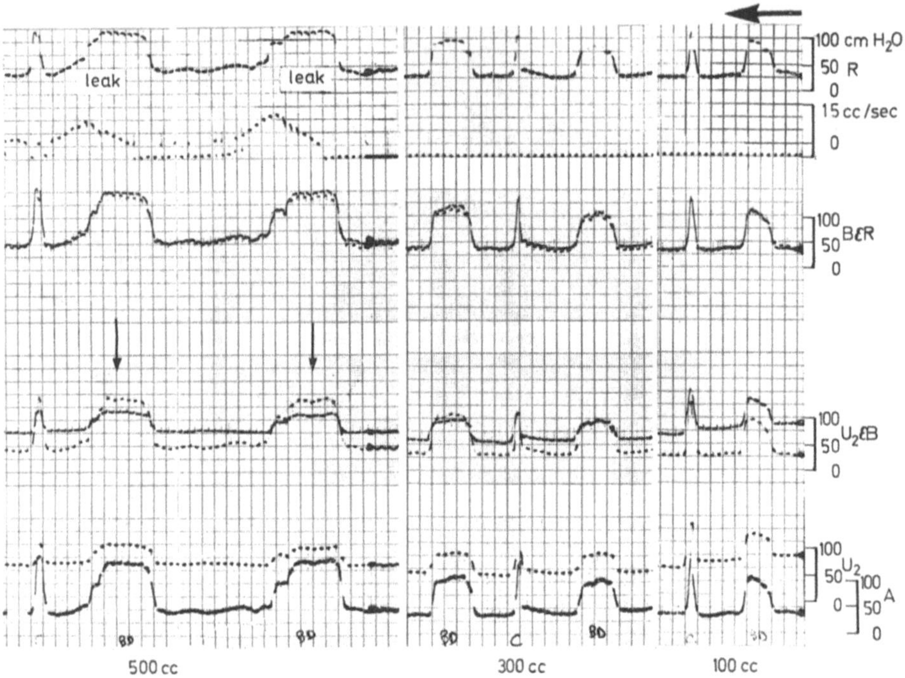

Fig. 21. Urethral pressure profile at right. At left is shown effect of stimulation of trigger zone (vertical arrows), which excited marked detrusor response and increased pressure in striated external sphincter and anal sphincter

in females, it can be persistent and severe enough to lead to detrusor overactivity, hypertrophy, and may end in decompensation.

In functional overactivity, proximal urethral pressure is normal or slightly higher because of overactivity of the striated component. Midurethral pressure is extremely high because of the spasticity of the external sphincter, which is usually part of a diffuse functional spasticity of the pelvic floor but is occasionally also associated with overactivity of the anal sphincter or the sphincter vaginae. Functional overactivity is intermittent in relation to the voiding act. Generally, voiding is functionally obstructed by spasm of the striated external sphincter, but occasionally normal voiding occurs because of momentary relaxation of the sphincter.

2. Neurogenic overactivity is commonly seen with upper motor neuron lesions. Spasm of the pelvic floor, including the striated external sphincter, occurs. In some instances, pudendal neurectomy or local sphincterotomy may be indicated to relieve this spasticity. Neurogenic external sphinter spasticity is always a part of a generalized spasticity; its degree depends on the extent of the neurogenetic lesion. Neurogenic overactivity can be determined by obtaining urethral pressure profiles before and after pudendal nerve block (Fig. 20). Not infrequently the bladder neck has been incriminated as the site of increased outflow resistance in neurogenic cases and needless or even harmful transurethral resection has been done, only to discover later that the real site of high resistance is in the region of the external sphincter.

Smooth Muscle Weakness. Smooth muscle weakness occurs secondary to general weakness of the pelvic floor and striated external sphincter. The support of the bladder base and vesicourethral junction is weakened, indirectly weakening the closure efficiency of the smooth musculature of the proximal urethra, though there is no actual damage to the smooth muscle. The pattern of a urethral pressure profile remains normal, but the pressure level itself is much lower than is normal. The vesicourethral relationship is altered, changing the orientation of the muscles as they proceed from the bladder to the urethra and weakening their closure efficiency. The vesicourethral junction is funneled while at rest, whereas it should be so only at the initiation of voiding. This leads to loss of sphincteric closure in the proximal urethra and markedly reduces the closure pressure throughout the rest of the urethra.

The functional (sphincteric) length of the urethra is therefore shortened by the loss of the very first segment of the urethra, which is the most critical segment for maintaining urethral closure, especially under stress. A marked degree of stress incontinence and its effect on the functional (sphincteric) length of the urethra is seen in Fig. 21. The functional length of the urethra is shortened even more with progressive bladder filling because of the progressive widening and pulling on the vesicourethral junction. This phenomenon explains the observation that, although closure efficiency may be adequate with minimal bladder filling, it gets weaker with progressive distention. In some cases, both the anatomic and the functional length are shortened with very low closure pressure.

In the male, weakness of the smooth muscular element is also frequently seen as a result of surgical intervention, commonly transurethral resection for prostatic hyperplasia in which resection extends outside the adenoma, destroying the smooth sphincteric element with or without associated damage to the striated sphincter. The smooth sphincteric element that should be preserved is at the level of the internal meatus, in the compressed prostatic capsule and at the apex of the prostate below the level of the verumontanum. There should be an attempt to preserve all these components, though not all of them are essential to the preservation of continence. Loss of all of them, however, definitely results in complete incontinence, regardless of whether the striated sphincter is preserved. If the latter is intact, it might overact to compensate for the weakened or lost smooth spincteric mechanism, but is usually not uccessful.

Smooth Muscle Overactivity. Smooth muscle overactivity has never been encountered as a primary change in the urethral smooth sphincteric mechanism. Slightly higher proximal urethral resistance (chiefly a function of the smooth muscular sphincteric element) is occasionally encountered in association with marked hypertrophy of the detrusor or the proximal urethra and has been seen in spastic neurogenic bladders. The raised proximal urethral resistance here may be a secondary effect of the associated spastic external sphincter, which alone can increase the efficiency of even the normal proximal urethral smooth sphincteric element. It is also conceivable, however, that the activity of the smooth muscular sphincteric element itself may increase as the entire detrusor hypertrophies as a result of distal obstruction or neurogenic dysfunction.

We have never been able to demonstrate increased smooth muscular sphincter resistance with a normal or moderately hypertrophied detrusor, thus we feel the smooth musculature cannot be responsible for obstruction at the level of the bladder neck. Smooth muscle overactivity is of no pathologic significance unless there is marked detrusor hypertrophy involving the internal meatus and proximal urethral musculature and even then it is hardly ever of any functional significance.

Summary

The bladder and its urethral sphincteric mechanism form a complex structure. They are interdependent and function in harmony. The urethral sphincteric mechanism comprises 1. a smooth muscular element, which is not essential for continence and control and which is rarely obstructive, and 2. a striated voluntary muscular element which complements the sphincteric action of the smooth element, but which can be overactive and produce significant obstructive pathology.

References

Bors, E.: Neurogenic bladder. Urol. Surv. **7**, 177–250 (1957).

Bors, E., Comarr, A. E.: Neurological disturbances of sexual function with special reference to 529 patients with cord injury. Urol. Surv. **10**, 191–222 (1960).

Caine, M., Edwards, D.: The peripheral control of micturition: A cineradiographic study. Brit. J. Urol. **30**, 34–42 (1958).

Cass, A. S., Hinman, F., Jr.: Constant urethral flow in female dog. I. Normal vesical and urethral pressures and effect of muscle relaxant. J. Urol. (Baltimore) **99**, 442–446 (1968).

Claridge, M., Shuttleworth, K. E. D.: The dynamics of obstructed micturition. Invest. Urol. **2**, 188–199. (1964).

Clegg, E. J.: The musculature of the human prostatic urethra. J. Anat. (Lond.) **91**, 345–351 (1957).

Denny-Brown, D. E.: Nervous disturbances of vesical sphincter. New Engl. J. Med. **215**, 647–652 (1936).

Disse, J.: Handbuch der Anatomie, Vol. 7. Jena: Gustav Fischer 1902.

Elliot, T. R.: The innervation of the bladder and urethra. J. Physiol. (Lond.) **35**, 367–445 (1907).

Enhörning, G.: Simultaneous recording of intravesical and intraurethral pressure. Acta chir. scand. (Suppl.) 276, 1961.

Fisher, R. E., Tanagho, E. A., Lyon, R. P., Tooley, W. H.: Urethral calibration in newborn girls. J. Urol. (Baltimore) **102**, 67–69 (1969).

Gleason, D. R., Lattimer, J. K.: The pressure-flow study: A method for measuring bladder neck resistance. J. Urol. (Baltimore) **87**, 844–852 (1962).

Green, L. F., Kelalis, P. P.: Retrograde ejaculation of semen due to diabetic neuropathy. J. Urol. (Baltimore) **98**, 693–696 (1967).

Homsy, G. E.: The dynamics of the ureterovesical and vesicourethral junctions. II. Effect of autonomic and somatic excitation on the resistance to flow of the distal ureter and lower urinary tract. Invest. Urol. **4**, 408–418 (1967).

Kalischer, O.: Die Urogenitalmuskulatur des Dammes. Berlin: S. Karger 1900.

Kennedy, W. T.: The muscle of micturition: Its role in the sphincter mechanism with reference to incontinence in the female. Amer. J. Obstet. Gynec. **52**, 206–217 (1946).

Kohlrausch, O.: Zur Anatomie und Physiologie der Beckenorgane. Leipzig: S. Hirzel 1854.

Krahn, H. P., Morales, P. A.: The effect of pudendal nerve anesthesis on urinary continence after prostatectomy. J. Urol. (Baltimore) **94**, 282–285 (1965).

Langworthy, O. R., Murphy, E. L.: Nerve endings in the urinary bladder. J. comp. Neurol. **71**, 487–505 (1939).
Langworthy, O. R., Kolb, L. C., Lewis, L. G.: Physiology of micturition. Baltimore: Williams & Wilkins Co. 1940.
Lapides, J.: Structure and function of the internal vesical sphincter. J. Urol. (Baltimore) **80**, 341–353 (1958).
Lapides, J., Ajemian, E. P., Stewart, B. H., Breakey, B. A., Lichtwardt, J. R.: Further observations on the kinetics of the urethrovesical sphincter. J. Urol. (Baltimore) **84**, 86–94 (1960).
Lapides, J., Hodgson, N. B., Boyd, R. E., Shook, E. L., Lichtwardt, J. R.: Further observations on pharmacologic reactions of the bladder. J. Urol. (Baltimore) **79**, 707–713 (1958).
Lapides, J., Sweet, R. B., Lewis, L. W.: Role of striated muscle in urination. J. Urol. (Baltimore) **77**, 247–250 (1957).
Le Gross, C. S.: Some remarks on the anatomy and physiology of the urinary bladder and the sphincter of the rectum. J. anat. Physiol. **17**, 442 (1883).
Learmonth, J. E.: A contribution to the neurophysiology of the urinary bladder in man. Brain **54**, 147–176 (1931).
Lowsley, O. S.: The development of the human prostate gland with reference to the development of other structures at the neck of the urinary bladder. Amer. J. Anat. **13**, 299–349 (1912).
Lowsley, O. S.: Embryology, anatomy and surgery of the prostate gland. Amer. J. Surg. **8**, 526–541 (1930).
Lyon, R. P., Tanagho, E. A.: Distal urethral stenosis in little girls. J. Urol. (Baltimore) **93**, 379–387 (1965).
Macalpine, J. B.: The musculature of the bladder-neck of the male in health and disease. Proc. roy. Soc. Med. **28**, 39–56 (1934).
Macleod, D.: The trigonal muscle, with special reference to its composition and urinary function. Brit. J. Urol. **13**, 135–148 (1941).
Macleod, D.: The posterior wall of the prostate gland. Brit. J. Urol. **14**, 153–171 (1942).
McCrea, E. D. A.: Musculature of the bladder. Proc. roy. Soc. Med. **19**, 35–43 (1926).
Muellner, S. R.: Lack of special urethral lesion in exertional urinary incontinence. New Engl. J. Med. **234**, 400–402 (1946).
Muellner, S. R., Fleischner, F. G.: Normal and abnormal micturition; study of bladder behavior by means of the fluoroscope. J. Urol. (Baltimore) **61**, 233–241 (1949).
Muellner, S. R.: The etiology of stress incontinence. Surg. Gynec. Obstet. **88**, 237–242 (1949).
Muellner, S. R.: The physiologic components of urinary bladder; their clinical significance. New Engl. J. Med. **241**, 769–772 (1949).
Muellner, S. R.: The physiology of micturition. J. Urol. (Baltimore) **65**, 805–810 (1951).
Murphy, J. J., Schoenberg, H. W.: Observations on intravesical pressure changes during micturition. J. Urol. (Baltimore) **84**, 106–110 (1960).
Nemoy, N. J., Govan, D. E.: Urinary continence in the absence of an intact external sphincter. J. Urol. (Baltimore) **102**, 200–204 (1969).
Potts, I. F.: The mechanism of ejaculation. Med. J. Aust. **1**, 495–497 (1957).
Rose, S. S.: An investigation into sterility after lumbar ganglionectomy. Brit. med. J. 1953/I, 247–250.
Scher, S.: Some observations on the anatomy of the bladder neck and posterior urethra with reference to prostatic obstruction. Brit. J. Urol. **22**, 116–124 (1950).
Scott, J. E. S., Clayton, C. B., Dee, P. M., Simpson, W.: A study of the hydrodynamics of the female urethra. I. Experiments with models. J. Urol. (Baltimore) **96**, 763–769 (1966).
Scott, J. E. S., Clayton, C. B., Dee, P. M., Simpson, W.: A study of the hydrodynamics of the female urethra. II. Static pressure profile measurements. J. Urol. (Baltimore) **96**, 770–778 (1966).
Scott, R., Jr., McIlhaney, J. S.: The voiding rates in normal male children. J. Urol. (Baltimore) **82**, 224–230 (1959).

SIMEONE, F. A.: A neuromuscular mechanism in the ductus epididymis and its impairment by sympathetic denervation. Amer. J. Physiol. **103**, 582–591 (1933).

SMITH, J. C.: Some theoretical aspects of urethral resistance. Invest. Urol. **1**, 477–481 (1964).

SMITH, J. C.: Urethral resistance to micturition. Brit. J. Urol. **40**, 125–156 (1968).

SMITH, J. C., EDWARDS, D., BRYANT, G. H.: A practical method of measuring urethral resistance to micturition. Brit. J. Urol. **38**, 542–546 (1966).

TANAGHO, E. A.: Spastic striated external sphincter. Brit. J. Urol. **43**, 69–82 (1971).

TANAGHO, E. A., McCURRY, E.: Pressure and flow rate as related to lumen caliber and entrance configuration. J. Urol. (Baltimore) **105**, 583–585 (1971).

TANAGHO, E. A., MEYERS, F. H.: The "internal sphincter": Is it under sympathetic control? Invest. Urol. **7**, 79–89 (1969).

TANAGHO, E. A., MEYERS, F. H., SMITH, D. R.: Urethral resistance: Its components and implications. II. Striated muscle component. Invest. Urol. **7**, 195–205 (1969).

TANAGHO, E. A., MILLER, E. R.: Initiation of voiding. Brit. J. Urol. **42**, 175–183 (1970).

TANAGHO, E. A., MILLER, E. R., MEYERS, F. H., CORBETT, R. K.: Observations on the dynamics of the bladder neck. Brit. J. Urol. **38**, 72–84 (1966).

TANAGHO, E. A., SMITH, D. R.: The anatomy and function of the bladder neck. Brit. J. Urol. **38**, 54–71 (1966).

TANAGHO, E. A., SMITH, D. R.: Mechanism of urinary continence. I. Embryologic, anatomic and pathologic considerations. J. Urol. (Baltimore) **100**, 640–646 (1968).

TANAGHO, E. A., SMITH, D. R., MEYERS, F. H.: The trigone: Anatomical and physiological considerations. 2. In relation to the bladder neck. J. Urol. (Baltimore) **100**, 633–639 (1968).

TANAGHO, E. A., SMITH, D. R., MEYERS, F. H., FISHER, R.: Mechanism of urinary continence. II. Technique for surgical correction of incontinence. J. Urol. (Baltimore) **101**, 305–313 (1969).

UHLENHUTH, E., HUNTER, DE W. T., JR., LOECHEL, W. E.: Problems in the anatomy of the pelvis. Philadelphia: J. B. Lippincott & Co. 1952.

WALKER, J. W. T.: On the surgical anatomy of the prostate. J. anat. Physiol. **40**, 189–209 (1906).

WESSON, M. B.: Anatomical, embryological and physiological studies of the trigone and neck of the bladder. J. Urol. (Baltimore) **4**, 279–315 (1920).

WOODBURNE, R. T.: Structure and function of the urinary bladder. J. Urol. (Baltimore) **84**, 79–85 (1960).

ZINNER, N. R., PAQUIN, A. J., JR.: Clinical urodynamics. I. Studies of intravesical pressure in normal human female subjects. J. Urol. (Baltimore) **90**, 719–730 (1963).

Clinical Problems Associated with Urodynamic Abnormalities with Special Reference to the Value of Synchronous Cine/Pressure/Flow Cystography and the Clinical Importance of Detrusor Function Studies

R. T. Turner-Warwick*

Many patients attending a routine urological clinic present with symptoms which are the result of voiding imbalance. In some of these the urodynamic nature of the problem is self-evident, in others it may be less obvious and even obscured by some widely accepted concepts relating to aetiology and management.

The evaluation of a urodynamic status by history, routine clinical examination, static X-ray films, endoscopic appearances and deduction is often considerably less reliable than is generally appreciated.

Even under the best circumstances the bladder itself may be an unreliable witness, recording to the patient at one time a sensation of acute distension when it is empty and, at another, of emptiness when it is distended; the problem becomes compounded if the patient's reliability as a witness is also in question; recognising that the bladder is potentially an inherent liar we are often in need of a truth machine to define the situation objectively.

In his contribution to this Symposium Dr. Graham Whiteside has presented the technique of synchronous cine/pressure/flow cystourothrography that he and Dr. Patrick Bates have developed at Middlesex Hospital from the original research technique devised by Dr. Earl Miller; ours is designed as a practical routine for the investigation of patients referred to the urodynamic clinic in our Department of Urology.

The estimation of the detrusor pressure by electronic subtraction both during the filling and the voiding phases is often important to clinical practice. We use this investigation specifically for the study of those cases in which detrusor function has proved to be important to the identification and the management of the underlying condition and for those in which the synchronous record of the data is helpful to the evaluation of a particular condition, but, even when the synchronous record is not so important, it may be clinically more convenient than the separate collation of the

* The Middleesex Hospital, St. Peter's Hospital and The Institute of Urology, London, Great Britain.

relevant data since it can be completed within about 20 minutes once the routine is established and this is a relatively economic expenditure of patient's, physician's technician's and instrument time.

From the patient's point of view the instrumentation involved is perhaps comparable to a cystoscopic examination and not infrequently the information it provides may be considerably more relevant than routine endoscopy and occasionally it renders this irrelevant.

We all welcome the opportunity of this timely Symposium to discuss and re-evaluate some of our concepts relating to urodynamic function. We are severally interested in different aspects; how things work normally, why they go wrong, techniques for investigation and procedures correcting the clinical problems and the results.

In this paper I am specifically contemplating the effects of urodynamic dysfunction, and their clinical significance and management. The indications for cine/pressure/flow cystography in our unit over the past of 20 months are enumerated thus:

Non prostatic outlow obstruction in males	170
Recurrent urine infection in females	220
Post cystoplasty follow-up	55
Neuropathic bladder (non paraplegic)	173
Diabetic bladder dysfunction	32
Voiding dysfunction after pelvic surgery	48
Post prostatectomy symptoms	74
Stress and urge incontinence	473
Other indications	394
	1639

In each group there are certain basic queries relating to specific functions that require to be answered.

Re-evaluation of Precepts as a Result of Urodynamic Findings

Detrusor Function: An accurate knowledge of a patient's detrusor function is fundamental to the diagnosis and treatment of many problems presenting in routine urological practice, and particularly to the symptoms of frequency, urgency and leaking in relation to the conditions summarised. Estimation of detrusor function on a clinical basis is often grossly inaccurate: a clinical complaint of urgency or "painful bladder spasms" is for instance more frequently the result of hypersensitive abnormalities than of detrusor dysfunction; somewhat surprisingly patients with acute cystitis and interstitial cystitis are able to inhibit their detrusor contractions effectively within the capacity limited by hypersensitivity and are not therefore recorded as unstable systolic bladders (Bates, 1970). Patients suffering from uninhibited detrusor contractions of normal dimensions are often only aware of it as a result of frequency or leaking when a relative degree of sphincter weakness renders the control balance precarious, as in some cases of stress incontinence associated with unstable bladders and certain neuropathic bladders.

We must also re-evaluate some widely accepted criteria for the estimation of detrusor function for instance, the endoscopic appearance of bladder wall trabeculation is often considered as an indication of bladder muscle hypertrophy, bladder wall thickness, and, supposedly, an elevated detrusor pressure; in the male this may be broadly true, but in the female the appearance of trabeculation is of little significance and indeed it is frequently associated with an unusually thin walled bladder and a low normal or absent detrusor function. This is only one of many interesting and significant differences between male and female bladder and sphincter function (TURNER-WARWICK, 1970).

	Male	Female
Infrequent voiding syndrome	Unusual	Not uncommon; voiding imbalance may develop.
Detrusor power	Usually strong	Often relatively weak.
Trabeculation	Usually a fair indication of bladder wall thickness and raised detrusor pressure.	Doubtful significance; often associated with thin bladder wall and low detrusor pressure.
Bladder neck incompetence	Rarely spontaneous: usual result after prostatectomy.	Not uncommon, with or without vaginal prolapse.
Distal urethral sphincter mechanism	Efficient: unless compromised surgically.	Relatively inefficient.
Clinical result of bladder neck incompetence	Continence (retrograde ejaculation).	Stress continence dependent on distal mechanism.
Bladder neck obstruction	Not usual; high destrusor pressures.	Remarkably rare; occasional opening failure associated with atonic detrusor.
Obstructing distal dyskinesia	Occasional cause of residual obstruction after bladder neck relaxation or prostatectomy.	Commonest site of outflow obstruction.
Distortion competent element in obstruction	Very rare	Urethral telescoping or kinking may be contributary, with or without vesicovaginal prolapse.

In evaluating abnormalities of detrusor function we recognise that the time honoured cystometrogram is useful only for the detection of relatively gross abnormalities under specific stimulus of filling because, even within the group of neuropathic bladder dysfunction, there are significant abnormalities of detrusor behavious which are not revealed under these conditions.

The term "atonic bladder" is a term commonly applied to patients with a large post-voiding residual urine in which no obvious mechanical cause can be identified endoscopically (bladder neck obstruction, prostatic enlargement or urethral stricture); in practice many of these patients prove to have normal detrusor functions associated with a urodynamically identifiable obstruction.

Bladder Outflow Obstruction

We cannot reasonably reassure a male patient that he has no significant bladder outflow obstruction on the basis of an absence of a residual urine on a post-voiding urographic film and symptoms that he considers to be mild; rather we should regard the presence of a residual urine as a relatively late stage of detrusor decompensation amounting to bladder failure and we would hope that treatment would anticipate this escalation.

Again, as will be discussed below, we can no longer accept the diagnosis of "obstructive bladder neck hypertrophy" based on endoscopic assessment, however senior the observor; and furthermore, although commonly "treated" true bladder neck obstruction in the female has proved to be a remarkably rare condition (Fig. 1).

These examples are contrary to some of our inherited urological precepts. It is clear that we must revise some of them in the light of urodynamic understanding.

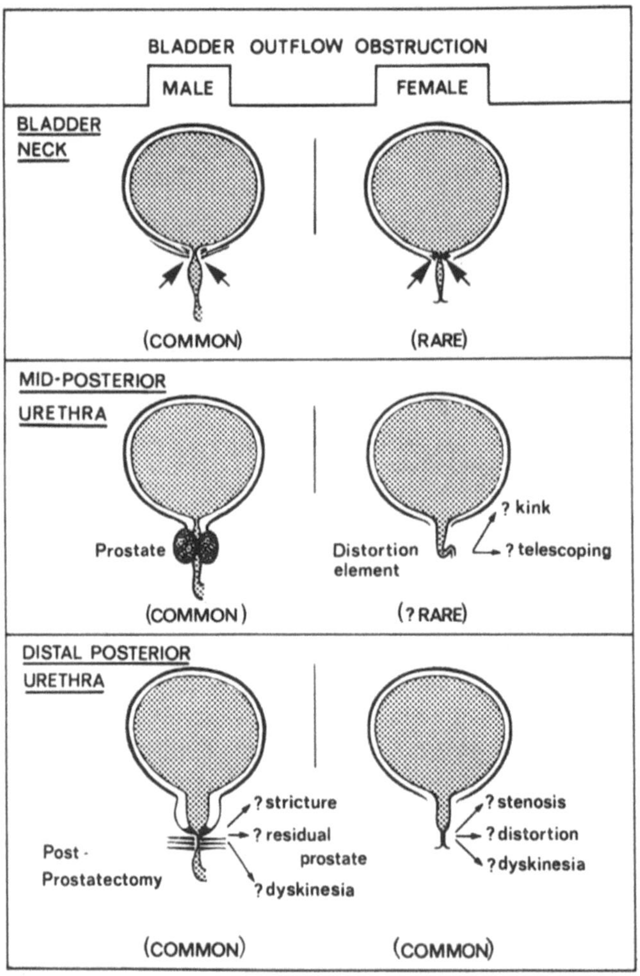

Fig. 1. Comparison of the sites of posterior urethral obstruction in the male and female

Terminology

The internal sphincter is better described as the bladder neck sphincter; it is formed by the embracing arrangement of the lower ends of the detrusor muscle fibre bundles and the normal bladder neck opening and the cough-competent closing mechanism which exist at this level must be the result of the functional arrangement of these fibres since the earlier concept of a concentric and separately innervated bladder neck muscle has long since been disproved by many authors.
The so-called "external sphincter" is clearly a composite and complicated mechanism; it is perhaps better referred to as the distal urethral mechanism and it is composed of the extrinsic striated muscle and the intrinsic urethral mechanism (Fig. 2).
The extrinsic urethral striated muscle relating to the pelvic floor or "urogenital diaphragm" are easy to demonstrate in the male. In the female, however, the extrinsic muscle system is relatively poorly defined.
The closing pressure of the intrinsic urethral mechanism is the result of elements of striated muscle, smooth muscle and tissue elasticity in the walls of the distal posterior urethra; the relative contribution of these individual components differ considerably from patient and to patient in the female there is frequently a superimposed closing pressure due to urethral distortion which may bear little relation to the apparent degree of vesico-urethro-vaginal prolapse.

The Study of Isolated Sphincter Mechanisms

It is fundamental to the understanding of urodynamic problems that although the terminology is the same there are marked differences between the male and female in the extent, location and relative efficiency of each element; some of these differences have been summarised.
Certain routine in surgical procedures result in sphincter destruction and some of these provide special opportunity for the functional study of individual sphincter mechanism in isolation (Fig. 2).

Analysis of Sphincter Function

The normal bladder neck is a cough-competent sphincter mechanism; it can retain urine so long as the patient is able to inhibit a detrusor contraction which would open it. This situation is readily demonstrable in patients whose distal sphincter mechanism (extrinsic muscles and intrinsic urethral mechanisms) have been destroyed by pelvic fracture and it is indeed remarkable to see on the screen of a synchronous cine/pressure/flow cystographic evaluation, the way in which such a patient can inhibit the detrusor mechanism to close the bladder neck in midstream. The same situation can be studied in the female with a traumatic proximal urethrovaginal fistula (a fistula due to vaginal marsupialisation a urethral diverticulum usually lies distal to part of the intrinsic urethral mechanism).

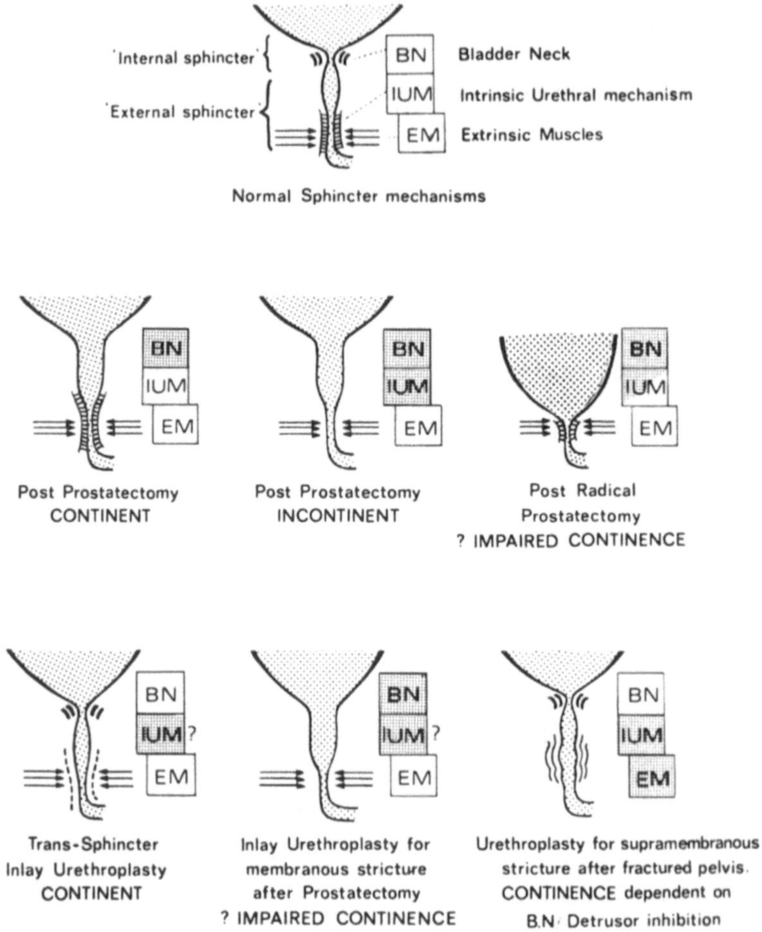

Fig. 2a

The intrinsic urethral mechanism can be readily demonstrated on a voiding cysto-urethrogram of a normal patient for when the stream is interrupted by the intrinsic muscles of the distal mechanism, this is immediately followed by retrograde emptying of the urethral contents through the bladder neck (Caine and Edwards 1958).

Retrograde posterior urethral emptying is achieved by the closing pressure of the intrinsic mechanism, it is rapid and usually complete within quarter of a second. Slow emptying may be the result of a weak intrinsic closing pressure but synchronous studies show that it may also result from a normal mechanism closing against a high voiding pressure of persistant detrusor contraction which maintains the bladder neck open. In some cases of obstructive dyskinetic failure of the bladder neck opening mechanism, the posterior urethra fails to empty efficiently in a retrograde manner as a result of premature bladder neck closure so that a small amount of urine may be trapped in the posterior urethra (Whiteside's sign) this may subsequently discharge distally as the external muscles relax; and is one of the causes a post-voiding drip.

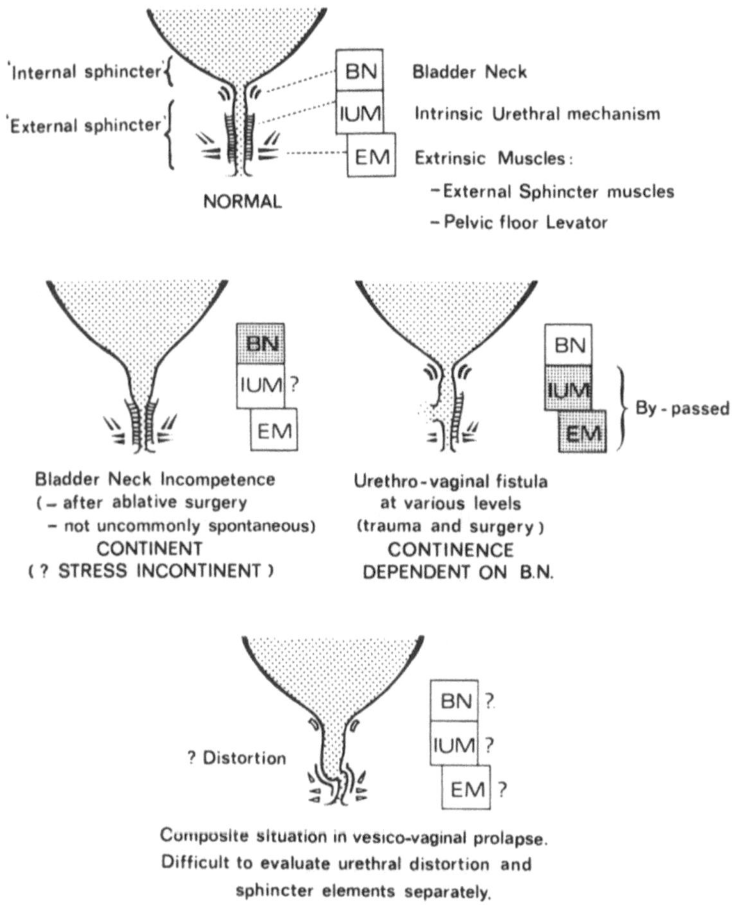

Fig. 2b

Fig. 2 a and b. a Isolated sphincter mechanism study after surgical procedures in the male. b Isolated sphincter mechanism study in the female

In order to be continent of urine under quiescent bladder conditions, the patient requires either i) a competent bladder neck mechanism or ii) a competent intrinsic urethral mechanism (Turner-Warwick, 1967); the extrinsic striated musculature surrounding the urethra (usually referred to as the "external sphincter") is apparently unable to maintain a reliably occlusive tone and hence patients who lack both the bladder neck mechanism and the intrinsic urethral mechanism are usually incontinent. This situation sometimes seen after prostatectomy as a result of damage to the intrinsic urethral mechanism within the apex of the prostate; it may apparently occur during a transurethral resection even if the loop has not strayed distally and, during a routine enucleation of a benign prostate, it is also at risk if the distal urethra is torn across with a finger or cut under tension while retracting the enucleated tissue upwards (Turner-Warwick, 1970). After a total prostatectomy the proximal part of the intrinsic urethral mechanism lying within the apex of the prostate is resected (Fig. 2) and patients who have had this procedure are naturally less reliably continent than those who had a subtotal prostatectomy.

The bladder neck is usually ablated during the procedure of prostatectomy; attempts to retain the bladder neck anatomy often lead to dysfunction in the sense that it tends towards either non-occlusion or frank obstruction as a result of secondary contracture. It is virtually impossible to damage the extrinsic urethral musculature inadvertently during a prostatectomy during either open or transurethral procedures.

A trans-sphincter repair of a stricture in the membranous urethra following a prostatectomy presents a precarious situation because urinary continence must necessarily be at risk since the bladder neck function was destroyed at the time of the ablative procedure and the intrinsic urethral mechanism is scarred as in the strictured area; it might be expected that all such patients would be incontinent but in practice the urethral closing pressure of the reconstructed urethral tissues sometimes proves to be sufficient to maintain continence.

The closing pressure at various levels of the urethal mechanisms can be measured objectively; not only may an overall pressure profile be obtained but the elastic tissue, smooth muscle and striated muscle a tissue elasticity components can be demonstrated experimentally by selective subtraction after pharmacological blocking. In the female, urethral distoriton (telescoping and kinking) may contribute to a closing pressure on straining which is significant in cases of detrusor weakness and relative outflow obstruction (Fig. 1), this factor although occasionally important, is difficult to measure objectively and accurately; it may or may not be associated with a degree of vesico-urethro-vaginal prolapse that may or may not be clinically impressive on clinical examination.

Urethral Length, Elongation and Angles

Contradictory opinions relating to the importance of the length of the urethral mechanism are a source of current confusion. It is important to recognise that it is the functional efficiency of the intrinsic urethral closing mechanism that we are interested in, and beyond a certain minimum this is not necessarily dependent upon the length of the tube. It behaves us to take an objective urodynamic view of the various surgical hypotheses.

It is well recognised that "elevation of the bladder base" in the stress incontinent female may succeed in restoring control. It cannot reasonably be argued that it is necessarily the increase in the actual urethral length that is the all important factor in these operative procedures. An anterior vaginal repair probably achieves little more than restore the urethra to a normal position and relieve the sphincter mechanism of a compromising distortion. The slight degree of urethral stretching involved in a vesico-urethral suspension may result in an increase in length but it is probably the concomitant reduction in the urethral lumen rather than the actual length that enables any residual urethral mechanism to achieve competent closure.

Reconstructive "urethral elongation" is apparently based on the same hypothesis but it represents a different situation and it is not surprising that reconstruction of a longer urethra by a tube flap frequently falls to recreate continence unless fortuitously or by hopeful design, its intrinsic structure succeeds in recreating a critical closing pressure.

Similarly geometrical measurement of the posterior bladder neck on lateral radiography was originally proposed by JEFFCOTE (1958) as an anatomical evaluation; he himself has repeatedly stressed that it cannot be assumed that a bladder neck which creates a potentially acceptable "angle" necessarily indicates that the bladder neck is functional, occlusive and cough-compentent; it is a matter of some personal embarrassment to him that his observations are frequently thus over-interpreted.

Essentially we should be interested in whether or not the bladder neck is actually cough-competent for as such it represents the primary and most reliable cough-leak preventer.

Dyskinetic Bladder Neck Obstruction

As previously mentioned, bladder outflow obstruction at the bladder neck level is a remarkably rare condition in the female, in the male however it is by no means uncommon (Fig. 1), its lesser degrees are difficult to identify without detailed urodynamic study, yet it represents a commoner cause of various clinical presentations than is sometimes appreciated.

The Natural History of Dyskinetic Bladder Neck Obstruction in the Male

The more severe degrees of Bladder Neck Obstruction naturally present in childhood; milder examples frequently escape diagnosis until later in life. A careful history from such a patient will however usually indicate that the condition has been present as long as the patient can remember although this is often masked since the patient is likely to accept his lifelong tendency towards a slightly hesitant "non-competitive" stream as normal to him and may even deny it until he is able to compare it exuberantly with the post-operative result after outflow readjustment.

For this reason many patients do not present until they develop urine infection, "prostatitis" or increasing obstruction due to minimal superimposed obstruction due to early prostatic enlargement; others present with troublesome dysuric symptoms and often a history of having been previously reassured of their normality as a result of a supposedly full urological investigation, (rectal and endoscopic examination having revealed no abnormality and the absence of a residual urine being erroneously accepted as excluding significant obstruction).

Many patients with mild dyskinetic bladder neck obstruction habitually show slight hesitation in starting, this also tends to make them more than averagely prone to an episode of urinary retention after an intercurrent operation such as a hernia repair; others have a peculiar difficulty voiding in a public urinal and some of these are even referred to a psychiatrist on this account. Per contra, the over-anxious patient may focus his anxiety on insignificant urinary symptoms; and a careful urodynamic evaluation is particularly important for the evaluation of urinary symptoms when there is an overt potential overlay for it is essential to differentiate those who are likely to be relieved by a selective sphincterotomy from those who are not, and may even be made worse by it for other reasons.

It must be appreciated that the patterns of bladder neck behaviour vary from the overtly normal to the overtly abnormal and it is not clinically helpful to attempt an artificially exact definition of the dividing line.

If a marginal flow pattern is associated with an average detrusor pressure it may be of no significance but the same flow pattern associated with a greatly elevated detrusor pressure may represent a very significant obstruction. Often the distinction between these voiding states cannot be made on the basis of symptoms; a given status of voiding imbalance may be associated with moderately severe symptoms in one patient and relatively few in another. It is fundamental to appreciate that absence of residual urine by no means excludes a significant and potentially serious bladder outflow obstruction: the residual urine is evidence that the bladder has actually failed.

The Mechanism of Dysynergic Bladder Neck Obstruction

The later stages of bladder neck obstruction may be recognised as a result of conventional routine urological investigation (Figs. 3–5) and are referred to by such ill conceived terms as bladder neck contracture, median bar, sclerose du col etc. It is not however the bladder neck hypertrophy that causes the obstruction, a more logical urodynamic explanation is a dysynergic failure of the detrusor/bladder neck opening mechanism.

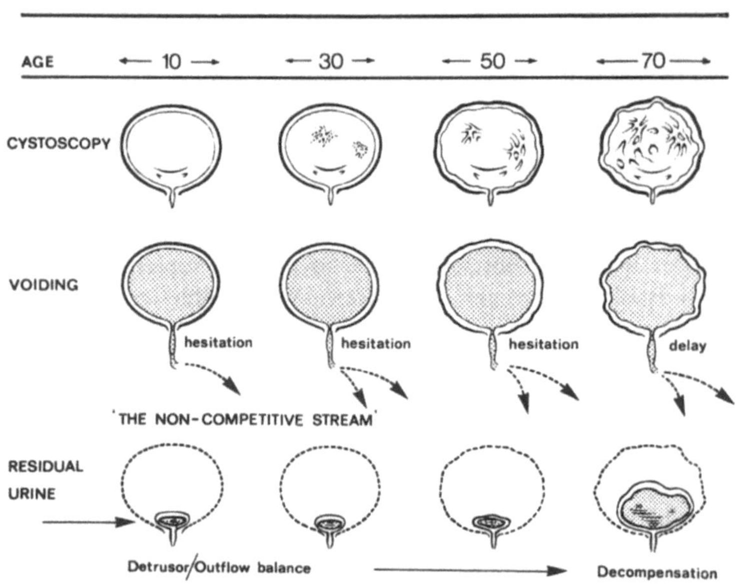

Fig. 3. The life history of dysynergic failure of bladder neck opening in the male

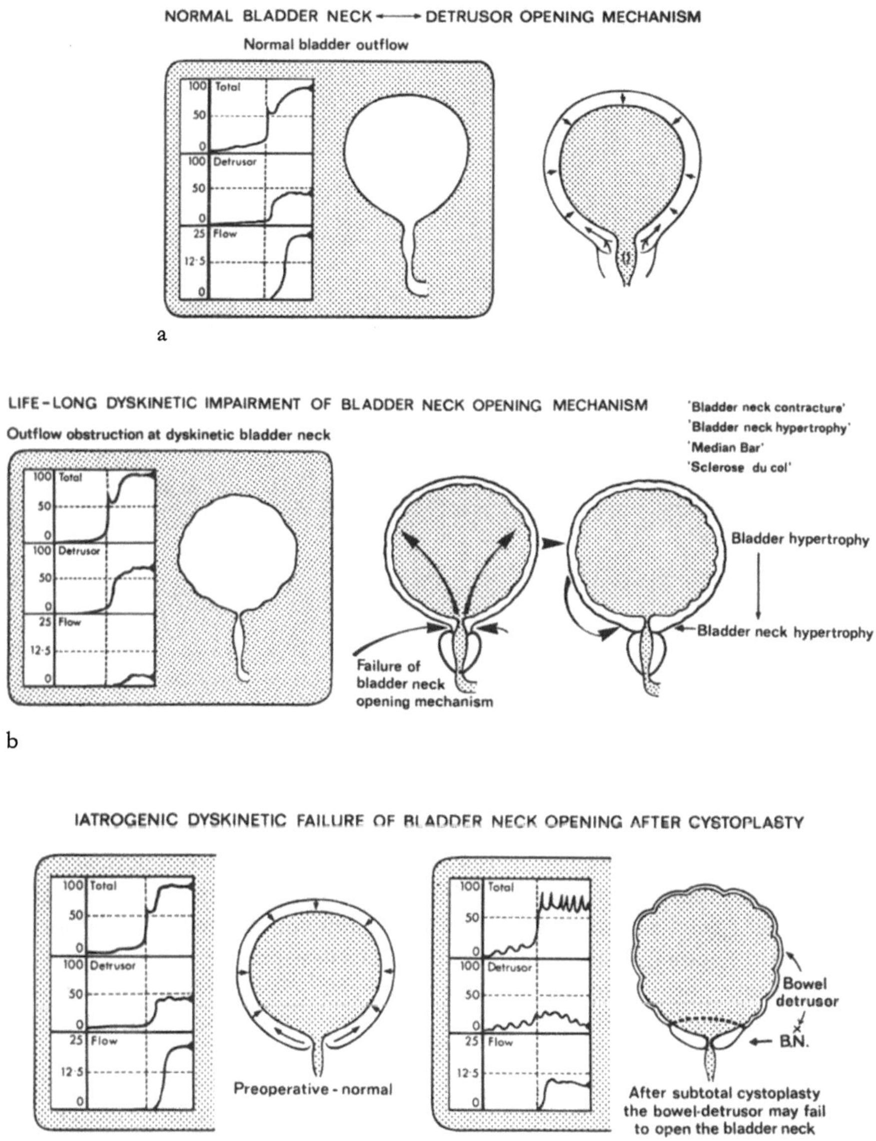

Fig. 4 a–c. a Normal bladder neck opening. b Dyskinetic bladder neck obstruction. c Failure of bladder neck opening mechanism after bowel-detrusor replacement

The normal detrusor bladder neck opening effect is shown in Fig. 4 together with a single frame of a four parameter cine pressure flow cystourethrographic evaluation. Bladder neck opening is preceded by detrusor contraction and the appearances contrast with those of dysynergic bladder neck obstruction (Fig. 4b) in which the failure of bladder neck opening seen on the screen is associated with a poor flow and

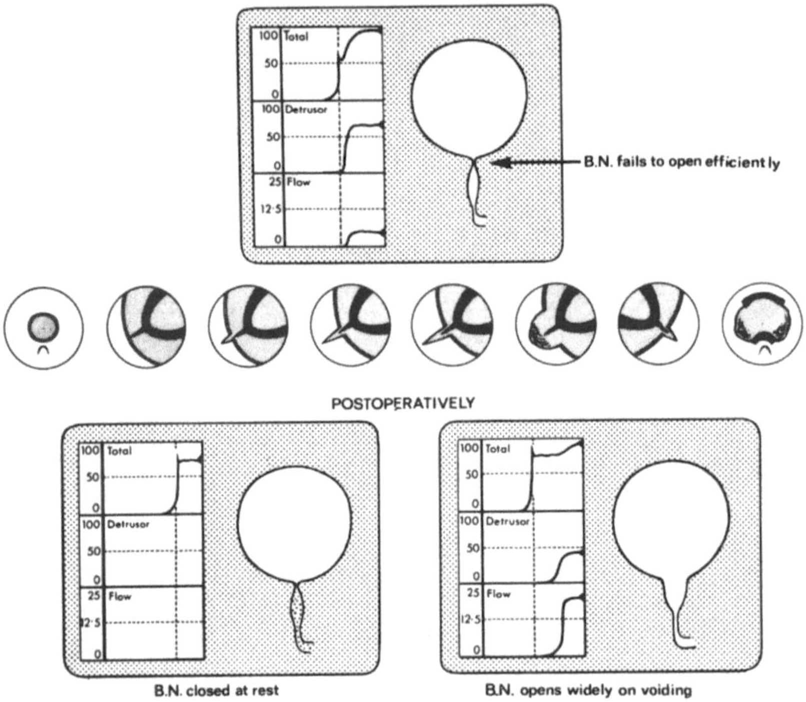

Fig. 5. The treatment of bladder neck obstruction by selective endoscopic diathermy incision

an elevated detrusor pressure. The logical result of a dysynergic bladder neck obstruction is secondary hypertrophy of the bladder wall and the bladder neck hypertrophy is merely the fortuitous result of this as it is essentially composed of the same detrusor muscle. Again, as a natural result of the identity of this muscle it is not surprising that the voiding balance should remain in its near-balanced situation for years, detrusor contraction continuing to achieve complete evacuation so that a residual urine does not develop until, at a relatively late stage in the natural history, it finally decompensates.

It is clear from the natural history (Fig. 3) that the dysynergic failure of bladder neck opening is a lifelong condition and the logical explanation of this would be some dysfunctional arrangement of the detrusor fibres around the bladder neck area but the actual mechanism of bladder neck function is clearly complex and is likely to present us with an academic challenge for some time to come.

Diagnosis

In its earlier stages the diagnosis of dysynergic bladder neck obstruction is therefore entirely dependent on urodynamic evaluation. No amount of endoscopic gazing or guesswork at this stage will indicate whether a particular bladder neck will open or remain closed during a voiding detrusor contraction. In its later stages the condition may be suspected as a result of the conventional endoscopic appearances of bladder and bladder neck hypertrophy and the multitude of descriptive terms previously referred to (bladder neck contracture, bladder neck hypertrophy, median bar and sclerose du col) bear witness to the confused state of aetiological conception. It is most important to appreciate, however, that although the endoscopic appearances and bladder and bladder neck hypertrophy may lead one to suspect dysynergic bladder neck obstruction they may be indistinguishable from the bladder and bladder neck hypertrophy due, for instance to a distal urethral stricture; in such a case it must on no account be assumed that hypertrophy of a previously normally functioning bladder/bladder neck mechanism will result in dysynergic obstruction: failure to appreciate this simple urodynamic concept results in unnecessary surgical ablation of a synergic bladder neck hypertrophy secondary to a distal stricture which seriously compromises urinary continence after a urethroplasty (Fig. 2, TURNER-WARWICK 1970). In such cases if the bladder neck is hypertrophied or secondarily affected by inflammation it is most important to check its functional state by careful urodynamic study before proceeding to treatment.

The Iatrogenic Bladder Neck Obstruction Following Cystoplasty

The importance of detrusor function in bladder neck opening can be further illustrated by the occasional development of iatrogenic dysynergic failure of bladder reck opening following the extensive detrusor muscle resection involved in a subtotal bowel cystoplasty (Fig. 4c). In such cases the combination of the small residual bladder base muscle and the bowel detrusor may be insufficient to open the previously normally functioning bladder neck. The characteristic urodynamic voiding pattern in these patients and their management by selective sphincterotomy has been previously discussed elsewhere (TURNER-WARWICK 1967, 1970) and is summarised below (Fig. 8).

Recurrent Urine Infection in the Male

Recurrent urine infection in the male is a relatively common presenting symptom of dysynergic bladder neck obstruction; as previously mentioned the patient may be unaware of his poor stream.

"Chronic Prostatitis"

Among the patients diagnosed as suffering from "chronic prostatitis" are a few whose main problem is one of dysynergic voiding imbalance and whose symptoms can be resolved by its adjustment. The main reason for this diagnostic difficulty is that the clinical syndrome of prostatitis is difficult to define and among its symptoms are a number in common with the equally ill-defined manifestations of voiding imbalance; in both groups of patients overt infection may or may not be a feature. It is therefore worthwhile checking the voiding efficiency of all patients who labour under the diagnosis of "chronic prostatitis" for bladder neck obstruction is a definitively treatable condition.

It is of interest that the results of transurethral resection in "chronic prostatitis" are notoriously variable. It would not be surprising if patients with an undiagnosed underlying bladder neck obstruction were among those who did best, for even an extensive posterior transurethral resection may fail to improve a patient with posterior prostato-vesiculitis and for these patients we ourselves prefer a posterior dissection prostatectomy (TURNER-WARWICK 1970).
It is obviously wise to investigate patients with episodic infection at a time when the inflammatory process is minimal and, as previously mentioned, borderline voiding flow curves may require more detailed urodynamic evaluation.

The "Small Fibrous Prostate"

The diagnosis of a "small fibrous prostate" is emotive both to the endoscopic resectionist and the open enucleationist; the one because the development of secondary bladder neck obstruction due to post-operative sclerosis following endoscopic

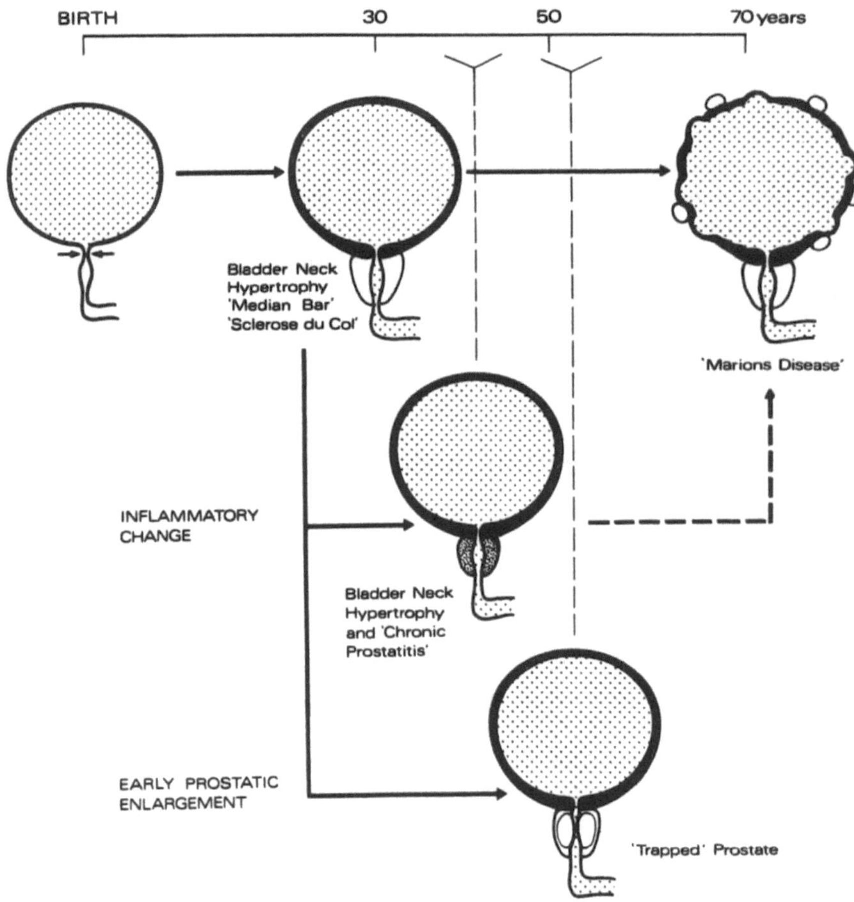

Fig. 6. The life history of bladder neck obstruction with superimposed infection and early prostatic enlargement

loop resection is a well recognised hazard, and the other because enucleation is virtually impossible. In point of fact urodynamic evaluation indicates that the prostate itself rarely represents the obstructive element in these cases. Again if bladder outflow obstruction is proven and urodynamically located to the bladder neck we prefer to achieve selective bladder neck relaxation by endoscopic diathermy knife incision in preference to resection (Fig. 5).

Endoscopic Bladder Neck Incision

We use a small flat insulated resectoscope diathermy knife developed from COLLINS panendoscope knife. We have usually locating the incision in the 4 o'clock and 8 o'clock position. The longitudinal extent of the incision is from the bladder base to the verumontanum, the incision is intended to divide the thickness of the a dyskinetic bladder neck and the depth of this naturally varies considerably according to the degree of hypertrophy. In order to be certain of relieving a bladder neck obstruction the incision is carried down through the bladder neck muscle until the muscle fibre pattern begins to "lattice"; at this stage the deeply V-shaped incision often appears to "open out" as if several loopfuls of tissue had been removed (Fig. 5).

Following incisions at the 40 o'clock and 8 o'clock position there appears to be a midline "hump" somewhat resembling a middle lobe; although this may tempt the finger of a loop resectionist, it is best left intact for, in spite of preconceptions of endoscopic appearances of obstruction it does not in fact cause outflow obstruction and it may contribute to the prevention of retrograde ejaculation. The two intact urothelial ridges between the lateral incisions (Fig 5) probably also contribute to the prevention of secondary bladder neck contraction. The post-operative urodynamic evaluation of this procedure regularly indicates the establishment of a satisfactory flow pattern with reduction of the detrusor pressure: somewhat unexpectedly, and in contrast to the usual appearances of a transurethral loop resection, after a bladder neck incision the bladder neck is usually closed at rest but opens quite unexpectedly widely during voiding. We have not yet seen secondary bladder neck sclerosis following this procedure and the relaxation appears to be effective over periods of years.

In view of the relatively early age at which dyskinetic bladder neck obstruction can be made as a result of urodynamic investigation the problem of reflux ejaculation becomes a relatively important consideration in the treatment of the case. Retrograde ejaculation seems to be relatively unusual after bladder neck incision compared with loop resection but if treatment cannot be postponed until parenthood is achieved we have occasionally modified the procedure using a single incision and limiting its depth planning a urodynamic re-evaluation after an interval of several weeks with a view to repeating the procedure if indicated.

In general it is a simple matter to achieve bladder neck relaxation by endoscopic incision where residual urinary control at this level is of no importance in view of the elasticity and the great variations of thickness of bladder neck musculature that can be associated with dyskinetic obstruction it is relatively difficult to achieve a well controlled reduction of the voiding resistance of this mechanism without risking its occlusive function and when this is important it seems best to plan on a conservative incision with urodynamic evaluation after a healing interval.

The initial cut of a loop resection destroys more than half the circumference of the urothelium and we ourselves have abandoned it for the treatment of bladder neck obstruction expecting to reduce the incidence of secondary bladder neck sclerosis and we reserve open VY-plasty for those patients who are referred with secondary contracture.

The Trapped Prostate

The patient presenting in his fifties, with significant outflow obstruction but with relatively early prostatic enlargement, presents an interesting urodynamic problem. Quite frequently the patient's symptoms are not only apparently disproportionate to the relatively minor degree of prostatic-urethral occlusion visualised endoscopically, but a careful history may reveal lifelong symptoms suggestive of mild dyskinetic bladder neck obstruction. It seems likely that in some of these patients the outflow obstruction is the result of summation of the two potential obstructing elements neither of which alone would have been particularly troublesome; it is possible also that in the coexistence the early expansion of the prostate may also be to some extent confined by the bladder neck hypertrophy (Fig. 6). (Turner-Warwick and Whiteside, 1969, Turner-Warwick et al., 1973.)

In some cases its treatment borders upon that previously discussed under "small fibrous prostate"; it is commonly treated by transurethral loop resection but some patients however specifically request a form of treatment which offers the best chance of maintaining their fertility for a number of years. While a partial sub-bladder neck loop resection may sometimes achieve the desired result for a limited period there is some risk of secondary bladder neck contracture. We have had occasion to try the urodynamically interesting alternative of relieving the bladder neck element of the obstruction by deep endoscopic incisions in the 4 o'clock and 8 o'clock position without any attempt at prostatic resection and it is our impression that this relatively simple procedure offers a good chance of urodynamic benefit, at least until the prostate enlarges definitively, with a reduced risk of reflux ejaculation.

Post-Prostatectomy Symptoms

We are all familiar with patients who complain of frequency, urgency, poor stream and difficulty in urinary control after the relief of prostatic or bladder neck obstruction and recognise that they are in fact more numerous than we would hope, assuming we regard sphincter impairment and residual mechanical obstruction due to secondary bladder neck sclerosis, residual prostatic tissue and secondary urothelial stricture as potentially preventable surgical complications.

We have already discussed the problem in so far as it is related to the sphincter mechanism (Fig. 1).

Careful urodynamic investigation is particularly important in those post prostatectomy patients whose symptoms cannot be related to persistent urine infection or an endoscopically recognisable residual obstruction. Some complaints prove to be entirely unfounded, indeed, a bi-product of the four parameter cine-pressure-flow study is that it may successfully resolve the patient's symptoms merely as a result of the convincing demonstration to him of their fallaciousness.

Some patients prove to have a bladder of small capacity associated with an inability to inhibit persistent relatively high pressure detrusor contractions. It is an interesting fact that male patients with bladder outflow obstruction may show detrusor instability by bladder contractions on the filling cystometrogram; these are by no means necessarily indicative of an underlying neuropathic abnormality for in the majority

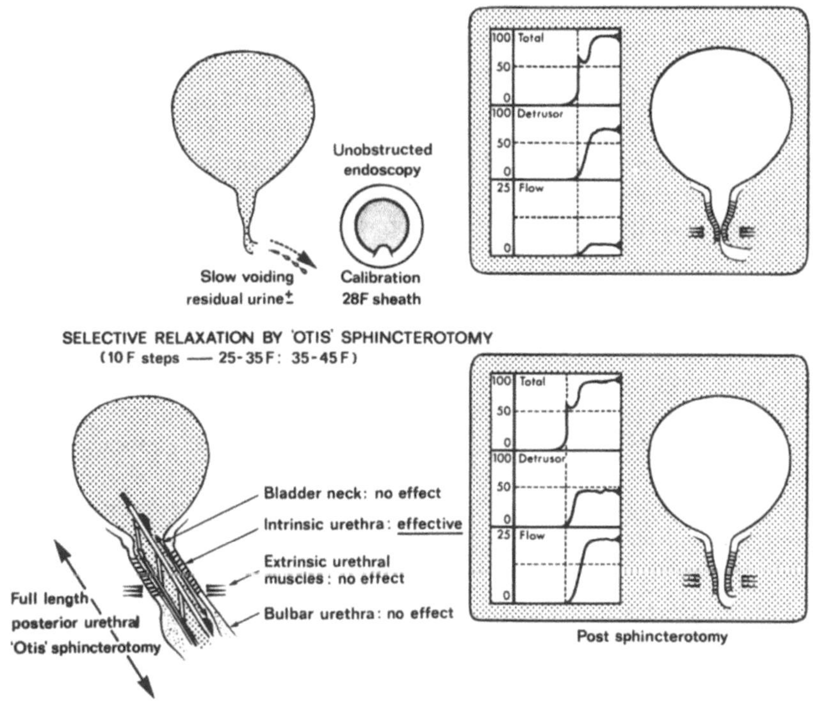

Fig. 7. The treatment of post-prostatectomy outflow obstruction due to hypertonic distal sphincter mechanism (stricture and residual prostatic tissue excluded)

of cases the filling and voiding behaviour of the detrusor returns to normality after the relief of obstruction. In a few cases the unstable bladder contraction persist, causing symptoms of frequency and urgency; this situation is of importance for it may be mistaken for a sphincter weakness, unfortunately it is also difficult to treat effectively.

A considerable number of patients with persistent symptoms after prostatectomy prove to have a relatively poor voiding flow curve in spite of the fact that their urethra calibrates to 28F and no residual prostatic tissue or bladder neck scarring is visualised endoscopically. In such patients there is obviously a significant voiding imbalance and synchronous cine/pressure/flow cystourethrography is likely to show a relative outflow obstruction located in the distal urethral mechanism probably as a result of a dyskinetic intrinsic urethral mechanism; it is our impression that this situation seems to occur more frequently in those patients in whom perhaps a parallel dyskinetic bladder neck mechanism played a significant role in the original obstructive aetiology.

Detrusor pressure studies are of particular interest in this group of patients. In some the detrusor pressure is abnormally high in association with bladder hypertrophy resulting from previous prostatic hypertrophy and in some of these there is persistence of the obstructive type of detrusor instability which is hardly surprising in view of the persistent distal obstruction: in others, objective measurement of the detrusor pressures show abnormally low values enabling the bladder to be accurately referred to as "atonic". The degree of obstruction is clearly a matter of relativity to the voiding force.

The treatment of such patients is usually relatively simple in the male once the situation is definitively diagnosed for selective relaxation of the intrinsic urethral mechanism can usually be achieved by the Otis instrument as indicated in Fig. 7. The passage of urethral sounds is of course ineffective since the dyskinetic urethral mechanism will readily calibrate to the maximum diameter that the anterior urethra can accept.

We prefer to recalibrate the intrinsic urethral mechanism cautiously in 10F steps with re-evaluation of the effect of each after a healing period before proceeding to the next; this seems to give good warning of a critical sphincter deficiency which might otherwise lead to incontinence.

If the 10F recalibration can be achieved without using the sliding knife of the Otis instrument this may be preferable. If the incision of the intrinsic urethral mechanism is required in order to achieve the recalibration its direction is probably immaterial; we have however usually avoided the 6 o'clock position (especially in the female) preferring the 12 o'clock and additionally the 3 o'clock and 9 o'clock. The proximal and distal extent of the incision on either side of the intrinsic mechanism is immaterial for, as indicated in Fig. 7 a shallow incision in the bladder neck does not affect its bulky and elastic mechanism (in the present context the bladder neck function is usually already destroyed by the previous prostatectomy). The extrinsic urethral muscles are similarly bulky and unaffected and an incision in the bulbar urethra is of course immaterial. The fact that a long, Otis sphincterotomy provides in effect a selective relaxation in the intrinsic urethral mechanism greatly facilitates the procedure for, although it can be accurately calibrated, the actual location of the incision in this particular area would not always be easy.

The Urodynamics of Bowel Cystoplasty Function

Among the clinical problems which initially stimulated our interest in urodynamic study were those related to the functional results of cystoplasty (Turner-Warwick and Handley Ashken, 1967). It was clear that many of the reported late complications were due to failure to re-establish efficient voiding in an unbalanced system resulting from the replacement of a relatively effective bladder detrusor by an inefficient and non-occlusive bowel detrusor; a system of selective sphincterotomy was developed to re-establish the voiding balance.

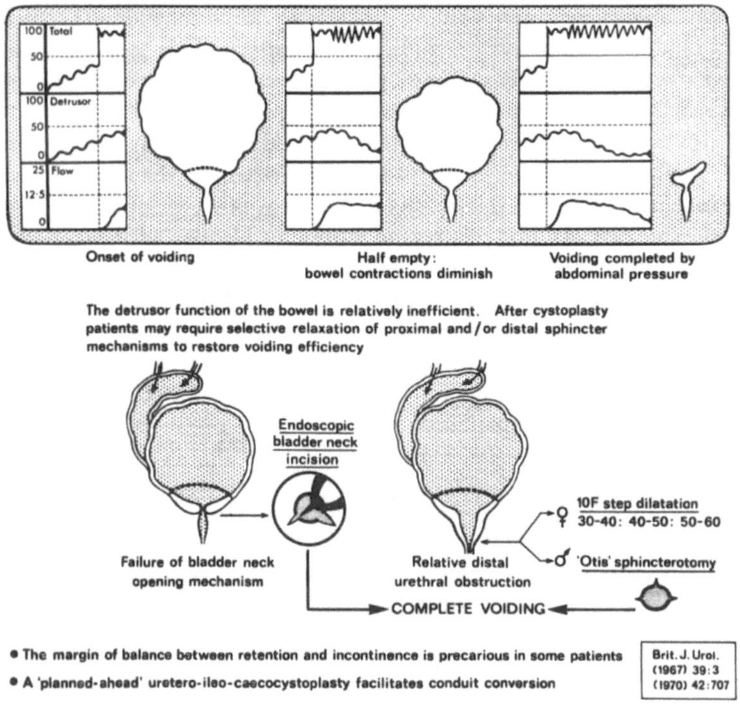

Fig. 8. Voiding efficiency after cystoplasty and its restoration by selective sphincterotomy

The urodynamic characteristics are essentially similar whether the bowel replacement is ileal, caecal or colonic; for technical reasons we generally prefered the caecal or ileocaecal unit. The general urodynamic behaviour of a cystoplasty is summarised in Fig. 8; the progressive filling cystometrogram shows the development of typical bowel contractions which cannot of course be inhibited by the patient, these contractions diminish equally progressively during voiding so that when the bladder contents are partially evacuated they no longer present a significant evacuating force and the patient has to learn to complete voiding by abdominal straining.

We have stressed the importance of routine urodynamic follow-up for such patients in order to check the voiding efficiency. Failure to achieve this results in a large residual urine, urine infection, mucus retention, stone formation and pyelitis. Simple voiding cystourethrography is our investigation of choice for this and we prefer to see the video cystographic record ourselves because we feel we can judge the situation better from this than from a still film record (it is the surgeon who bears responsibility for his selective sphincterotomy). The type of sphincterotomy performed is of course dependent upon whether the site of the relative outflow obstruction is at the bladder neck or in the intrinsic urethral mechanism (Fig. 8). It is also useful for the patient to observe the monitor screen during voiding as they find it is helpful in learning the new technique of voiding. We have performed a large number of synchronous cine/pressure/flow records after cystoplasty but the pressure studies tend to be so uniform that we no longer find them important except in those cases

in which the balance between voiding efficiency and continence is precarious and difficult to establish; similarly the voiding flow pattern tends to be slow and intermittent even in the efficiently voiding cystoplasty and this alone does not therefore provide a reliable check on voiding efficiency (TURNER-WARWICK, 1973).
As previously mentioned a normally opening bladder neck may fail to open after a subtota lcystoplasty due to the small bladder base detrusor remnant and ineffective bowel replacement (Fig. 4); this results in an iatrogenic type of dysynergic bladder neck obstruction (Fig. 4). For this reason, although bladder neck relaxation is very rarely required in the female on account of naturally occurring bladder neck obstruction it is not infrequently required after cystoplasty but, owing to the unreliable efficiency of the distal urethral mechanism in the female we do not advise prophylactic relaxation of the bladder neck in the female (as we do in the male) preferring to wait until we have objective post-operative urodynamic evidence for it only proves necessary in the minority of patients.

Stress Incontinence

The urodynamic factors involved in the clinical problem of stress incontinence are perhaps considerably more complex and important than is generally recognised.
We have previously noted the prime importance of the bladder neck in guarding against a stress leak in the female for it is normally cough-competent and should

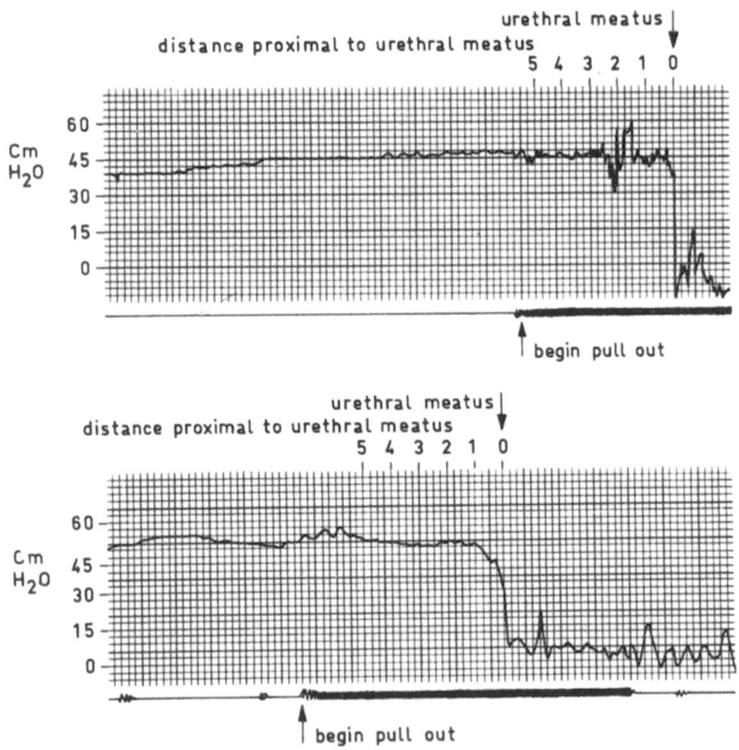

Fig. 9. Cysto-urethrographic diagram showing types of bladder neck and distal sphincter mechanism dysfunction in stress incontinence in the female

never be surgically violated without objective proof of the remarkably rare necessity. This is especially important for, unlike the male the distal mechanism is by no means reliably cough-competent irrespective of any vesico-vaginal prolapse or previous pregnancy.

This situation is summarised in the three columns of Fig. 9. The first column shows a cough-competent bladder neck with efficient voiding, the second a cough-incompetent bladder neck with efficient distal urethral compensation and the column an incompletely compensating distal urethral mechanism resulting in stress incontinence but with efficient voiding.

Most patients with stress incontinence void efficiently or even super-efficiently and this may be factor in their relative freedom from recurrent urine infection, there is however a group who have bladder neck incompetence and who void slowly and incompletely due to demonstrable distal urethral holdup (Fig. 9, Column 4).

These represent an important group which are relatively difficult both to evaluate and to treat; and the situation can perhaps be summarised in a number of generalisations: –

1. Urethral calibration by instrumentation is very poorly related to urodynamic calibration and voiding efficiency.
2. Urethral recalibration (by over-dilatation or by Otis instrumentation with interval reassessment) may or may not restore voiding efficiency even when carried to 60 F+.
3. Urethral recalibration may restore voiding efficiency but reveal a latent stress incontinence due to a cough-incompetent bladder neck which was in effect previously overcompensated by the distal mechanism. In such cases it is perhaps not entirely fair to attribute the stress incontinence to the Otis recalibration of the urethra but rather to the underlying bladder neck inefficiency that has been revealed thereby (clinical situation which is best anticipated and discussed).

There remains the problem of the distortional obstructive factor associated with a prolapsing female urethra which seems to have two main components of varying individual importance in any one case, a simple telescoping action and a sheering urethral kink. We find the distortional factor extremely difficult to evaluate objectively and impossible to estimate clinically for it is certainly by no means proportional to the degree of vesico-vaginal prolapse; an apparently normal degree of vault prolapse can occasionally cause considerable trouble especially after distal urethral following anterior vaginal wall urethral fixation following anterior vaginal repair while a severe degree of vesico-vaginal prolapse may or may not be associated with a distortional element.

The Relationship between Vesico-Vaginal Prolapse and Stress Incontinence

It is a matter of general agreement that there is no direct relationship between stress incontinence and the presence or the degree of prolapse. They often occur independently of each other but when they co-exist there is a tendency to assume that the prolapse is the cause of the stress incontinence: the clinical results of repair operations suggest that this is not so in about one third of cases; objective urodynamic evaluation of the results in the same group of patients indicates that less than half are actually stress continence, the difference being made up by patient appreciation of relative improvement.

The Relationship of Stress Incontinence and Bladder Instability

In addition to the question of simple sphincteric inefficiency in stress incontinence, synchronous cine/pressure/flow cystourethrography has focussed attention upon the problem of the unstable bladder (HODGKINSON, 1970, BATES, WHITESIDE and TURNER-WARWICK, 1970). The unstable bladder is one in which the patient is unable to inhibit a detrusor contraction; in severe cases this is manifest on the filling cystometrogram and may be associated with overt neuropathic abnormality, in the milder forms at present under consideration the uninhibited contractions are seen as a result of the stimulus of standing up or coughing.

The rise in the total bladder pressure produced by such contractions may not in itself be very significant since the effective detrusor pressure in females is often low; the fact that the detrusor contraction results in bladder neck opening (and therefore affects its cough competance) may however be very relevant since it naturally results in stress incontinence if the distal mechanism is inefficient. Under these circumstances involuntary voiding may be virtually complete if the patient is unable to inhibit or control the subsequent progressive detrusor contraction. The significance of cough stimulated bladder instability stress incontinence is currently under special study in our Unit but preliminary evaluation of both a retrospective and a prospective series (BATES, 1971) suggests that the incidence of bladder instability in these patients who fail to respond to routine precedures for stress incontinence is significantly high.

The Treatment of Stress Incontinence

The treatment of stress incontinence is therefore by no means always a simple matter. We believe that there must be a closer co-operative study between Urological and Gynaecological Departments and have set up a prospective protocol in our Urodynamic Clinic to compare the results of Anterior Colporrhaphy, Vesico Urethral suspension and anterolateral colposuspension and partial denervation.

Giggle incontinence: Giggle incontinence, in which the female patient voids involuntarily and uncontrollably during an episode of giggling or laughing is not particularly unusual in young girls; occasionally it persists into adult life. We have had the opportunity to evaluate several of these patients by cine/pressure/flow cystography and in none have we been able to detect any obvious abnormality and specifically no bladder instability on filling, standing or coughing. The detrusor contraction is presumably provoked at cortical level as a result of stimulation of or inability to inhibit. We have no specific treatment to offer (other than frivolously suggesting the possibility of excising the sense of humour). One of the main problems in these cases is however the potential social embarrassment and it is often very helpful to point out that the timely spilling of a glass of water may provide an acceptable cover story.

The "Atonic" Bladder

Patients who have a persistent residual urine must, in effect, have a degree of relative outflow obstruction since it results from the situation in which the component forces of the bladder outflow resistance exceed the component forces of bladder emptying.

Most female patients who have a large residual urine tend to be labelled, faut de mieux, as having "atonic bladders"; the same is true of males when the residual urine cannot be attributed to clinically obvious outflow obstruction due to the bladder neck, prostate, or a stricture; in many of these patients objective investigation in fact demonstrate that the detrusor function is quite acceptable.

The wide variations in detrusor pressure and outflow resistance that are found between the male and the female, and between one apparently normal female and another apparently normal female, make it virtually impossible, and clinically somewhat meaningless, to try and establish a "normal" range for either; satisfactory function in any individual depends upon the existence or the establishment of a satisfactory balance between the urethral sphincter closing pressure and the effective voiding pressure.

Whether or not the detrusor is functioning effectively it is often possible to re-establish efficient voiding in these cases by selective relaxation of the appropriate sphincter mechanism; in most cases the situation is essentially comparable to the voiding imbalance that may result from cystoplasty and there is a good chance that efficient voiding can be re-established on the basis of abdominal pressure.

It is however not always as simple as this especially in the female.

An inactive detrusor may result in failure of bladder neck opening and a relative bladder neck obstruction comparable to the iatrogenic failure resulting from cystoplasty (Fig. 4c).

If the bladder neck is relaxed in such a case and becomes incompetent the result may be incontinence if the closing mechanism of the distal urethra is inefficient. If however the distal urethra closing pressure still represents an outflow obstruction in spite of progressive step dilatation to 70F one is on occasions faced with the situation that a finger can be inserted into the urethra without obstruction or detectable closing pressure and yet the full bladder cannot be expressed. Under such circumstances the possibility of a distortion factor is obvious but if this is indeed a factor it is not necessarily overcome by clinical correction of a clinically apparent prolapse by anterior colporrhaphy or the clinically more effective anterior colposuspension (TURNER-WARWICK, 1969) but nevertheless critical placement of a relatively slight finger pressure on the anterior vaginal wall may facilitate voiding and we are currently evaluating the possibility of using the Bonnar Incontinence appliance for a "reversed effect" in these patients.

The "Cameloid" or Infrequent Voiding State

A significant proportion of females have a tendency towards infrequent voiding. Provided the situation remains in balance, as it does in the great majority of patients, all remains well but there appears to be a certain tendency for it to develop into episodes of recurrent urine infection and it is also this type of female patient who seems more prone to develop episodes of difficult voiding and of post-operative retention following intercurrent operations. The male equivalent of borderline abnormality would perhaps be the subclinical incipient bladder neck dysynergia but the mechanism of these two conditions is significantly different for the relative outflow obstruction in the female does not occur at bladder neck level but distally and although the bladders of these patients are characteristically somewhat trabeculated they are, quite unlike the male, often unusually thin-walled.

If voiding imbalance supervenes in a female patient with infrequent voiding syndrom it can often be treated by progressive 10F. step dilatation; bladder neck relaxation is very rarely indicated and should only be undertaken after careful objective proof of the necessity since it carries the aforementioned risk of stress incontinence. In the male the majority of patients with marginal voiding imbalance require a bladder neck relaxation in order to re-establish voiding efficiency for Otis urethrotomy is relatively ineffective treatment for the dyskinetic bladder neck (Fig. 7). Voiding imbalance in the male due to an isolated dysynergic distal urethral mechanism (as opposed to being associated with a dysynergic bladder neck mechanism) is in our experience an even rarer condition than bladder neck obstruction in the female, we have seen two examples and both resparded to progressive otis sphincterotomy.

Voiding Imbalance Following Pelvic Surgery

It is well recognised that voiding disabilities tend to occur after pelvic surgery of varying extent and these are commonly ascribed to the development of an "atonic" bladder; in many cases this represents not only an over-simplification of the urodynamic findings because the distortional factor and the sphincter mechanism may be equally important, but even if the detrusor pressures are found to be very low, a comparison with the pre-operative figures is essential in order to be certain that it was not a pre-existing state (as it not infrequently is in the female).

These patients merit particularly careful urodynamic evaluation because even if the voiding imbalance is disturbed as a result of detrusor weakening a carefully controlled selective sphincterotomy or correction of a distortion component is likely to offer a better solution than attempts at pharmacological juggling, partial cystectomy or longterm catheter drainage; we are in other words seeking a urodynamic solution rather a medical or surgical treatment.

The Neuropathic Bladder

The urodynamic problems presented by neuropathic bladder dysfunction are too numerous to discuss in the limited remaining time and the fundamental importance of detailed urodynamic evaluation with special reference to detrusor function is self-evident. Within this subject are two somewhat different groups of patients, those with severe spinal injury and those with relatively mild neuropathic dysfunction; there is of course a considerable overlap between the two but the clinical problems involved in trying to re-establish reasonably normal micturition in the latter group merit separate discussions; occasionally a simple selective sphincterotomy is all that is required to achieve this.

It would perhaps not be out of place to comment upon a problem which has been a particular study interest for us, that of dominant neuropathic motor instability: the bladder which develops high pressure systolic contractions when it is distended and which the sphincter mechanisms are unable to control, but yet which fade away when the bladder is half empty resulting in a large residual urine since the sphincter resistance is too great to be overcome by abdominal pressure. This may occur in association with a relatively stable underlying neuropathy, and a potentially promising degree of preservation of bladder sensation and sphincter control and function; in such patients the therapeutic value of creating an atonic bladder is self-evident for provided it can be achieved without impairing bladder sensation or the residual sphincter function there is a chance of readjusting the sphincter mechanism to balance and control the low pressure voiding system. We have previously reported a technique for supra-trigonal denervation (TURNER-WARWICK and HANDLEY ASHKEN, 1967) and are currently evaluating our limited success with this and other forms of motor denervation; the need for long-term follow-up of these patients with objective urodynamic evaluation was one of the projects which stimulated the establishment of the technique of cine/pressure/flow cystourethrography in our Unit (BATES, 1971).

Treatment of Interstitial Cystitis by Sensory Denervation

Finally I would like to mention a technique for the treatment of interstitial cystitis by supratrigonal sensory denervation with which we have had some encouraging results. It is designed for the treatment of those patients with interstitial cystitis in whom the bladder has not yet developed a fibrotic contracture, in other words those in which the bladder can be distended to a reasonable capacity under anaesthetic.

The principle is to achieve a bladder-dome sensory denervation which will allow the bladder to distend without causing the patient acute discomfort.

The bladder is explored through a suprapubic incision (Fig. 10) isolated down to the trigonal level; it is often a technical convenience to leave the adherent area of the peritoneum attached to the dome of the bladder. It is important not to disturb the uretero-vesical junctions unnecessarily in order to avoid increasing the risk of uretero-vesical reflux (which naturally exists as a result of the interstitial cystitis itself) but we usually extend the dissection in the mid-line between them behind the trigonal area. The dome of the bladder is elevated and sutured to the back of the antero-abdominal wall to prevent it contracting into the pelvis and becoming secondarily incarcerated by perivesical adhesion.

Follow-up of these patients over a period of years has been encouraging, the majority have lost their bladder discomfort and have maintained a reasonably normal bladder capacity without further treatment (WORTH and TURNER-WARWICK, 1973).

This technique is designed for the treatment of interstitial cystitis in its relatively early stages before bladder contracture has supervened. Quite apart from the fact that the progress of the condition is remarkably variable the diagnosis of the condition is difficult to substantiate objectively. In addition to careful investigation and bladder biopsy the diagnosis should be reasonably "time proven" before proceeding to sensory denervation; on the other hand the procedure is relatively simple to per-

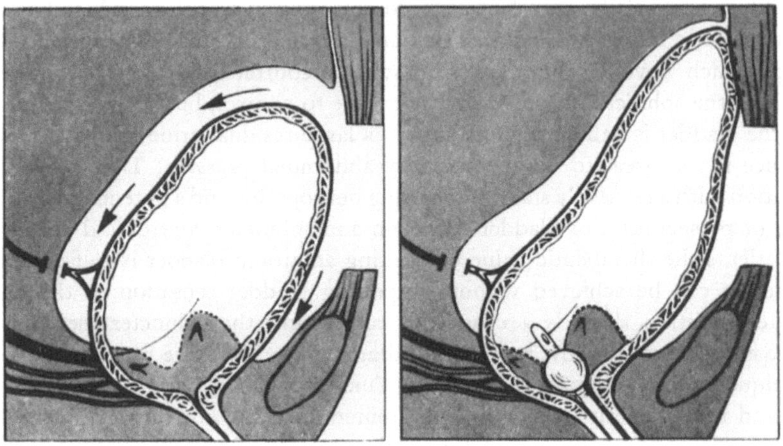

Fig. 10. The treatment of early interstitial cystitis (bladder distensible to acceptable capacity under anaesthesia) by supra-trigonal cystcolysis and anterior vesical suspension

form, and causes fewer potential hazards than some other forms of treatment including steroid therapy. It is important that the bladder base should remain undisturbed so that the patient remains aware of their bladder state and so that it is available for subtotal bowel cystoplasty if satisfactory bladder function is not re-established by denervation.

Conclusion

In conclusion I have tried to indicate the extent to which objective urodynamic evaluation of many groups of patients in our Urological Clinic have altered our concepts and management of their conditions; the wide range of these has necessarily reduced the depth to which any one aspect could be considered. This is not intended to be a review of various opinions but rather a summary of our present thoughts on various problems acknowledging that many of them are neither new nor original. They are set out as a basis for discussion and further work.

With any fundamental re-evaluation some apparently complex problems become simple and in some cases relatively easy to manage, others, which we previously considered to be simple are revealed as complex and while this does not necessarily immediately improve the management problem it may at least direct it along a more rational pathway. There are a considerable number of routinely occurring clinical problems which we would no longer feel able to manage efficiently without the relatively simple, yet relatively sophisticated facilities we use for their urodynamic evaluation; some of these problems we can now recognise on the basis of the clinical presentation but the recording uroflow meter is essential as a basic screening investigation for urodynamic problems.

We have in fact, established within our Urological Unit, a special urodynamic clinic which is based upon the closest co-operation between our clinical service, our urological radiology headed by Dr. Graham Whiteside, and our clinical investigation.

We have derived great mutual benefit from the session at which we jointly review the results of all patients referred for cine/pressure/flow cystourethrography during the previous week and find that these provide not only a useful post-graduate discussion but also the stimulus for departure from quite a number of accepted precepts.

References

BATES, C. P.: Continence and Incontinence. Ann. Rog. Coll. Surg. **49**, 18–35 (1971).

BATES, C. P., WHITESIDE, C. G., TURNER-WARWICK, R. T.: Synchronous cine/pressure/flow cystourethrography with special reference to stress and urge incontinence. Brit. J. Urol. **43**, 714 (1970).

CAINE, M., EDWARDS, D.: The peripheral control of micheration: a cure radiographic study. Brit. J. Urol. **40**, 747 (1958).

HODGKINSON, C. P.: Stress urinary incontinence. Am. J. Obst. Gynec. **108**, 1141–1168 (1970).

JEFFCOATE, T. N. A.: The principles governing the treatment of stress incontinence of Urine in the Female. Brit. J. Urol. **37**, 633 (1965).

TURNER-WARWICK, R. T.: The repair of urethral strictures in the region of the membraneous urethra. J. Urol. (Baltimore) **100**, 303 (1967).

TURNER-WARWICK, R. T., HANDLEY ASHKEN, M.: The functional results of partial, subtotal and total cystoplasty with special reference to ureterocaecocystoplasty, selective sphincterotomy and cysto-cystoplasty. Brit. J. Urol. **39**, 3 (1967).

TURNER-WARWICK, R. T.: Benign enlargement of the Prostate. London Med. Soc. (1969).

TURNER-WARWICK, R. T., WHITESIDE, C. G.: Investigation and management of bladder neck dysfunction. Modern Trends in Urol. 3rd Ed.: Sir ERIC RICHES, 295–311 (1969).

TURNER-WARWICK, R. T.: Colocystoplasty. Discussion. Brit. J. Urol. **42**, 706 (1970).

TURNER-WARWICK, R. T.: Cystoplasty. Chapter in "Urology": Ed. Blandy, Oxford Univ. Press 1973.

TURNER-WARWICK, R. T., WHITESIDE, C. G. W., WORTH, P. H. L., MILROY, E. J. G., BATES, C. P.: A Urodynamic View of the Clinical Problems Associated with Bladder Neck Dysfunction. Brit. Jour. Urol. **45**, 44–59 (1973).

TURNER-WARWICK, R. T., WHITESIDE, C. G. W., ARNOLD, E. P., BATES, C. P., WORTH, P. H. L., MILROY, E. J. G. WEBSTER, J. R., WIER, J.: A Urodynamic View of Prostatic Obstruction and the Results of Prostatectomy. Brit. Jour. Urol. **45**, (1973).

WORTH, P. H. L., TURNER-WARWICK, R. T.: The Treatment of Interstitial Cystolysis. Brit. Jour. Urol., **45**, 65–71 (1973).

Static and Dynamic Pressure Parameters in the Closure of the Bladder

P. Graber*

Continence means the ability of the bladder to retain urine without leakage at any capacity and even during muscular exercise. This definition signifies a negative pressure gradient between the bladder and extracorporal area, this being maintained under conditions of rest as well as under increased static or dynamic stress. The problem of continence is thus a question of the local effective pressure or of the pressure profile.

At rest, the urethra can be compared with an elastic pipe with a splitshaped lumen in which zones of higher and lower wall tension alternate. Immediately at the outlet of the bladder and in its uppermost part there is a low pressure zone. Here it is quite often impossible to register a pressure change as compared with the bladder. This finding is not only surprising in view of the "Internal sphincter" postulated at this site, but also from the point of view of anatomy: Apparently the sturdy semicircular muscle plate situated at the bladder outlet, the "baseplate", does not create any intraluminar pressure worth mentioning. A gradual pressure increase begins only 2–3 cm distal to the internal meatus; it is continuous in the female urethra and stepwise in the male due to the interposition of the prostate. The highest pressure in both sexes is found at the point of insertion of the urethra in to the pelvic floor, i.e. in the area of the so-called external sphincter.

Thus, under conditions of rest it is possible to determine a negative pressure gradient between bladder and urethra as the pressure in the bladder is lower than the highest tension measured in the urethra. If we analyze the profile critically, however, it is found that the point where bladder closure really takes place is not identical to the area of maximal pressure in the urethra. Radiologically, it can easily be demonstrated that the contrast medium never penetrates beyond the bladder neck into the urethra and thus it cannot possibly come into contact with the high pressure zone.

The problem of the bladder-closing apparatus cannot be understood on the basis of the classic theory, for this organ is always described as a system consisting of a pressure chamber and a drainage tube with an enclosed valve. But the apparently paradoxical behaviour can be understood when one abandons the idea of a valve in favour of a system where the bladder functions together with the pelvic floor as an integrated unit. Thus the bladder, being influenced by the intra-abdominal pressure,

* Urological Clinic (Chief: Prof. G. Rutishauser). Surgical Department (Director: Prof. M. Allgöwer). University of Basle, Switzerland

can be compared with a planimetric load lying on an elastic buttress, the pelvic floor. The urethra passes through this buttress as a slack tube with only a little tonus of its own and dependent on its surroundings for pressure. The behaviour of this system can be studied by means of a simple hydraulic model (Fig. 1). In this model the bladder consists of an elastic sphere whose internal pressure does not alter greatly when its volume changes. This sphere lies upon a waterfilled bucket of constant volume with initially rigid side walls and base. This system is transcoursed by the "urethra", a tube with little wall tension of its own. The pressure changes in the model can be studied by simultaneous recording from the "bladder" and "urethra". If the "bladder" pressure rises due to external force it is seen that the internal tension of the "urethra" increases accordingly. The explanation of this is simple; the transmitted force expands equally on all sides in the bucket, thus producing a closing force directed towards the centre, i.e. towards the "urethra". As no positive pressure gradient occurs, the model remains continent under all circumstances.

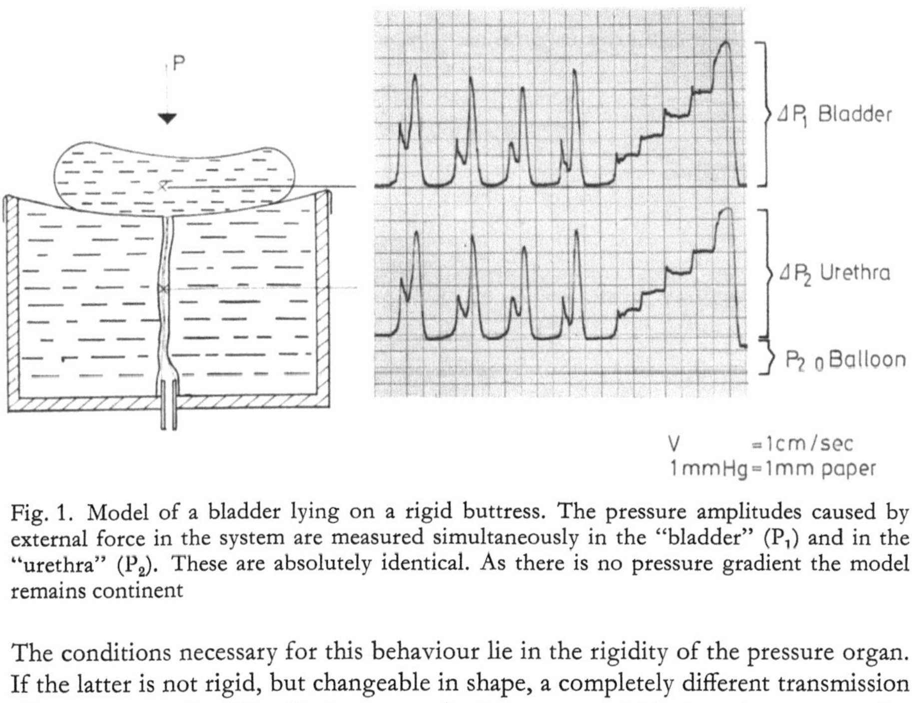

Fig. 1. Model of a bladder lying on a rigid buttress. The pressure amplitudes caused by external force in the system are measured simultaneously in the "bladder" (P_1) and in the "urethra" (P_2). These are absolutely identical. As there is no pressure gradient the model remains continent

The conditions necessary for this behaviour lie in the rigidity of the pressure organ. If the latter is not rigid, but changeable in shape, a completely different transmission of pressure results (Fig. 2). Stress results in an appreciable loss in pressure due to the deformation of the bucket with an elastic base. This is a sign that the action moment towards the centre decreases so that the model becomes incontinent.

Using the same methods of measurement we examined a group of continent and incontinent female patients and found the results corresponded quite well with those we achieved on the sphincterless model. The pressure recordings were made through a conventional balloon catheter, the urethral wall pressure being recorded by means of the indwelling balloon and the bladder pressure through the open end of the catheter (Fig. 3). In continent female patients the base of the pelvis is found to be an elastic basic element, constant in shape, which is able to convert an

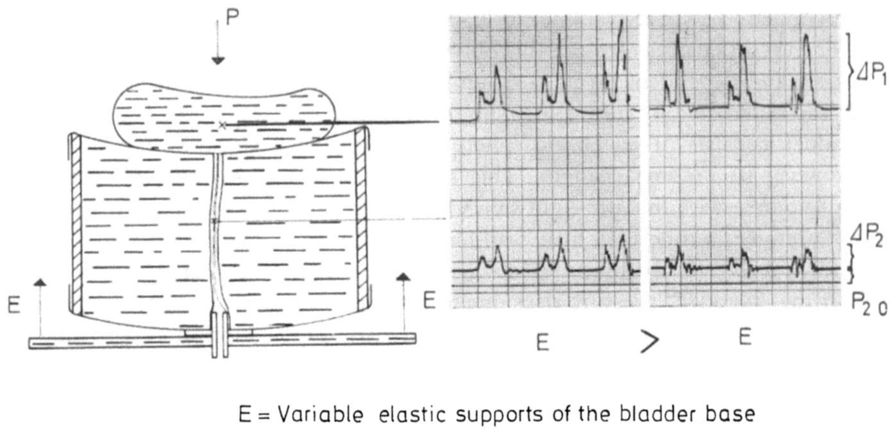

Fig. 2. Results of the simultaneous pressure registered in a deformable buttress. With increased slackness of the "bladder base" the pressure amplitude of the "urethra" becomes smaller than that of the "bladder", indicating the loss of intrinsic pressure in the bucket. The model becomes incontinent

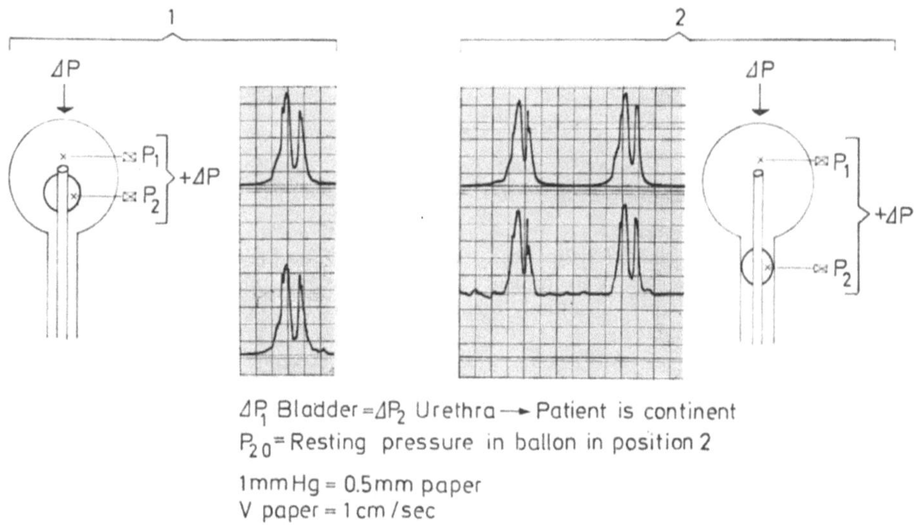

Fig. 3. Experimental measurement of the dynamic reaction of the pelvic floor in female patients. 1. Gauge. Both recording points are in the bladder. The signals are electronically coordinated until the pressure test (coughing) produces identical curves over both canals. 2. After withdrawal of the balloon into the urethra, its initial pressure rises to P_{20}. The pressure curves ΔP_1 and ΔP_2 show the load on the bladder and respectively that of the urethra by coughing. The resulting amplitudes are the same in this case so that the patient is continent

outside force into an equivalent counter-pressure and thereby build up a closing moment to the urethra. The pressure amplitudes transmitted through the urethra and bladder during coughing stress are thus fully identical. Female patients with stress incontinence showed a flattening-out of the curve for pressure in the urethra. In an inadequate pelvic floor the energy transmitted produces deformation of the buttress and thereby the internal closing pressure becomes inadequate. This change in shape can be diagnosed radiologically as a cystocele.

It is known from electromyographic examinations of the pelvic muscles that abdominal stress reflexly produces an increase in activity. In our experimental procedure, an additional contraction is recognized by the fact that the pressure amplitude of the urethra during coughing impulse exceeds that of the bladder. In fact, this type of reaction can occasionally be found and is even the rule in paraplegics, showing a certain detrusor-sphincter dyssynergy. In these cases it can be shown convincingly that the components active in bladder closure normally play only an insignificant part. By eliminating the additional muscular contraction by bilateral blockage of the pudendal nerves it is possible to achieve identical pressure amplitudes from bladder and urethra on repetition of the coughing test. Such patients remain continent.

Any model is a considerable simplification of the reality. It is not surprising to find additional factors which could not be demonstrated on the models shown. On observation of the pressure profile of a urethra recorded under high static load, it is found that the increased bladder pressure obtained by stress is not fully compensated by a corresponding increase in tonus in the area of the external sphincter. Instead of the expected pressure step, a continuous decrease in the wall tension along the urethra is found. This observation can be compared with two well-known technical facts: the distribution and the transformation of energy. We know that pressure under planimetric loads is characteristically divided into vertical and horizontal tensions which differ numerically only with the elasticity constants of the tissues subjected to pressure. As a result of this distribution a reduction of tensions occurs with increasing tissue depths.

Such curves coincide remarkably well with pressure profiles of the urethra under static load. This would also explain why no incontinence sets in here either, despite an apparent positive pressure gradient.

The conversion of pressure or destruction of energy in the technical world is brought to our notice daily in car crashes on the roads. Attempts at optimal passive security have resulted in constructions in which the resistant driver's compartment is protected by a pliable construction in front of it. This alternating effect of hard and pliable elements can also be recognized in the closure apparatus of the bladder. A rigid, stable bladder base plate primarily prevents the funnel shaping of the outlet under pressure and thus the admission of urine into the urethra. The proximal part of the urethra with its poor wall tonus functions as a progressive zone of deformation, absorbing part of the energy which is finally transmitted on to the muscle plate of the urogenital diaphragm which has resumed its rigidity.

The comparison with crashing cars can be further studied upon, for it gives a simplified illustration of the basic possibilities of impaired continence: the change in shape, or the conversion of energy planned by the constructor cannot take place if the deformable anterior part is inadequate or if the passenger's compartment is not built as a solid buttress. The first case can be compared with male post operative in-

continence, where the elastic zone essential to continence has been too radically removed. The second case can be compared with female incontinence, where the stability of the pelvic floor is inadequate.
All attempts to isolate anatomically one or more sphincter muscles have led to artificial separation of a system which can only be understood and accepted as a functional unity.

Discussion

The Function of the Bladder Neck

Urine incontinence is present when there is a positive pressure difference between bladder and urethra (Boyarsky, Hinman, Tanagho, Turner-Warwick). Under physiological conditions, the pressure increases to the point where bladder and urethra unite in the area of the orificium urethrae internum. This increase of pressure markedly obstructs the bladder neck.
In the distal urethra there are segments where the pressure is much higher than in the region of the orificium internum. These segments represent another safety factor: increases of intravesical pressure in a stress situation, if higher than the pressure of the orificium internum, allow the urine to enter the posterior urethra without an involuntary voiding (Tanagho). Even a minimal increase of pressure in a small urethral segment over intravesical pressure is sufficient to guarantee continence (Boyarsky).
Under physiological conditions, one can divide the continence mechanism in man into three parts: bladder neck, posterior urethra and sphincter urethrae externus (Allen, Donker, Tanagho). To maintain continence under pathological conditions, however, all three factors are not necessary. Postoperative continence after radical prostatectomy is proof of this (Marberger). Concerning the morphological substrate which guarantees continence after a radical prostatectomy, there are different views:
1. Tanagho believes that, in the area of the new bladder neck, a sphincter mechanism develops after a radical prostatectomy, with a following cysto-urethro-neostomy. This mechanism consists of newly arranged muscle fibres of the detrusor. In some patients, one can observe the development of a kind of tubular urethra proximal to the diaphragma urogenitale, which also supports the continence mechanism. That is why Marberger suggests that in a radical prostatectomy the urethra be well isolated and pulled 1 to 2 cm into the minor pelvis.
On the basis of simultaneous cinematographic, electromanometric, and electromyographic investigations, Tanagho believes that the obstruction mechanism after a radical prostatectomy arises from many factors. The abrupt pressure increase in the region of the new bladder neck shows that there must be a new sphincter urethrae. Regular myopotentials in connection with increased activity are a sign of a resting period with reflex influences on the muscles of the pelvic floor.
2. Turner-Warwick has investigated sphincter mechanisms after prostatectomy. In most procedures the function of the bladder neck mechanism is, intentionally or unintentionally, ablated. He believes that the most important component of the

distal mechanism maintaining continence after prostatectomy is the intrinsic mechanism of the membranous urethra (lying within the extrinsic muscles) but also extending proximally into the apex of the prostate. If the supramembranous urethra is damaged during prostatectomy, the continence of the patient is at risk since the closing tonus of the extrinsic muscle alone seems insufficient. During radical prostatectomy the supramembranous part of the intrinsic distal urethral mechanism is removed and continence tends to be more precarious.
TURNER-WARWICK also notes that, when the entire distal mechanism is destroyed in association with a pelvic fracture, effective continence can be maintained if bladder neck mechanism remains intact after urethroplasty.
GIL VERNET also emphasises the importance of the distal sphincter mechanisms.
3. DONKER thinks that the smooth muscles of the distal urethra together with the longitudinal sphincter externus are the most important factors in continence after prostatectomy. Upon intravenous application of sympathicolytic agents after radical prostatectomy, the patient immediately becomes incontinent after having previously shown complete continence.
4. ZINNER regards the specific passive tissue properties as the most important criteria for continence and incontinence of urine. Not only the quantity of the smooth muscles but the pattern of muscular, elastic and collagenic parts of the tissue, their physical characteristics and their geometric arrangement are important for the functional result.

The Measurements of the Urethrovesical Angle and its Importance for the Function of the Bladder Neck

B. Terhorst, H. Melchior, H.-P. Bastian, and F. K. Lynen*

The function of the bladder neck is to act as a valve; complete bladder obstruction must be guaranteed during the resting- or filling-period and complete voiding of the bladder at micturition. The ansiform arrangement of the detrusor muscles at the bladder neck and the fan-shaped structure of the muscles in the region of the base plate with the trigonum are very important to unimpaired functioning of this valve mechanism (Tanagho and Smith, 1966; Woodbourne, 1967). Disorders of the valve mechanism cause incontinence and dysuria.
The measurement of residual urine, cystography and urethro-cystoscopy are very useful in the differential diagnosis of incontinence and dysuria as is a detailed medical history. Later functional investigations are made by cysto- and sphincteromano-metry as well as by the measurement of the urethrovesical angle during the resting period and during micturition.

Methods

The patient is positioned on a tilting table. A catheter (No. 10 French) is inserted. The bladder is filled with x-ray contrast medium (Conray FL:water = 1:1) until 300 ml or until the patient has the desire to urinate. The catheter is plugged and fixed. The tilting table is erected and the vesico-urethral angle is visualized on the monitor. For the radiogram taken laterally the x-ray tube is centred on the trochanter maior. The radiogram is taken with the catheter in place during the resting period and during micturition after removal of the catheter.
The angle between the urethra and the base of the bladder dorsal to the urethra (urethro-vesical angle taken laterally) is measured (Fig. 1).
The investigations were performed on 60 patients, mainly male. The case material was not selected, and we examined healthy persons as well as patients with dysuria and incontinence.

* Abteilung Urologie (Director: Prof. Dr. W. Lutzeyer) und Chirurgie (Director: Prof. Dr. M. Reifferscheid) der Medizinischen Fakultät der Rheinisch-Westfälischen Technischen Hochschule, Aachen, Germany

Results

a) Healthy Persons

The urethrovesical angle taken laterally was measured during the resting period and during micturition in 23 patients with normal bladder function. During the resting period it was 119° ± 3,6° and during micturition 138° ± 3,0°. The individual variation in the urethrovesical angle between resting period and micturition was 19° ± 3° (Fig. 1).

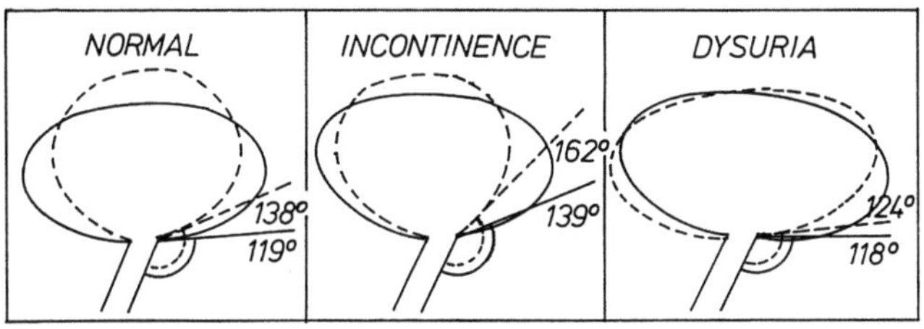

Fig. 1. The urethrovesical angle of normal persons and patients with incontinence or dysuria during the resting period and during the micturition

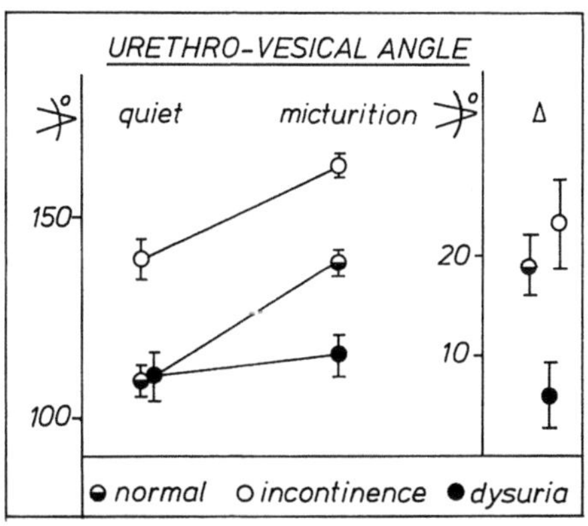

Fig. 2. Statistical findings of the urethrovesical angle in normal persons and patients with incontinence or dysuria

b) Incontinence of Urine

The urethrovesical angle taken laterally was measured in 25 patients with incontinence. In the resting period we measured an urethrovesical angle of 139° ± 4,9° and during micturition one of 162° ± 2,5°. In these cases the individual change of the urethrovesical angle between resting period and micturition was 23° ± 4° (Fig. 2).

c) Dysuria

In 12 patients with dysuria the urethrovesical angle taken laterally was measured during the resting period at 118° ± 3,6° and during micturition at 124° + 4,7°. The individual angular aperture was 6° ± 3,4° (Fig. 2).

Discussion

Differential diagnosis of incontinence and dysuria may be aided by the medical history, by clinical investigations with measurement of the residual urine and cystoscopy and by functional exploration examination using cysto-sphinctero-manometry and radiological determination of the urethro-vesical angle taken laterally. The measurement of the urethro-vesical angle taken laterally is performed during the resting period and during micturition. In contrast to Kaufmann and Ikle (1956) and Jeffcoate and Roberts (1952), we do not take the radiogram at high intraabdominal pressure. We believe that a radiogram of this type is only important in women with prolapse and cystocele.

According to our investigations, the urethrovesical angle during the resting period is 119°. During micturition, there is a physiological opening of the bladder neck in association with an extension of the posterior labium up to 138° (Fig. 1). In comparison with this the posterior urethrovesical angle of patients with incontinence is 139° even during the resting period, which corresponds to the angle of normal patients during micturition. In patients with very pronounced incontinence of urine the posterior urethrovesical angle is often almost obtuse. It is interesting that in patients with incontinence the angle is not only enlarged during the resting period, but also becomes wider during micturition (Fig. 1 and 2). In patients with dysuria the urethrovesical angle during the resting period does not differ from that of the healthy persons. But it is an important characteristic of this group that the posterior labium of the bladder neck does not extend during the micturition. This is why the change in the urethrovesical angle is only minimal (Fig. 1 and 2).

The measurement of the urethrovesical angle is familiar to gynecologists (Kaufmann and Ikle, 1956; Jeffcoate and Roberts, 1952; Hartl, 1953; Lochmüller and Zimmer, 1968). In healthy women the angle is given as 100° (Jeffcoate and Roberts, 1952; Lochmüller and Zimmer, 1968; von Rütte, 1971). In contrast to this, we found a more obtuse angle of 119°. This is probably because most of our patients were men. The anatomic conditions in the area of the posterior bladder neck are different in man because of the prostate and the seminal vesicles.

For trouble-free micturition the extension of the posterior labium of the bladder neck is an important parameter. This is pointed out by the test results in patients with dysuria. In patients with incontinence the urethrovesical angle is almost obtuse. This has important therapeutic consequences: the functional incontinence of urine caused by an enlarged urethrovesical angle can be treated by suspension plastic operations. Zoedler and co-workers (1970) were able to correct functional incontinence in almost 100 percent of cases by lifting the bladder neck with a nylon net ribbon. The suspension plastic operation of Marshall-Marchetti (1949), with antefixation of the urethra and the bladder neck, is based on the same mechanism. On the other hand, functional dysuria can be treated by antefixation of the detrusor vesicae at the anterior abdominal wall in association with extension of the urethrovesical angle after Rochet. This is why preventive antefixation of the detrusor is recommended whenever exenteration of the minor pelvis, as or extirpation of the rectum is performed. The object of this procedure is to prevent the bladder from falling back into the pelvic cavity, and to produce flexion in the area of the bladder neck.

Summary

The measurement of the urethrovesical angle taken laterally yields information usefully in the differential diagnosis of incontinence and dysuria. Our investigations show that the extension of the posterior labium of the bladder neck is important for trouble-free micturition.

References

HARTL, H.: Die funktionelle Harninkontinenz der Frau. Stuttgart: Enke 1953.
JEFFCOATE, T. N. A., ROBERTS, H.: J. Obstet. Gynaec. Brit. Emp. **5**, 685 (1952).
KAUFMANN, P., IKLE, A.: Beitrag zur Beurteilung der relativen Urininkontinenz mit der lateralen Zystourethrographie. Geburtsh. u. Frauenheilk. **16**, 29–43 (1956).
LOCHMÜLLER, H., ZIMMER, F.: Die Urethrozystographie. Münch. med. Wschr. **110**, 2076 (1968).
MARSHALL, V. F., MARCHETTI, A. A., KRANTZ, E.: Surg. Gynec. Obstetr. **88**, 509 (1949).
VON RÜTTE: Persönliche Mitteilung.
TANAGHO, E. A., SMITH, D. R.: The anatomy and function of the bladder neck. Brit. J. Urol. **38**, 54–71 (1966).
WOODBOURNE, R. T.: Anatomy of the bladder. In: BOYARSKY, S.: The neurogenic bladder. Baltimore: The Williams & Wilkins Comp. (1967).
ZOEDLER, D.: Die operative Behandlung der weiblichen Harninkontinenz mit dem Kunststoff-Netz-Band. actuelle urol. **1**, 28–34 (1970).

The Velocity Profile of the Human Urethra: Measurement and Significance

N. R. Zinner*, A. M. Sterling**, R. C. Ritter***, J. H. Harris, F. B. Gessner, J. M. Reid, J. Hedges, D. Chow, and F. E. Barber****

Pressure, flow rate, fluid velocity and urethral diameter are closely interdependent variables. The bladder pressure represents the energy input into the system and is generated by the detrusor alone or in combination with superposed abdominal muscular straining. The velocity, or more precisely, the kinetic energy of the external stream represents the energy out of the system. If the urethra-fluid system were completely lossless, the bladder pressure would equal the kinetic energy of the external stream and from a measurement of either bladder pressure *or* exit velocity, one could easily calculate the other. However, we have previously shown that the female urethra is not lossless but that 30 per cent or more of the energy generated by the detrusor is lost by frictional dissipation within the urethra (Ritter *et al.*). What are the sources of this frictional loss? In part it depends upon the dynamic and static geometry of the outflow system but related, also, to the fluid velocity itself. While considerable investigation and discussion has centered on the bladder pressure, the flow rate and considerations of the frictional losses, little attention has been paid to the way the fluid velocity varies either along or across the urethra. Yet, measurement of this variation may be one of the most important in providing an understanding of the mechanisms of urethral function and the role of the flowing stream in preventing or permitting outflow-tract disorders.

Local Velocities in a Flowing Stream

Local Average Axial Velocity

Fig. 1A shows an irregular channel and is used to depict how the average axial velocity varies at different locations along the channel even though the volume flow rate is the same. The wider sections would have lower mean axial velocities and the

* Department of Surgery (Urology), Charles R. Drew Postgraduate Medical School, Los Angeles, Cal., U.S.A.
** Afdeling Urologie, Academisch Ziekenhuis Leiden, Leiden, The Netherlands.
*** Department of Physics, University of Virginia, Charlottesville, VA., U.S.A.
**** Departments of Urology, Chemical Engineering, Electrical Engineering, Mechanical Engineering and Center for Bioengineering, University of Washington, Seattle, WA., U.S.A.

narrower sections, higher mean axial velocities. Measurement of such mean local axial velocities might provide information of the functional diameter of the urethra at the particular site of the measurement.

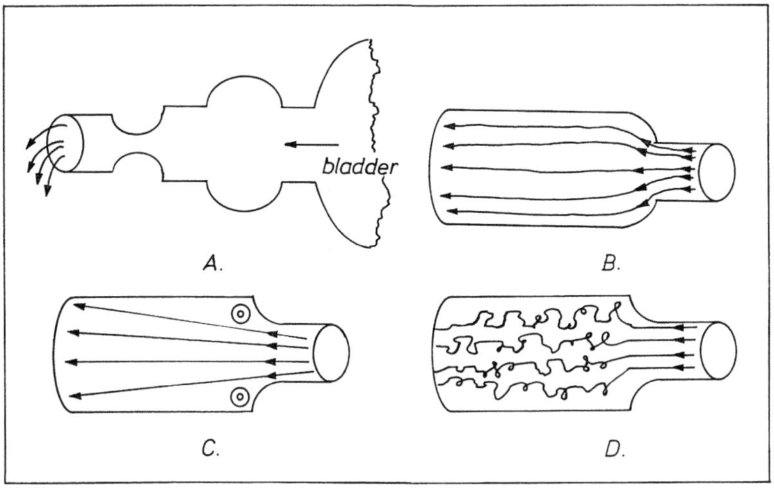

Fig. 1 A–D. A Consider relations between velocity and cross-sectional diameter in this hypothetical urethra. B Radial velocity within an expansion. C Separation bubbles in an abrupt expansion. D Turbulence in an expansion. See text for descriptions of flow properties depicted in Fig. 1 A–D

Radial Velocity within an Expansion

In addition to the velocity in the axial direction, there is a radial component of velocity in an expansion or contraction. As the urine stream expands to fill the wider section or contracts to pass through the narrower section, some of the velocity would be in a lateral or radial direction. Fig. 1 B shows the radial component of velocity within an expansion. We can see from this figure that such a velocity component might drive bacteria contained within the fluid into the walls of the urethra.

Velocity within a Separation Bubble

Fig. 1 C shows another tube with a more abrupt expansion than in Fig. 1 B. As the fluid moves into *this* expansion, a separation bubble develops (or, in other terms, a separation vortex) which differs from both the axial velocity and the radial velocity. It is important to distinguish this phenomenon from turbulence. While we have shown tubes with *expansions* to demonstrate radial velocities and separation bubbles, the same phenomena occur in *contractions* as well.

We can see from Fig. 1 C that bacteria contained within the fluid vortex can be recirculated locally and remain within the duct even though flow is occurring. Also, bacteria originally on the walls may not be rinsed by the stream.

Turbulence within an Expansion

Another consideration of local velocity is that of turbulence (Fig. 1D). In this case, the elements of fluid do not flow in a uniform direction. Rather, directions and speeds of the fluid elements are random. This random pattern results in an energy loss which is greater than the loss occurring from radial velocities or from stream separations. In addition to energy losses, which are greater than in laminar flow, turbulent flow has the property of providing increased mixing of the fluid which will increase the rate of diffusion of particulate matter such as bacteria to and from the urethral wall.

It should be emphasized that turbulence can develop distal to expansions and contractions even though the criterion which is used to characterize the overall flow would indicate the fluid motion to be laminar.

The major types of local velocities we have described can occur together or separately. If local velocity is very slow and the wall expansion or contraction gentle, the fluid can expand without separation. If the flow is more rapid or the expansion more abrupt, separation occurs and a separation bubble can develop. If the flow is still more rapid, local turbulence develops.

Clinical Significance of Local Velocity Measurements

Urethral Geometry

One reason to measure local velocity components is to understand the local geometry and from this, to understand the local urethral properties which may relate to bacterial rinsing or deposition. In general, regions of relatively high mean local axial velocity will indicate narrowings, regions of relatively low mean axial velocity will indicate expansion. If, in addition, the average local axial velocity *and* simultaneous volume flow rate are measured, the local effective diameter can be calculated. Variations of the local flow conditions such as radial velocities and circulation within separation bubbles will indicate grossly, the abruptness of change in urethral diameter. Changes in the turbulent structure of the flow may indicate geometry variations such as inapparent strictures which produce jetting of fluid and result in intensified turbulence.

Energy Loss

Another reason to measure local velocities is to understand how energy is lost within the system. Since bladder pressure represents the energy input into the system and the exit velocity represents the energy output, the difference between entrance and exit energy represents the energy consumed within the urethra. This may be related to the concept of a urethral resistance factor.

While the entrance and exit energy measurements provide information to show *how much* energy is lost within the entire urethral system, they do not tell *how* or *where* it is lost or *how much* is lost at each location. The "where", "how", and "how much" can also be important in understanding disease. Two different urethrae might register identical overall resistance factors yet each would require entirely different forms of treatment.

Fig. 2 shows two canals, each with the same exit velocity and flow rate. Since the system with the irregular tube consumes more energy, the bladder pressure in this system would have to be higher than the bladder pressure in the system with the more regular tube. Accordingly, the resistance factor does not locate the source(s) of resistance, describe whether they are represented by curves within the urethra, expansions, contractions, or areas of wall roughness and the like. It is necessary to measure local velocities to understand how and where energy is lost.

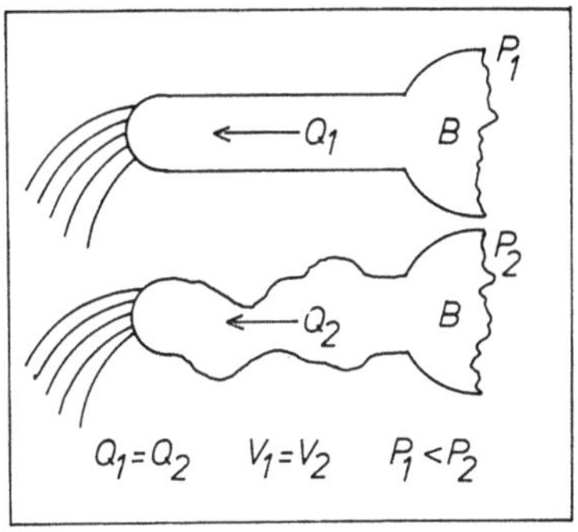

Fig. 2. Relations among properties of flow (Q), exit velocity (V), and bladder pressure (P) are as shown. Measurement of any or all of these would not locate site(s) or character of the higher flow resistance locations in the more irregular "urethra"

Boundary Layer Conditions

To this point we have emphasized the large scale and easily demonstrated phenomena of variations in the local velocity near an obstruction. We now turn to the case in which obstruction is absent. We will first describe the well established phenomena for turbulent flow in a uniform pipe and then apply these concepts to a simplified model of the human urethra. This phenomenological approach will indicate the effects of urethral wall roughness and urethral length on important considerations of both energy loss and bacterial flux to and from the urethral walls.

Laminar Sublayer: It is customary to consider fully developed turbulent flow in terms of three zones; the laminar sublayer, the buffer zone and the turbulent core. The laminar sublayer is that layer immediately against the wall. In the urethra its minimum thicknss is about 25 microns (STERLING). In this region, flow is completely laminar despite the overall turbulence of the stream. At the wall itself, flow is zero but velocity increases towards the lumenal axis as with all laminar flow.

Buffer Zone: The buffer zone (transition zone) is a region of intermediate flow where most particle motion is axial but there is some turbulence. Here both turbulent and laminar effects are of equal importance. In a urethral cross section of about three millimeters, this layer is about 125 microns thick (STERLING).

Turbulent Core: This layer makes up the bulk of flow. The mean axial velocity is roughly uniform over the cross section of the core and laminar effects are negligible.

Clinical Significance of the Boundary Layer: Recall that the laminar sublayer is, at its thinnest, at least 25 microns thick. Also, recall that the size of a bacterium such as *E. Coli* is approximately two microns thick. Accordingly, an *E. Coli* bacterium on the urethral wall would be protected from the stream turbulence and lie well within the laminar sublayer.

The agitation effect of turbulence on bacteria at the urethral wall would not be present and there would be *no* scrubbing action on bacteria at the wall.

There are ways however, by which bacteria can be removed from the wall by the flowing stream. First, a few words about wall roughness.

If a wall protuberance does not extend beyond the laminar sublayer, the wall surface will behave as a hydraulically smooth surface; and surface roughness of this extent will have no effect on the flow. In view of this consideration then, one would expect protuberances of less than 25 microns to have no effect on the pressure loss in the urethra. From the observations of LLOYD-DAVIES, HAYES, HINMAN and from SILVERBLATT, the assumption of a hydraulically smooth tube appear to be reasonable. This consideration applies *only* to the *general overall* wall characteristics of the normal urethral surface and *not* to *specific locations* on the wall which may contain larger protuberances.

Concentration Gradient Across the Laminar Sublayer: Assuming that it is normal for bacteria to exist in greater concentrations at the urethral wall than in the bladder urine (Cox *et al.*), a diffusion gradient of bacteria develops within the laminar sublayer. If the laminar sublayer is thin, the gradient is steep; that is, there is a quick fall in the high bacterial concentration at the wall to the low concentration in the urine near the buffer zone. This then means that there is a large bacterial concentration gradient across the laminar sublayer. The higher the velocity of the urinary stream, the thinner the laminar sublayer and the larger the bacterial concentration gradient. Owing to this concentration gradient, bacteria will diffuse towards the buffer zone and from there, will be swept quickly to the turbulent core. From here, the bacteria will quickly be passed out of the urethra.

The reverse of this process may occur if the urine itself were infected and the urethral walls were relatively free of bacteria.

Entrance Effects: At the entrance to the urethra we have usually a smooth transition from a large vessel to a long tube. This area can be considered as a bellmouth entrance. Owing to the properties of flow at such a transition section, for all practical purposes, the fluid velocity is uniform almost to the wall itself.

Accordingly, the bacterial concentration gradient in the region near the wall of the urethral entrance would be greater than in any region along the urethral canal and bacteria may diffuse more readily into the core stream far more easily. This property of the bellmouth entrance flow may help explain why so few bacteria are normally found in the region of the bladder neck (Cox *et al.*).

The hydromechanics of development of the flow because of the entrance effects is more complex than presented here and other considerations may arise. The interested reader is referred to a more detailed description which is to be reported elsewhere (STERLING). To continue with this point however, owing to transitions of flow at entrances such as the bellmouth entrance, turbulence does not fully develop until some distance downstream. Within the transition region, complicated patterns develop which may either enhance or retard bacterial diffusion and enhance or retard frictional losses from what they might otherwise be in a region of fully developed turbulence. Our preliminary calculations indicate that the flow through the configuration of the female urethra might require as much as six centimeters to reach fully developed turbulence. Since the urethra is only about four centimeters long, many considerations of resistance and properties attributable to turbulence may require special considerations. In the male urethra, excepting for the proximal portions, fully developed turbulences can occur.

Local Velocity Measurements

In principle, local velocity measurements can be made both from inside and outside the urethra. At present there are instruments which can be inserted into the urethra to measure local velocity. Unfortunately, these instruments disturb flow patterns because they must be inserted into the canal. In addition to disturbances of normal flow patterns by the probe, there may be secondary physiological effects such as sphincter spasms and psychological effects resulting from inserted instruments which may produce abnormal urination in an otherwise normal individual. Despite these disadvantages, useful information can be obtained.

Measurements with Insertion-Type Instruments

The Pitot Tube : One type of velocitometer which can be applied to urethral measurements is a variation of the standard Pitot tube. It is a combination of two tubes, one with an opening on the end, one with a closed end and an opening on the side. The tube with the hole at the end measures the total pressure, that is, the sum of both the static and the dynamic pressure.

The tube with the hole in the side measures the static pressure only. Subtraction of the static pressure measurement allows one to determine the average local velocity. One problem with this instrument is that the measurement made at the side hole is very sensitive to variations in the angle of the tube with respect to the flow of the stream. Minor errors in tube position can produce gross errors in the velocity calculation.

Other types of instruments inserted into the canal may be less subject to instrumentation error but the insertion problems remain. These other instruments include thermisters and hot film anemometers.

Total Pressure Measurement (The Urethral Total Pressure Profile): The urethral total pressure profile measurement is not a measurement of velocity. Nonetheless, it is a related measurement and it appears useful to include it in the manuscript.

The total pressure within the urethra represents the sum of the dynamic and static pressure. Loss in total pressure along the urethral canal represents loss of energy which is irrevocably converted into heat.

The urethral total pressure profile is measured with a single tube with an opening at the end only. The impact pressure measured at the end opening represents both the static and dynamic pressures in the stream at the point of the measurement (GESSNER, GESSNER and ZINNER).

A tube of known length is inserted into the bladder *per urethram*. Prior to voiding, the tube is withdrawn in increments and the subject asked to cough after each

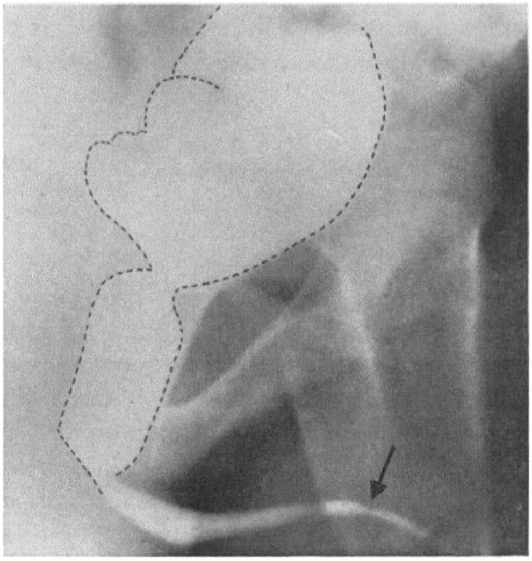

A

B

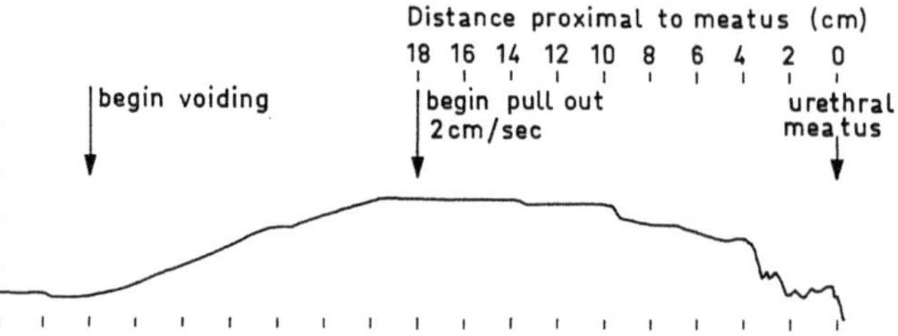

Fig. 3 A and B. A Voiding urethrogram of boy aged 13. Broken line drawn over bladder and proximal urethra to enhance visualization of these structures. Note seemingly minor narrowing in pendulous urethra at arrow. B Pressure traverse in same patient depicted in Fig. 3 A. Note absence of pressure drop in region of previous valve resection (about 12 cm. proximal to urethral meatus) despite abrupt narrowing in urethral caliber at this location. Pressure loss of about 50 cm H_2O occurs at about 3 cm proximal to urethral meatus and corresponds to region shown in arrow in Fig. 3 A

increment. In a continent patient, when the pressure impulse of the cough is no longer registered on the recorder, one can generally assume the bladder neck has been reached. Since overall catheter length is known (or possibly marked in measured increments) the urethral length can be calculated. The catheter is then inserted another few centimeters and the patient voids. During voiding, the tube is withdrawn at 2 cm per second with a motorized winch. The total pressure measurement is recorded continuously on a strip chart recorder which also marks time in seconds. The physician observes the instant the catheter tip passes the meatus and causes a mark to be made on the recording. Since insertion distance as well as speed and time duration of the traverse are known, one can relate any measured pressure drop to a particular location within the urethra.

Fig. 3 A and B show the voiding urethrogram and pressure traverse results in a thirteen-year-old Indian boy referred from Alaska. The boy had undergone resection

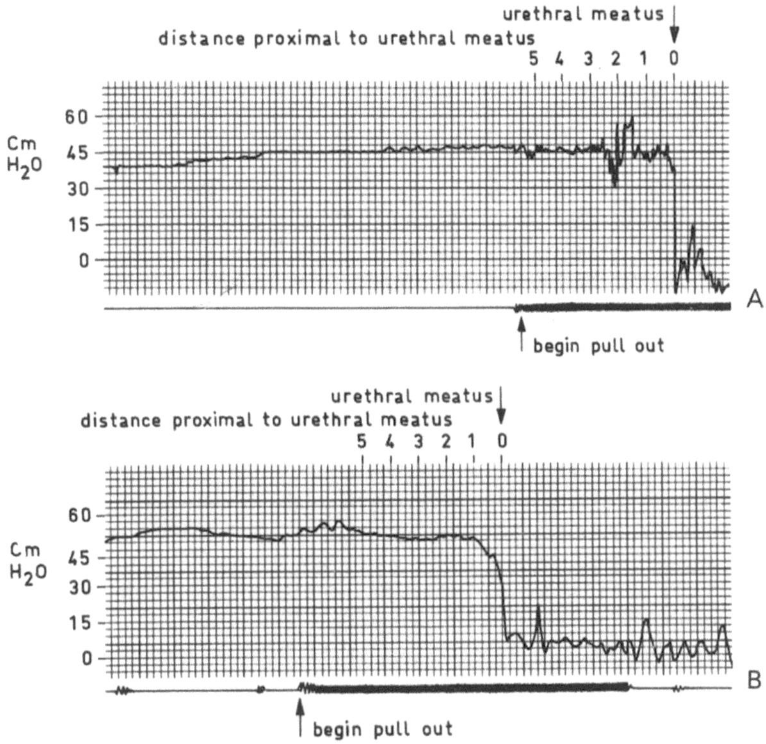

Fig. 4A and B. Pressure traverses in two female patients with normal urethrograms showing absence of significant pressure drop within urethra proximal to meatus

of urethral valves and bilateral ureteral reimplantation for reflux. Despite these procedures, reflux persisted and the boy had recurrent urinary tract infections. Mid-voiding bladder pressure plateaued at about 100 cm of water. The urethrographic appearance caused us to suspect the presence of an obstruction at or near the site of the valve resection. Cystourethroscopy was not helpful. The pressure traverse showed an unexpected loss of energy in the distal urethra about three centimeters proximal to the urethral meatus. This area was dilated and the boy's symptoms improved. Unfortunately, the patient has been lost to followup after his return to Alaska. Similar measurements made in 18 females have shown in every case that the entire pressure loss which occurs in the female urethra occurs at or near the urethral meatus (Figs. 4A and B). Measurement of energy remaining in the exit stream allows calculation of the extent of the loss at the meatus and can help in the diagnosis of meatal stenosis[1].

[1] These measurements assume that the bladder pressure is sustained at a plateau during the traverse. Ideally, a second pressure measuring catheter should be inserted into the bladder by suprapubic puncture. We have found however in most cases, that plateau pressure is sufficiently constant to allow meaningful measurement without the need for this second catheter. In some patients, fluctuations in mid-voiding pressure owing to straining or sphincter closures can be seen during the traverse and, in these cases, the second catheter will be necessary.

Measurements with Non-Insertion, Non-Interfering Type Instruments

Continuous Wave Doppler: The continuous wave Doppler processes signals scattered from particles within the fluid. From the back-scattered signals, calculations can be made of average local velocities and their variations (Albright and Harris). Using the CW Doppler, Albright, Harris and Zinner have measured local velocities within both artificial systems and the human male urethra (Albright *et al.*).

An accoustical transmitter is placed along the ventral surface of the penis and the patient urinates. Particulate matter in the urine causes a small portion of the incident energy to be scattered back to the monitor at frequencies determined by the Doppler principle.

Seeding the urine with oral ingestion of Sodium Bicarbonate or Acetazoliamide causes the precipitation of phosphates and enhances the ultrasonic scattering from the urine.

Fig. 5 shows the spectral peak and bandwidth of the analyzed recordings and their variation with position along a strictured mechanical flow system. Observe that the mean axial velocity increases in the region of the strictured segment as indicated by a shift of the spectral peak to higher frequencies, and that the turbulence, as indicated by the widening of the spectral bandwidth increases at a location beyond the strictured segment. The mean axial velocity within the stricture (location d) is 113.2 cm per second, and is 57.6 cm per second in the non-restricted portion of the tube (locations a and b).

From the above described method, we can see that both the local mean axial velocity and, to some extent, the degree of turbulence at that location can be determined. Fig. 6 represents the same type of analysis on a normal human male subject. Measurements were made at three locations as indicated in the figure legend. The mean axial velocities for these three locations are shown in Table 1. Note that the velocity within the urethra varies more than eight-fold from sites separated by as little as 1.5 cm. By measuring simultaneous external flow rate, the effective local diameter of these sites was calculated. These results are presented in Table 1 along with results of measurements taken in a total of four subjects.

While the measurements shown were made in males, miniaturization of the transducer should permit development of an instrument with transducer crystals placed in tandem for multiple simultaneous measurements along the urethral canal. Such an instrument could be inserted into the vagina for use in females since the entire urethra and possibly the bladder neck lie along the anterior vaginal wall separated from the vagina by a relatively thin septum.

Direct Accoustical Measurements: There are indications that an analysis of the noise generated by the flow itself will provide details of urethral geometry and mean velocity (Hedges *et al.*). The turbulent fluctuations in the flow give rise to pressure fluctuations at the urethral wall which responds by vibrating.

These urethral wall vibrations can be observed by lightly placing one's finger on the ventrum of the penis during urination[2]. The flow noise in the urethra could be re-

[2] The sensation one gets that urination is occurring is probably the result of urethral wall vibrations. During World War II, Air Force personnel in bombers had to urinate into an outlet in the rear section of the plane. Often, there was little or no light. We have been told that, if the stream was not visible, the airmen did not know when the flow had begun nor when it ended since the vibrations of the plane's engines masked the urethral vibrations during voiding

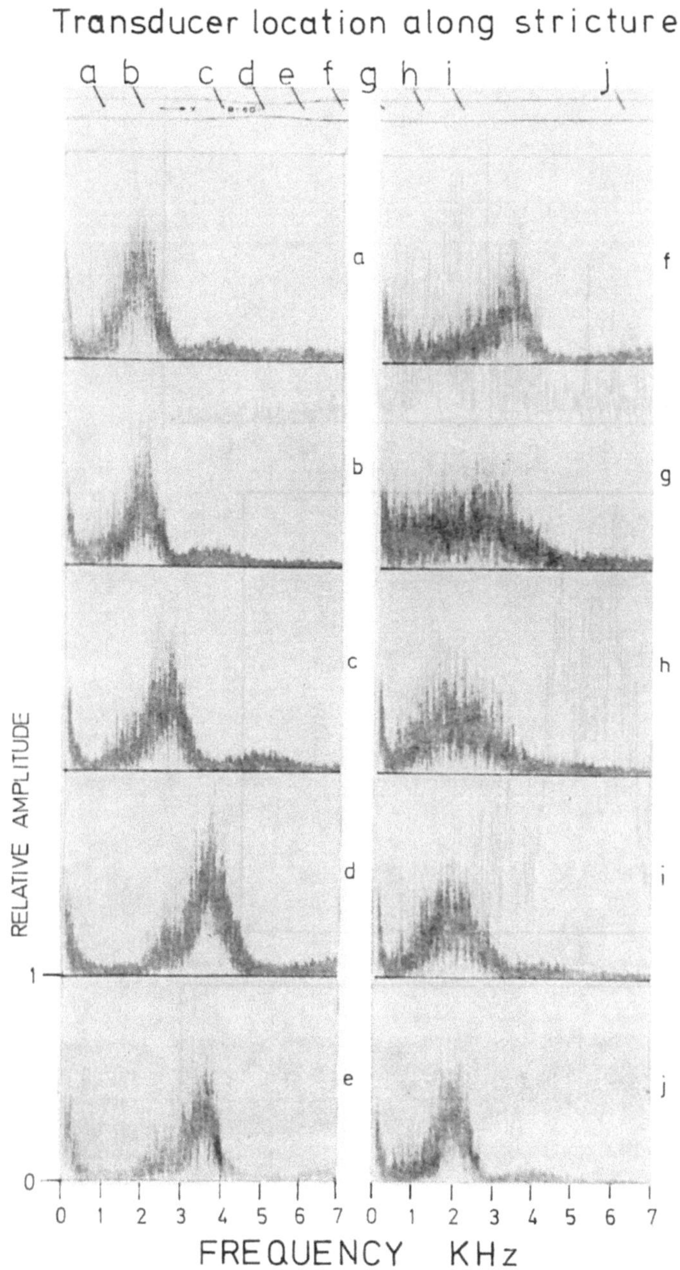

Fig. 5a–j. Spectral analysis of signals from C–W Doppler unit when used in model flow system shown at top of this illustration. Observe stricture at location D. Spectral analyses of measurements made at locations a–j are shown in correspondingly labelled tracings. Spectral peak indicates average axial flow velocity. Higher velocity produces proportionately higher frequency or return signals. Note that highest velocity is in region of strictured segment (location C) and that greatest spectral bandwidth (indicating greatest turbulence) is downstream of stricture

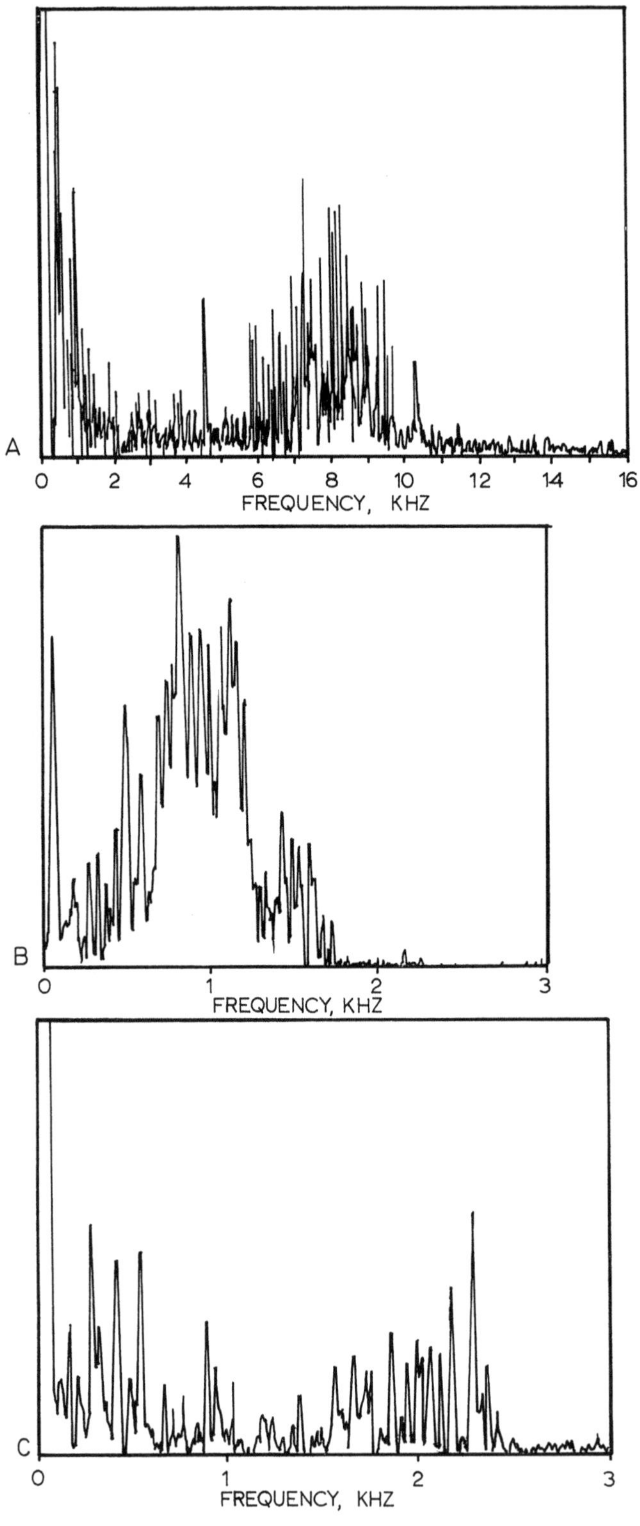

Fig. 6 A–C. Spectral patterns taken during mid-portion of normal male urination with Doppler transducer. A Transducer located at glans penis – about 1 cm proximal to urethral meatus. B Transducer located about 2.5cm proximal to meatus. C Transducer located midway between scrotum and glans. Note that scale for frequency is compressed in Fig. 6 A when compared to 6 B or C

corded by placing a microphone along the ventral side of the penis (or within the vagina) during urination.

A similar phenomenon which is more familiar to clinicians is the bruit observed over strictured vessels during stethoscopic examination. The analysis of the recorded signal is much like the analysis of the back-scattered signals from the Doppler in that the spectrum of the noise is calculated. In this case however, the spectrum is forced to fit a universal non-dimensional spectrum (FREDBURG *et al.*). This fitting procedure provides two constants from which the centerline velocity and urethral diameter can be calculated. The advantage of this measurement is the high degree of sensitivity to the diameter Theory indicates the intensity of the noise is proportional to the eighth power of the diameter.

Table 1. Urine flow velocities and urethral diameters for tests on single subject at different locations along penis as indicated in Table

Urethral Location	Transmitter Frequency MHz	Transducer Angle	Doppler Shift kHz	Flow Rate ml/sec	Velocity cm/sec	Effective Diameter cm
At glans (1 cm proximal to meatus)	5.18	45	8.3	8.7	170.0	0.26
Near glans (2.5 cm proximal to meatus)	5.18	45	1.0	5.3	20.4	0.57
Midway between scrotum and glans	5.18	45	2.3	7.8	47.0	0.46

We also expect that the change in the turbulent flow characteristics distal to an obstruction can be analyzed by this method to yield the size of the obstruction. The location of such an obstruction would be obvious from a traverse along the external urethra with one microphone or by multiple recordings taken simultaneously along the external urethra by a tandem arrangement of microphones.

We have not yet developed this technique to the point where quantitative results are available. We have, however, done some preliminary work on models made of Tygon tubing which contained an obstruction. The noise was recorded by a phonograph cardridge placed on the outside wall of the tubing.

Fig. 7 shows how the spectrum of the recorded signal changes at various locations about the obstruction. The similarity of this result to the spectrum of the backscattered Doppler signals confirms the merits of this approach.

Rotational Compound Scanner: This instrument employs three pulse echo crystals, each set at a different angle, to measure the geometry of internal structures directly (TOMEY and REID). These crystals are set on a rotating disc which rotates eight times per second. The disc itself is set within an enclosed probe which is placed on the skin (or other) surface overlying the structure of interest. In our case, the probe is placed on the ventral surface of the penis. The internal structure of the penis, with the urethra, is displayed continuously on a video tube. The display can be recorded with video tape, motion picture film or by individual photographs taken sequentially.

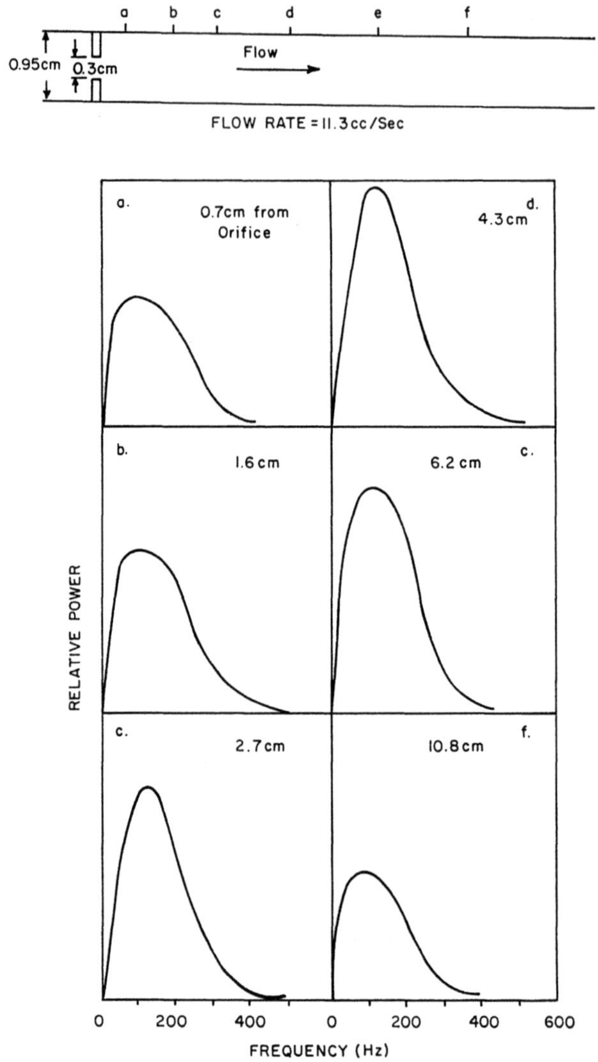

Fig. 7 a–f. Spectra of flow noise downstream from an obstruction in a soft plastic tube

At present, the resolution of the device is one – to two millimeters. Results of the measurements made at different locations along the urethra are shown in Figs. 8 A–C. Measurements made in a normal urethra in the mid-pendulous portion before, during and after urination are shown in Figs. 8 D–G. It is possible to miniaturize the instrument to permit insertion into the vagina for studies in females. The subject would then be able to wear the device as a tampon and urinate in a natural setting at will.

Measurements of the External Urinary Stream : The Drop Spectrometer : Because of the well-known difficulties of urethral velocity profiles, and the interference of most practical methods of measurement, we have done considerable study of the external urinary stream (ZINNER *et al.*, RITTER *et al.*). Based on physical information carried by drops of the urine stream, a Urine Drop Spectrometer has been developed which senses the drops by optical means. In the normal male voiding, which usually lasts from 10–30 seconds, 3000 to 6000 drops are recorded. The distribution of their temporal spacings has a mean of about 5 milliseconds and extends in an approximately Maxwellian-shaped curve from about one millisecond to beyond 10 milliseconds. The drop velocity distribution has a similar variability with a mean of about 150 cm per second. As estimated from the mean flow rate of 15 cc per second, the mean drop diameter is about 0.5 cm if the drops are assumed to be spherical. Actually, the drops have many different shapes, including quadrupole, octupole and higher moments which are normally used to specify liquid drop surfaces.

These drops contain the energy, angular momentum and geometric information imparted to the stream by the urethral canal. If the stream acts as a forcing function to a vibrational ring (e.g., a stricture) or to a more extended vibrating region whose "tone" is modified by disease, the vibrating walls provide a small modulation on the geometry which will affect the break-up pattern of the external stream. This principle has long been used to provide a form of hydraulic amplification (Sir BOYS), and indeed, the sensitivity of such modulation is without question.

Mean parameters of the breakup are affected in simpler ways. The mean drop size, velocity and spacing is affected directly by the flow reduction in specifically different ways depending on the location of the obstruction. An analogy to this is the difference in the jet stream of a garden hose whose flow has been throttled to a given value by constriction at the faucet or at the nozzle.

A major effort is needed to determine how much of this information, imparted to the stream by the canal, remains in useable form. We have performed auto- and cross-correlations, Fourier power spectra and coherence functions, and time interval distribution analysis to a limited extent on Drop Spectrometric data from models, patients and normal male subjects. Each type of analysis has shown particular sensitivity to certain properties of artificial streams (RITTER *et al.*). Moreover, the common obstructions of benign prostatic hypertrophy and meatal stenosis are clearly and sensitively differentiated with this empirical analysis.

A more fundamental study of the information of the drops is in progress. Although the human stream is more complex, certain aspects of the drop patterns are explained by the theory of capillary instability of liquid jets as developed by Lord RAYLEIGH, in 1879, advanced by NIELS BOHR in 1909 and by others since. This theory tells how to relate the mean values of the drop patterns from a simple laminar jet to the dimensional physical and dynamic variables of the initially unbroken stream. An outline of this theory is given in reference RITTER and STERLING and ZINNER and HARDING.

The fault of this theory is that it makes an excessively simple assumption about the initial conditions of the free cast stream just as it leaves the jet. Thereafter, the breakup pattern is set, and is limited to determination of only the mean drop size, spacing and velocity.

But what is of interest in the human is the variability of these quantities, i.e., their distributions which can be measured, and a mathematical way to connect them to the properties of the conduit casting the stream.

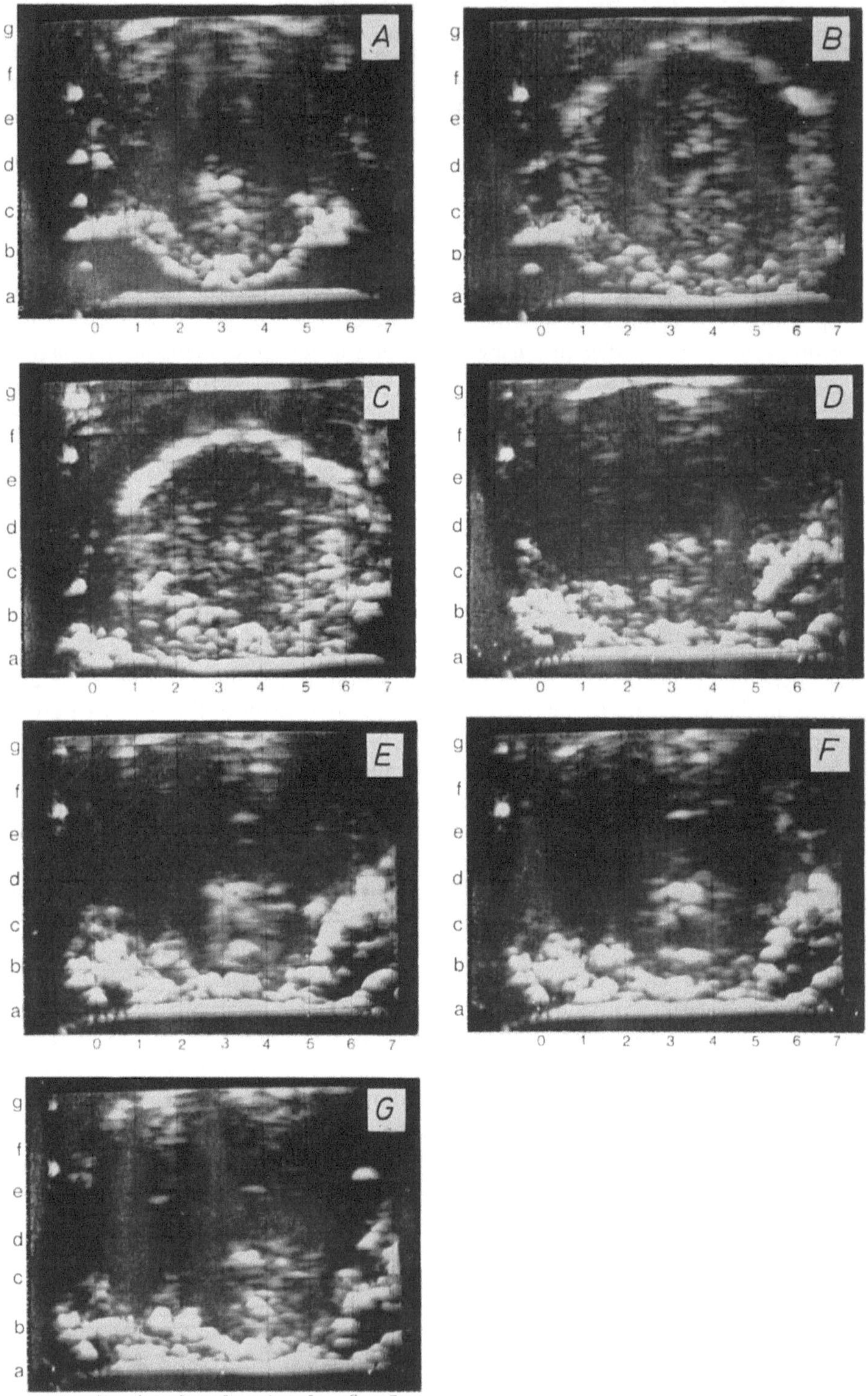

Fig. 8

The present theory treats the radius of the jet as a quantity which deviates from a constant value, a by a quantity η

$$r = a + \eta\ (z, \theta, t) = a + \eta_0\ Re\ \{e^{\beta t + i\,(kz + n\theta)}\}, \tag{1}$$

where standard cylindrical coordinates are used. The equations of continuity and motion are solved subject to three boundary conditions:

i) $$\left(\frac{\delta \varphi}{\delta r}\right)_{r=0} = 0, \tag{2}$$

ii) $$\left(\frac{\delta \varphi}{\delta r}\right)_{r=a} = \frac{\delta \eta}{\delta t} \tag{3}$$

iii) $$p = \sigma\left(\frac{1}{R_1} + \frac{1}{R_2}\right), \tag{4}$$

with notation as in reference Ritter and Sterling and Zinner and Harding.
In one new formulation we use new boundary conditions, evaluated at $z = 0$ and $t = 0$ as follows:

i) $$\eta\ (0, \theta, 0) = \sum_{nk} R_{nk} \cos\,(n\,\theta + \delta_{nk}), \tag{5}$$

ii) $$\left(\frac{\delta \varphi}{\delta r}\right)_{z=0,\,t=0} = \sum_{nk} V_{nk}\,(r) \cos\,(n\theta + \varkappa_{nk}), \tag{6}$$

iii) $$\left(\frac{\delta \varphi}{\delta z}\right)_{z=0,\,t=0} = \sum_{nk} W_{nk}\,(r) \cos\,(n\theta + \varepsilon_{nk}), \tag{7}$$

where we relax certain of the previous boundary conditions, as necessary.
In equation 5 the R_{nk} and δ_{nk} are given at the outset, and define the shape of the meatus. The V_{nk} and W_{nk} are components of the radial and axial velocity fields at $z = 0$ and are determined from solving the hydrodynamic problem of the fluid in the urethra or from models. If a time-varying urethral shape is taken to be the case, these quantities will also be function of t.
In principle, this analysis allows a complete connection ot the external stream to the internal one, assuming irrotational flow. A computer solution will, in most cases

---

Fig. 8. Cross sections of human penis at three different locations (A–C) and during different phases of urination at single location (D–G). Corpora cavernosa appear as black spaces and corpora spongiosum urethra appears as a region of white fluffy spots. Ventral surface of penis is indicated by heavy transverse white line. A Mid shaft of penis – not voiding. Note corpora spongiosum urethra in region (3—4, (b–c). Transverse line on left at level between b–c is artifact. Dorsum of penis shows at very top of figure. Compare urethral region with that in Fig. 8F. B Corona of penis – not voiding. Note increased mass of corpora spongiosum (2–5), (b–f) and decreased size of corpora cavernosa. C Mid glans penis – not voiding. Note that here, most of glans consists of corpora spongiosum urethra and the corpora cavernosa are represented by a small area dorsally. D Mid shaft of penis with voiding just starting. Compare to Fig. 8A. Note the urethra beginning to open (3–4), (b–c). E Mid shaft of penis with flow increasing. Note larger urethral cross-section. F Mid shaft of penis during maximum flow. Note urethral size now nearly fills space (3–4), (b–c). G Mid shaft of penis at end of urination. Note urethra is again closed

be necessary in order to provide this detailed basic connection of urethral properties to distributions of drop intervals, velocities, etc.

The significance in the theoretical interpretation of the cast human stream is twofold. First, we are freed from the purely empirical analysis presently being carried out, with the hazards of interpretation inherent in such empiricism. Second, the ability to understand effects of geometry on initial velocity profile at the meatus should lead to an understanding how to use mechanical devices with well specified interference effects of a minimal nature so that the method can be extended to the human female.

Summary and Conclusions

In this manuscript we have described some of the hydromechanical principles of local velocity, how it can be measured, and the significance of variations in local velocity in disease. Measurement of local flow velocities within the urethra may provide the information needed to understand the mechanisms of urethral function and the role of the flowing stream in preventing or permitting outflow-tract disorders.

References

Albright, R. J., Harris, J. H.: Ultrasonic Scattering by the Urethra from: Hydrodynamics of Micturation, Chapter 22, page 353. Hinman, F., Jr. (Ed.), Springfield, Ill.: Charles C. Thomas Publ. 1971.

Albright, R. J., Harris, J. H., Zinner, N. R.: Transcutaneous Measurement of the Urethral Lumen. Invest. Urol. **8**, 340 (1970).

Bohr, N.: Determination of the Surface Tension of Water by the Method of Jet Vibration. Phil. Trans. B. A **209**, 281 (1909).

Boys, Sir C. V.: Soap Bubbles and the Forces Which Mold Them, p. 161. Soc. for Promoting Christian Knowledge, London 1896 and Modern Paperback reprints.

Cox, C. E., Lacy, S. S., Hinman, F., Jr.: The Urethra and its Relationship to Urinary Tract Infection. II. The Urethral Flora of the Female with Recurrent Urinary Infection. J. Urol. (Baltimore) **99**, 632 (1968).

Fredburg, J. J., Lees, R. S., Dewey, D. F.: How to Listen to Arteries (Or What Your Doctor Would Hear if He Were a Fluid Dynamicist). Preprint, AIAA Eighth Aerospace Sciences Meeting 1970.

Gessner, F. B.: The Fluid Dynamics of Micturation. Invest. Urol. **8**, 324 (1970).

Gessner, F. B., Zinner, N. R.: On the Measurement of Local Urethral Resistance I. Analytical Considerations. Invest. Urol. **8**, 331 (1970).

Hedges, J. R., Sterling, A. M., Decher, R.: Jet Tones from Submerged Square Edged Orifices (in Preparation).

Lloyd-Davies, R. W., Hayes, T. L., Hinman, F., Jr.: Urothelial Micro-Contour. I. Scanning Electron Microscopy of Normal Resting and Stretched Urethra and Bladder. J. Urol. (Baltimore) **105**, 236 (1971).

Rayleigh, Lord: On the Capillary Phenomena of Jets. Proc. roy. Soc. Med. **29**, 71 (1879).

Reid, J. M., Barber, F. E., Nation, A. W. C., Mozerski, D. J., Strandness, D. E., Jr.: Ultrasonic Rotational Fast Scanner., Proc. of Conf. on Eng. in Med. and Biol. Las Vegas, Nevada, Oct–Nov. 1971.

Ritter, R. C., Zinner, N. R., Sterling, A. M.: Clinical Urodynamics VII: Statistical Analysis of Fine Properties of Urination (in Preparation).

Ritter, R. C., Zinner, N. R., Paquin, A. J., Jr.: Clinical Urodynamics II: Analysis of Pressure-Flow Relations in the Normal Female Urethra. J. Urol. (Baltimore) **91**, 161 (1964).

a) RITTER, R. C., STERLING, A. M.: Exposition of Pertinent Principles of Mechanics and Hydromechanics, from Hydrodynamics of Micturation, Chapter 9, Page 103. HINMAN, F., JR. (Ed.). Springfield, Ill.: Charles C. Thomas Publ. 1971.

b) ZINNER, N. R., HARDING, D. C.: Velocity of the Urinary Stream, from Hydrodynamics of Micturation. Chapter 16, page 186, HINMAN, F., JR. (Ed.). Springfield, Ill.: Charles C. Thomas Publ. 1971.

SILVERBLATT, F. J.: Personal Communication, 1970.

STERLING, A. M.: Consideration of Flow Development Within the Urethra (in Preparation).

TOMEY, G. F., REID, J. M.: Rotational Compound Scan. Chicago, Ill.: Proc. Eighth ICMBE, July 1960.

ZINNER, N. R., RITTER, R. C., STERLING, A. M., HARDING, D. C.: High Speed Cinematography of Human Urination. Invest. Urol. **6**, 605 (1969).

ZINNER, N. R., RITTER, R. C., STERLING, A. M., HARDING, D. C.: Drop Spectrometer: A Nonobstructive, Noninterfering Instrument for Analyzing Hydrodynamic Properties of Human Urination., J. Urol. (Baltimore) **100**, 914 (1969).

ZINNER, N. R., RITTER, R. C., STERLING, A. M., HARDING, D. C.: Clinical Applications of Drop Spectrometry as a Nonobstructive, Noninterfering Method for Analyzing Hydrodynamic Properties of Human Urination. J. Urol. (Baltimore) **102**, 485 (1969).

Correlation of Flow Rate Profile with Diseases of the Urethra in Man

F. B. SCOTT *

Uroflowmetry is a rapid, painless, physiologic means of obtaining informative, objective, inexpensive data on the patient with voiding difficulties. It serves as a screening test which saves the physician a great deal of time and provides a reference with which to follow the patient.

Materials and Methods

Patients: Studies were done on patients treated for various urologic problems at the St. Luke's Episcopal Hospital over the past three years. Uroflowmetry has been used routinely in all of the author's patients, but the bulk of the patients were those of some fifteen urologists who practice in this hospital. All patients were studied on written request by the attending physician. Each record was interpreted, and a reproduction of the actual recording was placed on the patient's chart. The author had the patient's sex, age, admission diagnosis, and the recording from which to render an interpretation which was subsequently correlated with findings of a complete urologic workup.

Equipment: The Scott Uroflowmeter[1] is an electromagnetic flowmeter based on Faraday's Law which states that an electrical potential will be developed across a conductor (in this case, urine) which is moving through a magnetic field (CARDUS *et al.*). The magnitude of the induced voltage is proportional to both the electromagnetic field strength and the velocity of the moving conductor. Since voltage is proportional to velocity rather than electrolyte concentration of the conductor, one need not worry about the variable concentration of urine in one patient versus another. The electromagnetic flow sensor is placed in the bottom of a toilet bowl located in the privacy of a small room and connected to a recording module in an adjacent room. A rectilinear, heat writing recorder provides an immediate graph which represents the flow curve, or flow rate profile which can be used for comparison to subsequent recordings in the same, or other person.

---

* This work is supported through contributions of the Albert B. Alkek, and the R. D. McDonald Funds, St. Luke's Episcopal Hospital

[1] The Scott Uroflowmeter, model 1001, LT Instruments, Inc., 5720 Greenash, P.O. Box 36221, Houston, TX 77036

For privacy the patient encloses himself in a small room which contains the toilet chair. The patient is instructed to void in his usual manner when he experiences an urge. He is instructed not to increase pressure by abdominal straining. Unless the first voiding event is obviously normal, the patient is requested to return two additional times since multiple determinations result in greater accuracy than a single observation.

Since uroflowmetry serves mainly as a screening test, some patients whose flow profiles are not clearly abnormal may undergo combined cystometry, uroflowmetry, electromyography, and cinecystography depending upon other findings (SCOTT *et al.*). Should additional complex urodynamic studies be required, however, the pre-existent flow data serves well for comparison to check on the credibility of measurements taken later when the pain and fear of instrumentation might keep the patient from voiding in his usual manner.

Results

During the past three years a total of 5,700 patients have had 9,230 uroflowmetry studies (Table 1).

Table 1. Voiding Flow Rates — St. Luke's Hospital

	1968	1969	1970
Number of determinations	2,754	3,018	3,463
Men	880	969	1,267
Women	717	851	933
Number of patients	1,597	1,920	2,190
Flows/Patient	1.7	1.5	1.6

During this interval of time the number of studies of the male have been slightly greater than of the female. The number has steadily increased each year and indicates that this test must be of value to the practicing urologist, or else he would not have ordered it on increasing numbers of patients. The number of flow profiles averages about 1.7 per patient.

The flow profiles of the same healthy person appear to belong to the same family of curves and the amplitudes are generally proportional to the volume (Fig. 1); however, there is considerable variance in digital data. Note, for example, that peak flows obtained from 63 voidings vary considerably at similar volumes. As one reviews the digital and graphic data in the same subject the similarity of flow rate profiles (Fig. 1) is more striking than digital information (Fig. 2). This was also true in studies obtained on a younger technician as well as on a group of healthy male controls. My impression is that although the flow appears to be proportional to the volume, the flow rate more closely relates to the degree of urgency which generally, but not always, parallels the volume. Uroflowmetry then, while an excellent screening test, is not an accurate tool with which to differentiate the normal from the abnormal. This is particularly so if one looks only at peak or mean flow measurements.

In addition to peak flow, mean flow, volume, and the interval required for voiding, the flow curve also provides an additional attribute. That attribute is the flow rate profile; the profile captures with a single image an entire action that may take several minutes to complete. The flow curve or flow rate profile, therefore, is truly a profile of an entire voiding event. The value of this becomes apparent as one looks at a number of profiles on various patients and accumulates experience and familiarity.

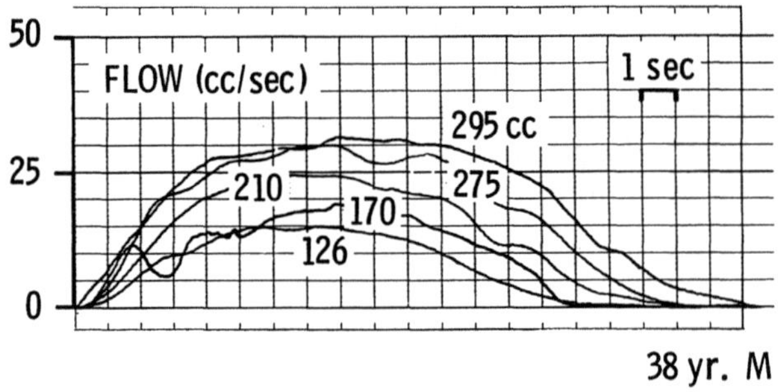

Fig. 1. Flow rate profiles in the same healthy person are strikingly similar. Peak flow appears to correlate well with volume. Note the variance, however, in the peak flows from the same person in Fig. 2

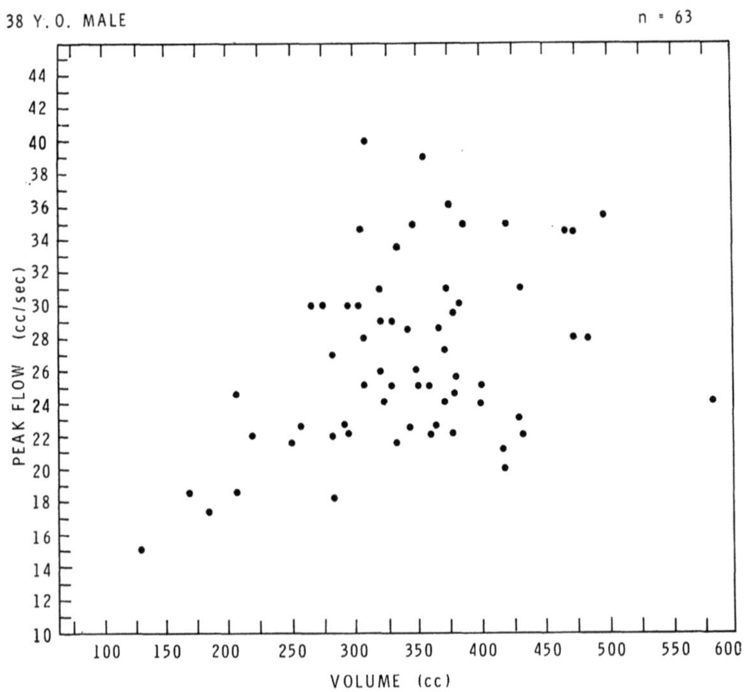

Fig. 2. The peak flows of sixty-three voiding events obtained on the same person whose profiles appear in Fig. 1 are quite variable, even at the same volume. For example, at a volume of 300 ccs., the peak flow ranges from 25 to 41 cc/sec

Obstruction Due to Prostatism

The peak flow of a patient with prostatism may fall within a normal range, particularly if the patient initially pinches the distal urethra or increases pressure by abdominal straining, but the flow rate profile will generally disclose the obstruction in spite of this maneuver The peak flow in patients with prostatism has been compared to healthy men in prior publications (SCOTT, SCOTT *et al.*). Though the flows are lower in the obstructed patient, there is considerable overlap with digital data of the normal. The flow rate profiles, on the other hand, are more diagnostic.
The flow rate is dramatically improved after a well done transurethral resection of the prostate. This improvement is usually apparent even on the day the catheter is removed.

Urethritis and Prostatitis

Review of the records of men with the clinical diagnosis of urethritis or prostatitis failed to disclose any typical flow rate profile characteristics. Indeed most patients had normal flows whereas some patients actually had superflows. There were isolated cases of men whose flow rate profiles were poor until the administration of antibiotics brought about improvement.

Urethral Stricture

The flow rate profile recorded from a patient with a small caliber stricture is typically prolonged and flat, usually without a distinctive peak flow (the maximal amplitude is constant). It is unusual to see a patient with prostatic obstruction produce a prolonged flat profile; such patients usually develop acute urinary retention before developing poor flow rates so commonly seen in stricture disease. Furthermore, the amplitude of the curve is fluctuant in prostatism rather than constant as it tends to be in stricture disease.
The 41-year-old patient whose flow rate profile is shown in Fig. 3 had had a poor urinary stream for several years. The improvement following dilation of an urethral stricture is dramatic, and serial subsequent flow rate profiles serve to illustrate re-

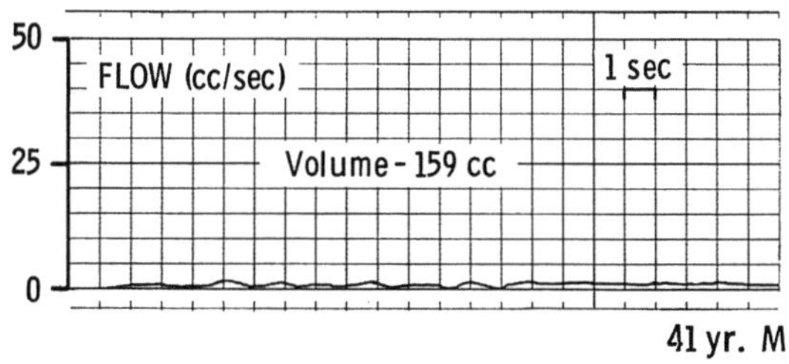

Fig. 3. The flow rate profile resulting from severe stricture is typically prolonged and flat. Almost three minutes were required for this patient to void only 159 cc

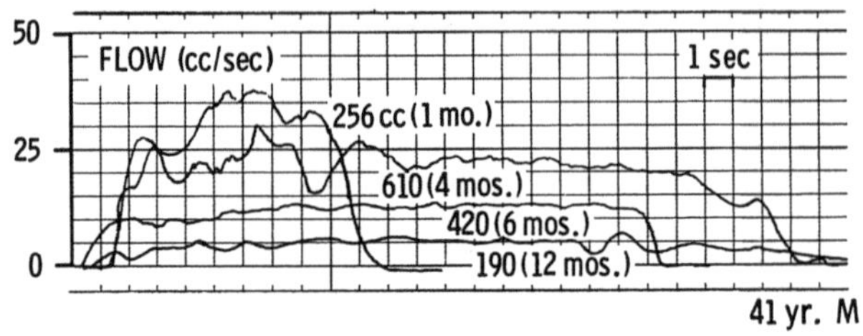

Fig. 4. Dilation of an urethral stricture brings about immediate improvement in the flow rate profile. Uroflowmetry at various times helps one to select the appropriate time for another dilation

gression of the stricture (Fig. 4). The flow rate profile is helpful in selecting the propitious time for subsequent dilation, rather than falling into a routine which results in unnecessary dilations. For example, the ideal time interval for dilation in this patient is about six months, at which time his flow has decreased to 14 cc/sec. If the patient is dilated routinely at three month intervals he will receive twice as many dilations as are necessary.

Unusual Problem Cases

Anterior Urethral Valve: M. S., a 9-year-old boy, developed an acute urinary tract infection as his presenting urologic complaint. Flow rates were distinctly abnormal. A voiding cine was normal (though in retrospect the entire urethra was not examined). Cystoscopy was normal except for bladder trabeculations. Pressure-flow studies showed significantly increased resistance (P/F^2) (Table 2).

Table 2. Micturition Study

M. S. 9 yrs.	P mm Hg	F cc/sec	P/F^2	Residual cc
	70	1.5	31.0	145
	48	1.0	48.0	100
	47	1.9	13.0	112

Electromyography showed appropriate relaxation of the sphincter coincident with voiding indicating that a spastic sphincter was not the cause of his obstruction. A repeat voiding cinecystogram with instructions to the examiner to observe the entire urethra during voiding, disclosed an anterior urethral valve. After open resection of the valve, the flow rate profile is that of a healthy boy. If one assumes the same bladder pressures ranging from 47 to 70 mm Hg. as found preoperatively, one can use postoperative peak flow of 26 cc/sec to calculate a resistance of 0.06 to 0.10, a value well within the normal range (Table 3).

Table 3. Postop Values Based on Assumed Pressures

M. S. 9 yrs.	P mm Hg	F cc/sec	P/F^2
	45	26	.06
	70	26	.10

Reflux Caused by Bladder Neck Obstruction : A 41-year-old man had lower obstructive symptoms of many years duration. In the two months before seeking medical care, enuresis occurred for the first time in his life. He recalled, from childhood, episodes of severe transient pain in his left side coincident with voiding. His post voiding residual was measured to be 400 ccs. The first sensation to distension appeared at a

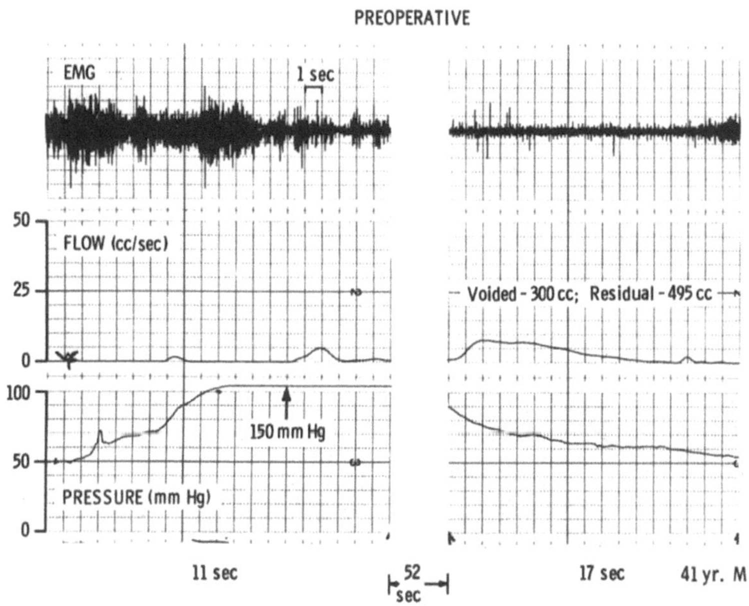

Fig. 5. A micturition study ruled out neurogenic bladder but revealed tremendously increased pressures which must have caused decompensation of the vesico-ureteral valve and hydronephrosis of the left kidney

volume of 350 ccs., and an urge to void occurred at 900 ccs. His flow rate profile was intermittent and of low amplitude, suggesting several possibilities: detrusor atrophy or megacystis syndrome, neurogenic bladder, or outlet obstruction. Cystoscopy revealed moderate trabeculations and possibly "median bar" hypertrophy. His voiding cystogram revealed prominent left reflux and transient right reflux. A micturition study (combined cystometry, uroflowmetry, and sphincter electromyography) (Fig. 5) revealed: 1. normal relaxation of the sphincter coincident with detrusor contraction, ruling out neurogenic disease, 2. maximum pressure ranges from 80 to 150 mm Hg (on the first event the patient strained, in his usual manner, causing a pressure of 225 mm Hg). These results clearly indicated bladder neck obstruction. Three days following transurethral resection of his bladder neck, a pressure-flow study revealed excellent postoperative results. Histologic studies showed "nodular smooth muscle hypertrophy". Pre and postoperative numerical data is impressive and shows postoperative resistance (P/F^2) to be well within normal range (Table 4).

Table 4

	Pre-Op			Post-Op	
	# 1	# 2	# 3	# 1	# 2
Pressure	225	150	80	64	66
Flow	9	9	10	37	32
P/F^2	2.7	2.8	0.7	0.4	0.6

The above problem case illustrates the need for measuring pressure as well as flow for accurate diagnosis. Furthermore, I would not have felt confident in resecting the bladder neck of this 41-year-old man had I not known that his sphincter functioned normally. A dyssynergic sphincter was eliminated as the cause of obstruction by combining electromyography with cystometry and uroflowmetry.

The opposite result was found in an 8-year-old boy who had had a successful hypospadius repair, but who proved to have enuresis and reflux with deteriorating renal function. A micturition study disclosed an explanation for why previous serial flows had been poor. Although cystoscopy showed no obstruction, the pressure-flow values did indicate obstruction; electromyography proved the sphincter to be the cause (Fig. 6). My interpretation of this phenomenon is that there is detrusor-sphincter dyssynergia probably secondary to a neurogenic disturbance.

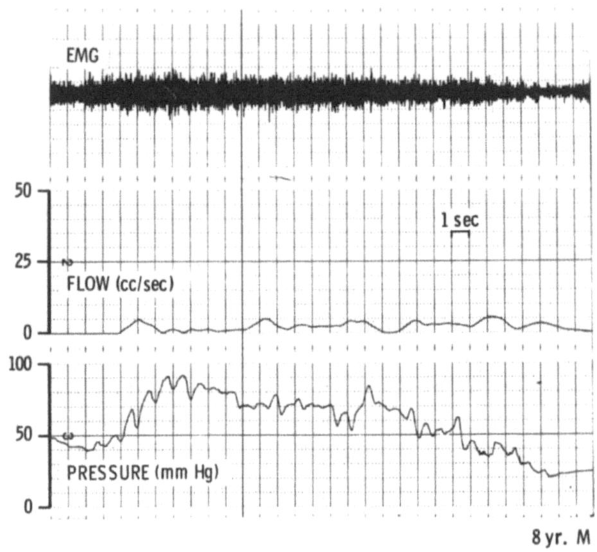

Fig. 6. Resistance to the flow of urine may be caused by sphincter spasm. This can be detected by combining electromyography of the sphincter with other urodynamic studies

Discussion

I have found the electromagnetic uroflowmeter to be an accurate and reliable method of determining the velocity profile of voiding. This method is not without its problems, however: a) There is a short timelag since the recording begins only after urine falls from the urethral meatus to the flow sensor rather than at the instant urine flows through the meatus. b) A "wag" artefact is created if the patient directs the stream alternately between the walls and the drain hole of the toilet bowl. This is not a significant artefact at high flow rates, however, at lower flow rates usually seen with very small volumes, particularly in children, this artefact is of greater concern. c) Urinary streams of great velocity may create foam and trap air so that it too passes through the flow sensor; this causes a very high "blip" on the flow profile

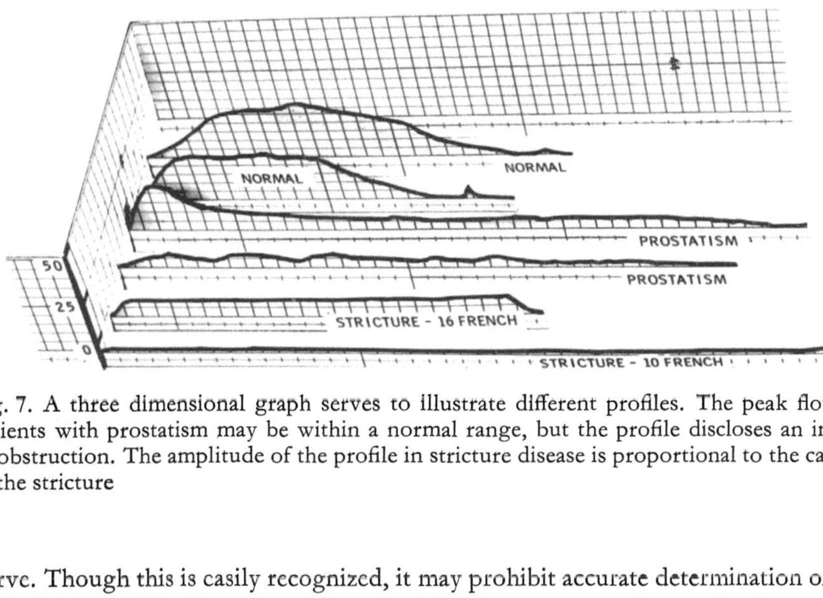

Fig. 7. A three dimensional graph serves to illustrate different profiles. The peak flow of patients with prostatism may be within a normal range, but the profile discloses an image of obstruction. The amplitude of the profile in stricture disease is proportional to the caliber of the stricture

curve. Though this is easily recognized, it may prohibit accurate determination of the peak flow. d) The flow sensor is a precision instrument and must be kept clean, otherwise abnormalities can creep into the system. However, the particular equipment now in use shows the most stable properties of any that we have tested, and calibration shows little or no change from day to day and week to week. The manufacturer (LT Instruments, Inc.) has attempted to improve on this flowmeter even to the point of employing entirely different physicochemical-electrical concepts and, as yet, has not been able to improve significantly on this flowmeter.

I have found the flow rate profile to be helpful increasingly in my own urologic practice; I have paid less attention to the numerical values of mean or peak flow presumably because I have learned to see more information in the profile. The flow rate profiles also assume similar characteristics for a given condition affecting the urethra (Fig. 7). This information is valuable in diagnosis and treatment of the patient, but the combined use of cystometry, uroflowmetry, and electromyography will more clearly differentiate various kinds of voiding disorders.

Summary

The electromagnetic uroflowmeter offers the advantages of convenience, objectivity, and accuracy in evaluating patients with voiding disorders. By the use of such equipment, a graphic display of instantaneous flow throughout voiding is obtained, giving as a single image the profile of an entire voiding event. This flow rate profile adds a new dimension to the value of uroflowmetry as a diagnostic tool.

References

Cardus, D., Quesada, E. M., Scott, Brantley, F.: The Use of an Electromagnetic Flowmeter for Urine Flow Measurements. J. appl. Physiol. **18**, **4**, pp. 845–847 (1963).

Scott, F., Brantley, Queseda, E. M., Cardus, D.: The Use of Combined Uroflowmetry and Cystometry and Electromyography in Evaluation of Neurogenic Bladder Dysfunction. Neurogenic Bladder, pp. 106-114, Ed.: S. Boyalsky. Baltimore: Williams & Wilkins Company 1967.

Scott, F. Brantley: Uroflowmetry in the Evaluation of Voiding Disorders. Med. Rec. (Houston) **60**, pp. 263–265 (1967).

Scott, F. Brantley, Cardus, David, Queseda, E. M.: Uroflowmetry Before and After Prostatectomy. J. S. Med. Ass. **60**, pp. 948–952 (1967).

Resistance Measurements in the Human Urethra

M. R. Bottaccini *, D. M. Gleason **, and J. C. Byrne ***

Introduction

A number of new experimental and clinical results have been obtained through the use of the voiding dynamometer. The investigators have adapted the rationale of the hydraulician and have discovered concepts which would otherwise have remained hidden. Since the vocabulary of the hydraulics engineer and the urodynamicist are often mutually incomprehensible it is necessary prior to the introduction of the results to present a review of the basic theory and the pertinent physical principles.

1. *Basic Concepts.* It has been accepted for a number of years that urodynamic computations involve the three universal principles of conservation of energy, mass and momentum. Such computations lead directly to the modified Bernouilli equation, to the classical undimensional equation of continuity and to the momentum theorem. The mere application of hydraulic equations is not particularly useful to the understanding of urethral dynamics unless the restrictions imposed by natural laws are understood qualitatively.

The study of resistance in micturition is a case in point, since "resistance" in engineering parlance is connected with *dissipation of energy* and with no other physical phenomenon. In urodynamics, unfortunately, a term for resistance has come into use which combines the effects of resistance and the effects of meatal valving in a single number. This computation leads to no conclusions since the quantity

$$R = \frac{P_B - E_R}{Q^2}$$

where:

R = improper resistance (dynes-sec/cm^5)
P_B = bladder pressure (dynes/cm^2)
E_R = energy per unit volume in voided stream (dynes/cm^2)
Q = flow rate (ml/sec)

does not lead to a characteristic number for a given subject.

* College of Engineering, Aerospace and Mechanical Engineering Department, University of Arizona, Tucson, AZ USA
** Bioengineering Division, Engin. Exper. Station, University of Arizona, Tucson, and the College of Medicine, University of Arizona, Tucson, AZ USA
*** Bioengineering Division, Engin. Exper. Station, University of Arizona, Tucson, and the Squier Urological Clinic, Columbia University College of Physicians and Surgeons, New York, NY USA

For example four consecutive tests on a single female patient (C. L. in Table 3) lead to values of R which differ by a factor of 100, whereas the actual dissipation of bladder energy is approximately 50% in all four micturitions (Table 1).

Table 1. Comparison of the traditional urodynamic resistance in four micturitions with energy dissipation

	R	L
1	0.449	0.48
2	0.0048	0.45
3	0.003	0.47
4	0.048	0.54

R = Traditional Resistance Parameter.
L = Fractional Loss.
(Test data taken from Table 3)

It is evident that the fractional loss for the subject is reasonably constant, whereas R varies drastically. The traditional resistance will not be discussed further and henceforth the loss will be tabulated in the form

$$L = \frac{P_B - E_R}{P_B}. \tag{1}$$

It will be shown that L is characteristic of the subject. Whenever it is required to write in terms of resistance, rather than loss, we can note that the Bernouilli equation for assumed steady flow has the form, (if the difference in elevation between the bladder and the meatus is included in P_B) (4)

$$P_B = E_R + K{}^1/_2\,\varrho\, V_E^2 \tag{2}$$

where:

K = resistance coefficient (dimensionless)
V_E = average velocity of voided stream (cm/sec)
ϱ = density of urine (gm/cm³)

Since $E_R = {}^1/_2\varrho\, V_E^2$ it follows immediately that

$$K = \frac{L}{1-L}. \tag{3}$$

Either K or L serve as a resistance parameter with the understanding that K represents resistance and L represents fractional energy lost. The word "lost" implies that the energy is no longer available to propel the urine and either has been dissipated or else has been used to generate turbulence and local eddies in the fluid.

2. It is obvious that the energy balance (equation 2) does not alone permit the computation of *flow*. If the bladder energy per unit volume P_B is established by the detrusor and by the pressure generated by the viscera and K is a parameter of the subject, then the *velocity* is established by

$$V = \sqrt{\frac{2P_B}{\varrho\,(1+K)}}. \tag{4}$$

The flow is obtained by the concept of conservation of mass (continuity)

$$Q = VA \tag{5}$$

where:

Q = flow rate (ml/sec)
A = crossectional area of the vena contracta (mm²).

(Note that A is approximately equal to the area of the meatal opening.) It follows from equations 4 and 5 that the flow rate depends on bladder pressure, crossectional distal area and resistance.

One point to note is that a change in area can generate a direct change of flow, whereas a much greater change of resistance is needed to effect an identical change of flow (BOTTACCINI and GLEASON). The effectiveness of distal area as compared with resistance is computed with equations 4 and 5 and shown in Table 2. Note that a 70% decrease in flow can occur with 70% decrease of area or with a 1000% increase in resistance.

Table 2.

Reduction in Flow	Area changes needed to cause flow change if resistance = 0	Resistance change needed if distal area is held fixed
%	%	%
10	10	75
50	50	271
70	70	948

3. Resistance is the result of viscosity and can be connected with the following causes:

a) Boundary shear: near a material-boundary, fluids move at the speed of the boundary. With a stationary boundary the fluid is at rest near the boundary. To bring the boundary fluid at rest, energy must be withdrawn from the moving stream, thereby giving rise to *wall shear resistance* (HANSEN). Such resistance depends on fluid viscosity, fluid average velocity, fluid density and wall roughness.

b) Separation: Whenever a stream changes direction or shape the fluid has difficulty conforming to the boundary. In such instances the liquid will *separate* from the boundary, thereby leaving eddies between the wall and the fluid boundary. The motion of such eddies is maintained by withdrawal of energy from the main stream. The more abrupt the geometry change the larger is the loss, and hence the larger the resistance (HANSEN).

4. With the preceding concepts in mind it is possible to establish a number of principles of urinary flow. These principles *cannot be violated*: any experiment in contradiction of these ideas must either be wrong or else misunderstood.

Principle No. 1. An interior change in crossectional area, that is, a change of crosssectional area anywhere except at the meatus, can affect the flow rate only by changing "resistance" and thus changing the flow energy. Moderate changes in internal geometry generate only minor flow fluctuations.

Principle No. 2. For a given *fixed flow energy* (which implies fixed internal geometry) established by the lower urinary tract, the flow rate will vary directly as the most distal crossectional area.

Principle No. 3. All bladder pressure, no matter from what source derived, is expended in distending tissues, overcoming resistance and in the maintenance of fluid velocity.

Equipment and Methods

The urethral dynamometer (Byrne *et al.*) (Fig. 1), a modification of the Whitaker and Johnston apparatus is used. The apparatus consists of a vertical disc mounted on a Statham UC3 force transducer and a UC3 Statham force detector attached to a Statham UL4 accessory in such a fashion as to weigh the urine continuously in a collection beaker. Electric differentiation is used to transform the weight output to flow rate. Bladder pressure is acquired through an indwelling catheter introduced suprapubically and connected to a Sanborn 400 mm Hg range 267A pressure transducer. All outputs are recorded on a Gould/Brush 220 six-channel stripchart recorder.

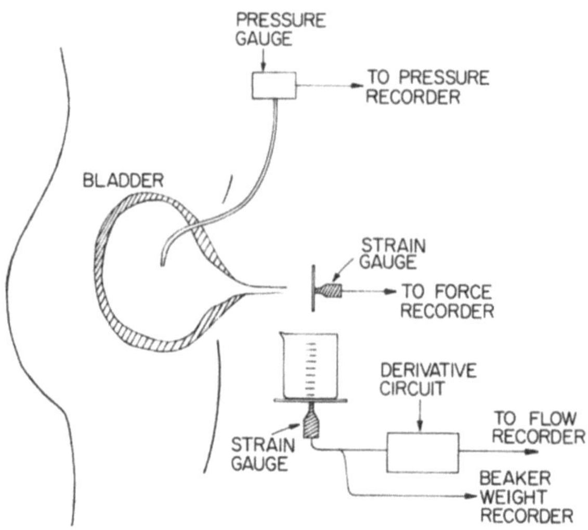

Fig. 1. Schematic representation of the dynamometer

A pressure reference datum is established at the height of the meatus after the patient is placed in voiding position. Male patients stand in front of the dynamometer with the distal urethra horizontal and the meatus approximately 5 cm from the disc. Females are seated on a modified "Spanish" labor chair so as to rotate the perineum in such a fashion that the DUS is horizontal. A plastic ring shaped to fit within the labiae, is inserted between the labiae minora to free the meatus from labial interference. The insertion of the speculum and the upward rotation tend to place the subject in an unnatural position and seems to affect the voiding pattern. After some experience, however, female subjects void normally, or at least generate consistent data from voiding to voiding.

Operating Theory of Dynamometer

The voided fluid strikes the disc and is deflected approximately 90° to the beaker. A force F is measured on the inpact disc and a flow rate Q is computed continuously from the beaker weight. Then from the momentum theorems of fluid mechanics we obtain:

$$F = \varrho Q V_E \tag{6}$$

where F = force in dynes. Since the residual energy is given by

$$E_R = {}^1/_2 \varrho V_E^2$$

it follows that

$$E_R = \frac{1}{2\varrho}\left(\frac{F}{Q}\right)^2 . \tag{7}$$

From the continuity equation we determine the crossectional area of the vena contracta

$$A = \varrho\left(\frac{Q^2}{F}\right) . \tag{8}$$

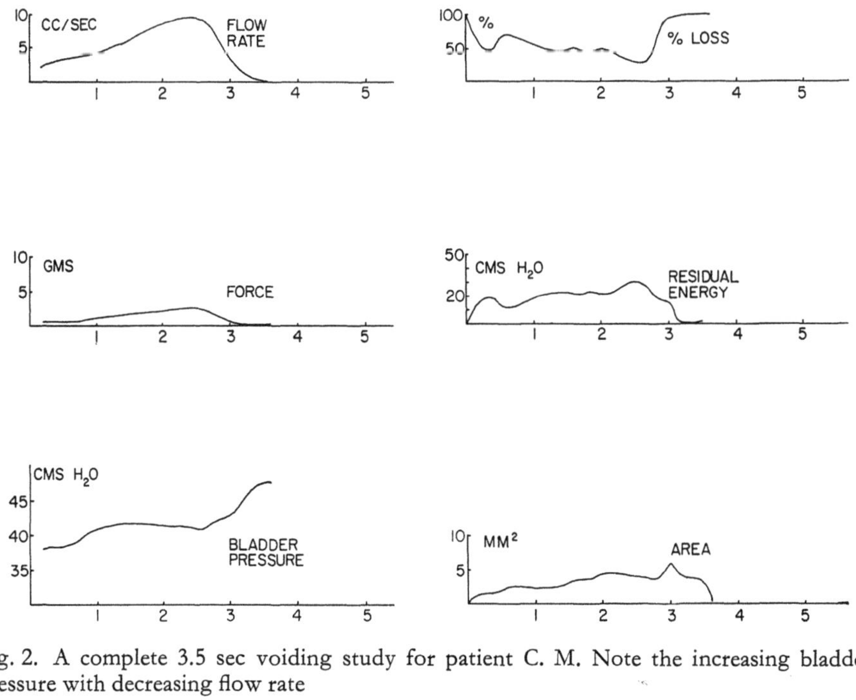

Fig. 2. A complete 3.5 sec voiding study for patient C. M. Note the increasing bladder pressure with decreasing flow rate

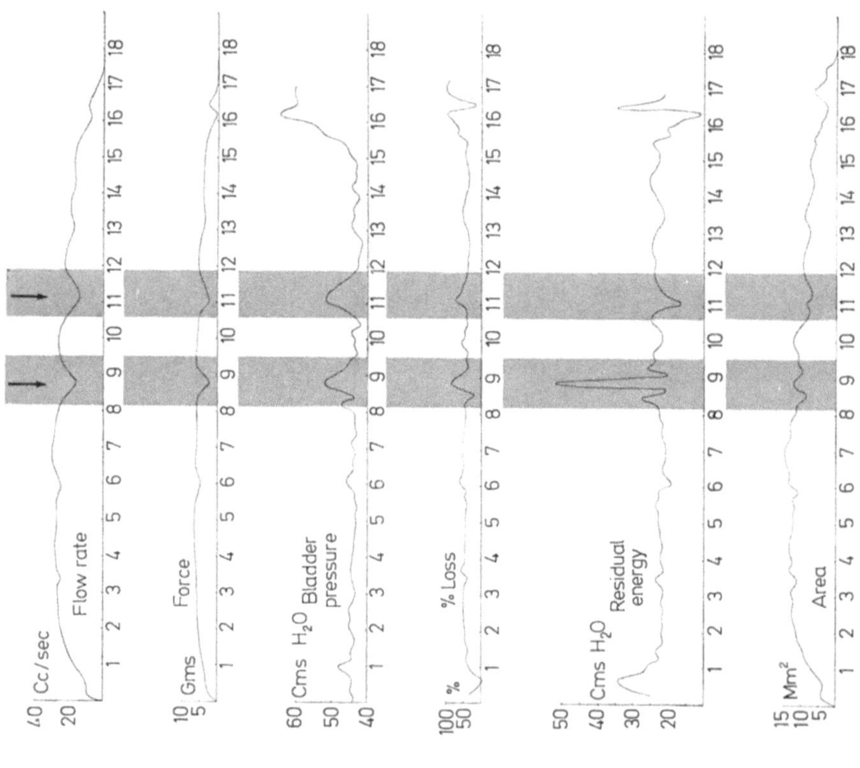

Fig. 3. Patient C. L. (2): a complete voiding lasting 18 sec. Shaded bands mark "events" which occur during micturition

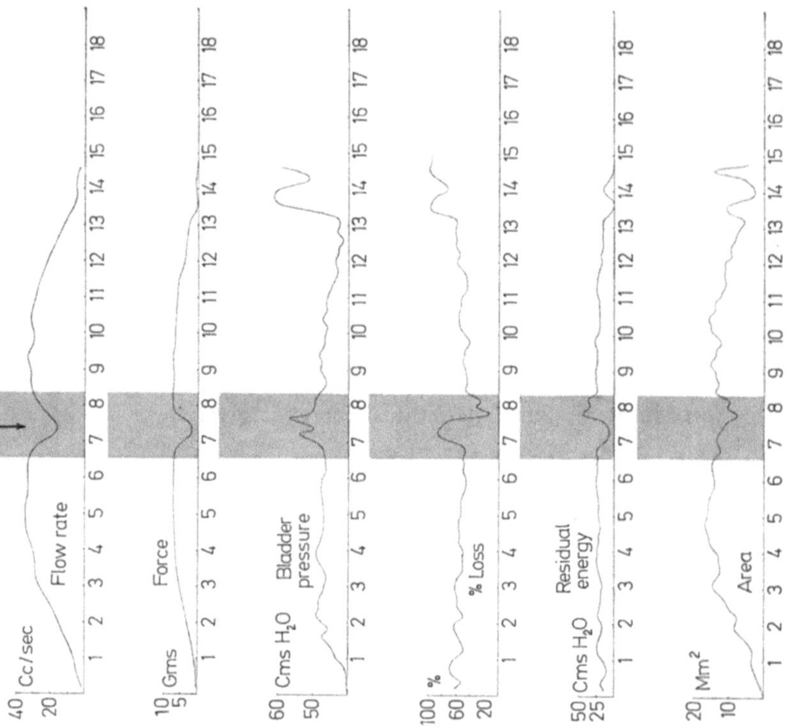

Fig. 4. Patient C. L. (3): a complete voiding lasting 15 sec. Shaded band marks an "event" which occurs during micturition

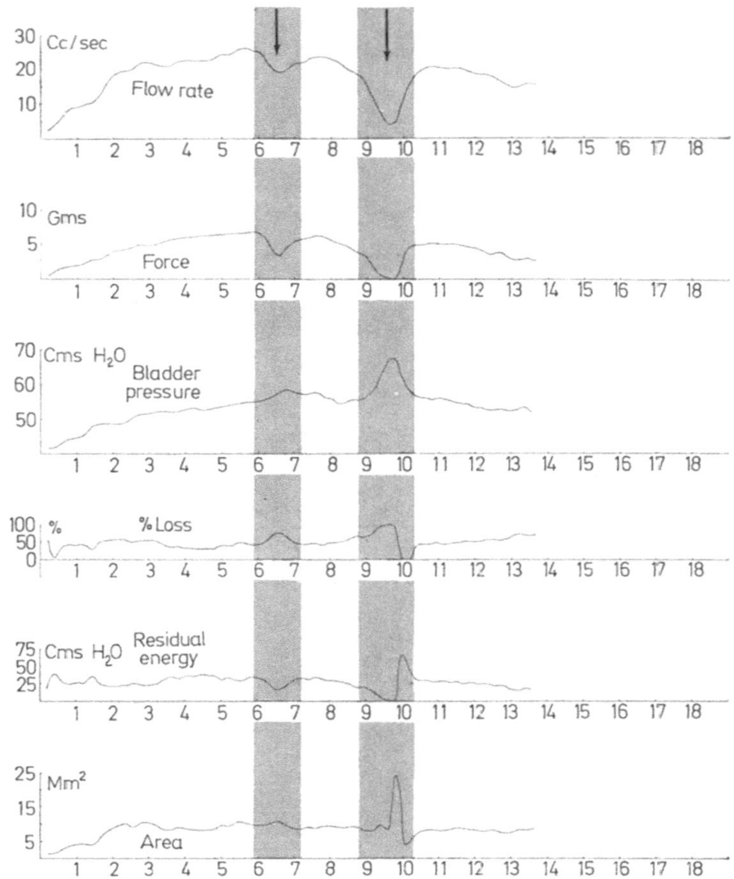

Fig. 5. Patient C. L. (4): a 15 sec portion of a voiding which lasted 24 sec. Shaded bands mark "events" which occured during micturition

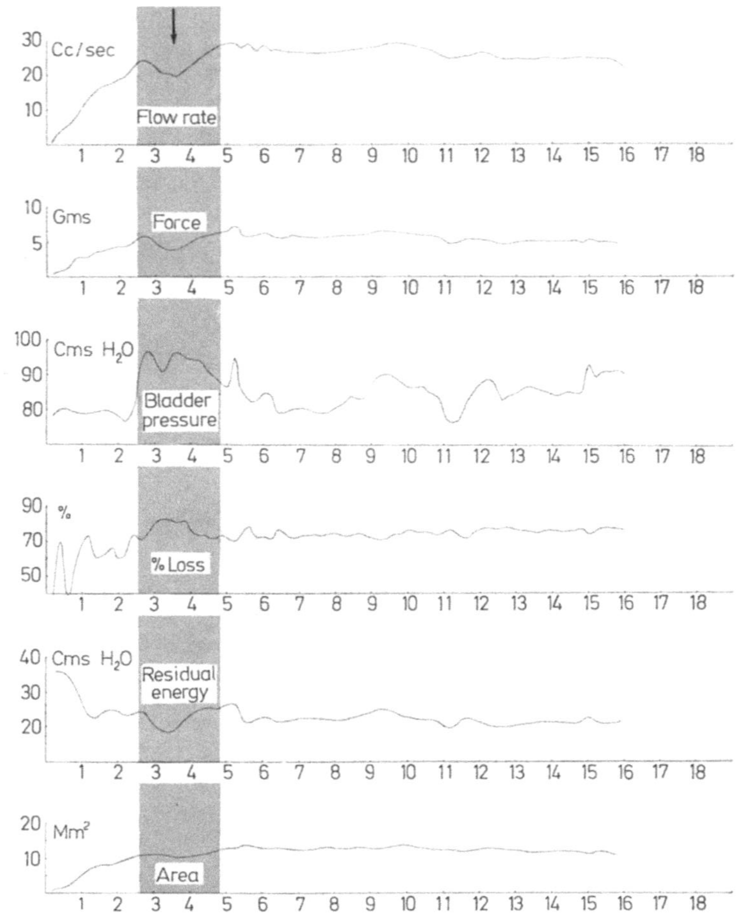

Fig. 6. Patient T. M. (1): a complete voiding lasting 16 sec. Band marks an "event" which occured during micturition

Equation 8, 9 yield all the information which can be extracted by the dynamometer. With the addition of the suprapubic puncture the loss parameter L can be computed by

$$L = \frac{p_B - E_R}{p_B}. \tag{9}$$

Nine patients, two females and 7 males, were tested with the results shown in Table 3. Tracings of the galvanometric records and of the computed data for all but one subject are presented in Figs. 2–9. The loss (in percentile form), the residual energy, and the meatal (vena contracta) crossectional area are presented together with the flow rate, disc force and bladder pressure. The complete voiding histories shown in Figs. 3–10 were obtained by manual reading of the galvanometer tracings and computation with a PDP 8 digital computer.

Fiftynine male subjects, patients and volunteers, were tested with the voiding dynamometer but without the bladder puncture. Flow rate and residual energy were plotted against each other (Fig. 10) and residual energy was plotted against age (Fig. 11).

Table 3

			Q cc/sec	p_B cms H_2O	E_R cms H_2O	A mm²	V cm/sec	L
C.L.	(1)	20 y.o. female	7	42	22	3.5	208	0.48
C.L.	(2)	asymptomatic	28	44	24	13	215	0.45
C.L.	(3)	one past episode	33	47	22	16	206	0.53
C.L.	(4)	of cystitis	20	54	28	8.5	236	0.48
M.H.		50 y.o. female distal urethral stenosis	20	81	39	7.0	286	0.52
F.M.	(1)	34 y.o. male bladder	27	81	21	13	208	0.74
F.M.	(2)	neck and meatal stenosis	13	89	15	7.5	172	0.83
J.G.		11 y.o. male frequency, dysuria and enuresis	40	90	31	16	250	0.66
W.H.	(1)	64 y.o. male retention	13	58	5.4	12	108	0.91
W.H.	(2)	post TURP	8	91	6.7	7.0	115	0.93
S.D.	(1)	57 y.o. male	17	157	2.0	27	63	0.99
S.D.	(2)	prostatism	8.8	161	1.6	10	88	0.99
S.D.	(3)		6.3	125	3.0	9	70	0.98
J.M.	(1)	47 y.o. male	3.9	94	3.9	4.3	90	0.96
J.M.	(2)	pendulous	5.0	108	4.8	5.0	103	0.96
J.M.	(3)	stricture	3.9	69	2.7	5.2	66	0.96
J.E.	(1)	15 y.o. male urethral	26	86	3.9	29	82	0.95
J.E.	(2)	valves	20	62	4.1	23	58	0.94
T.R.		36 y.o. male meatal stenosis chordee	8.6	90	30	3.5	245	0.67

5. Discussion. It is not possible, at this point to compare the, admittedly not normal, subject described in Table 3 with population norms. No measurements of loss (or resistance) appear in the current literature, although this difficulty shall soon be eliminated. The use of the urethral dynamometer, or of the hot wire anemometer (MELCHIOR *et al.*) or of a modification of the balloon pressure probe (DROUIN and MC CURRY) will soon yield large quantities of data and will define normal bladder pressures, residual energy and loss in the population. The best that can be done at the moment is to guess from such published experiments which are available.

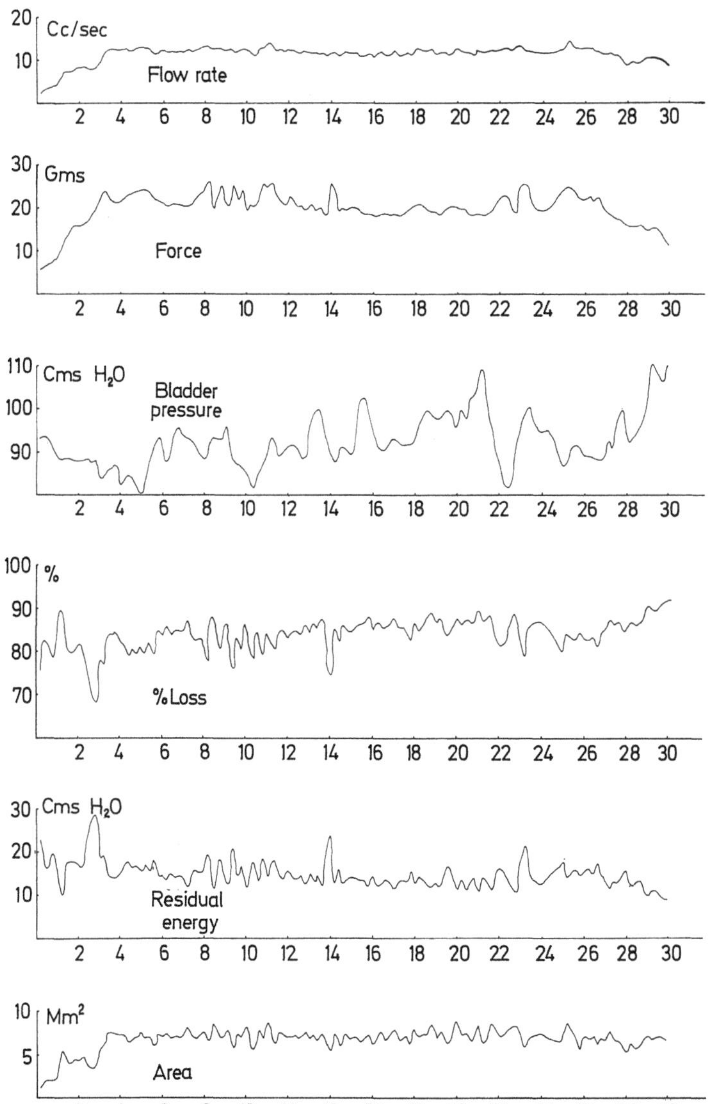

Fig. 7. Patient T. M. (2): a 30 sec portion of a 43 sec voiding. Note irregular pattern

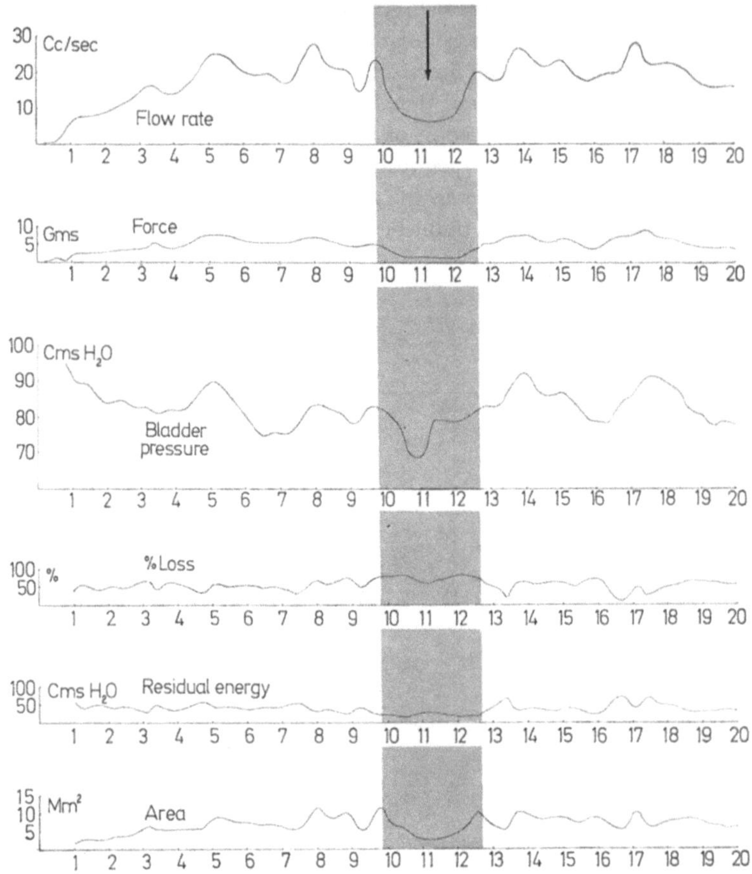

Fig. 8. Patient M. H.: a 20 sec. portion of a voiding which lasted 37 sec. Shaded band marks an "event" which occured during micturition

The authors are not familiar with any tests of resistance in the human female. Zinner *et al.* assumed that the female urethra is a narrow cylinder and decided that normal flow rate and normal bladder pressure are consistent with 30% loss. Gleason and Bottaccini computed the theoretical losses from voiding cystourethrograms. With the assumption of smooth walls they found that theoretically losses should not exceed 13%. Wise *et al.* found that computed losses from the VCU were quite small in the female. It must be concluded, therefore, that either the cystourethrogram does not give an hydraulically acceptable picture of the lumen or else that the urethral tissue is much rougher than assumed.

Cass and Hinman ran a series of resistance measurements on seven anesthetized female dogs. The authors were able to obtain the original data from the investigators and to compute the dissipation parameter. Losses appeared to be in the 30–60 percent range. Although we must be extremely careful in applying cross-species results, we must admit that the urethra of the female dog is hydrodynamically similar to the urethra of the female human and that a figure of 50% may be a median loss in the female. On this basis alone the two female subjects C.L. and M.H. appear to have normal loss. M.H. has an elevated bladder pressure and a distal area of 7.0 mm^2 (effective diameter of 3 mm). In a female the pressure is abnormally high and the area normal. This combination implies that the subject will generate an extremely high energy stream with a normal flow rate. Such a condition can be called *effective meatal stenosis* and can be detected only with the dynamometer or equivalent apparatus. The other female, C.L., is probably normal.

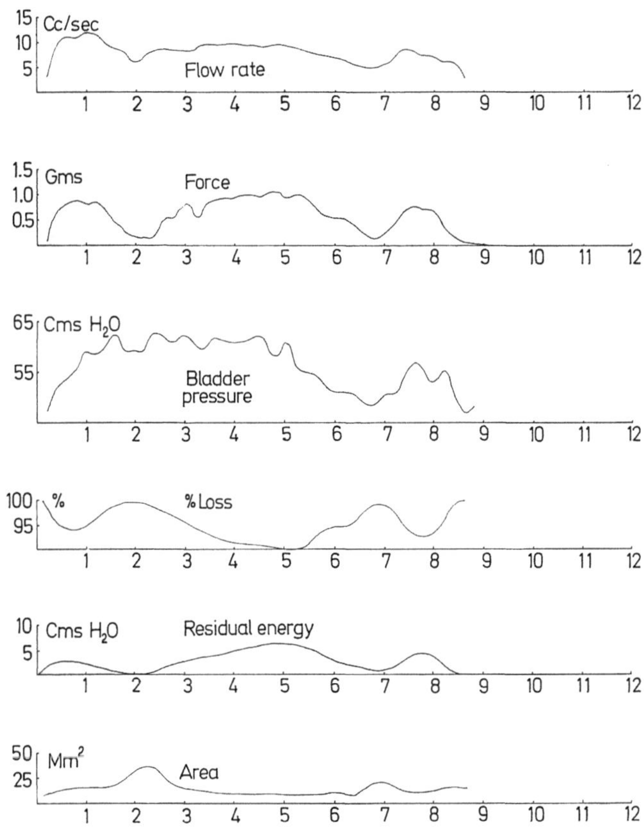

Fig. 9. Patient J. H.: complete voiding which lasted 9 sec

We note that C.L. seems to have a characteristic loss in spite of flow variations. This observation applies also to five male subjects (F.M., W.H., S.D., J.M., and J.E.). Fixed loss independent of bladder pressure and flow rate implies a fixed geometry.
a) All the males tested generated high bladder pressures. Pressures between 30–70 cm of water have been reported in the literature as normal for the human male (Gleason *et al.*). Wise *et al.* specifically showed in a study of 10 normal males that 30–75 cm appears to be a good range. In the subjects tested, the elevated bladder pressures may be considered pathologic. We note that most subjects lose over 90% of the bladder energy and that this loss appears to be the result of proximal constriction.

Since such high losses cannot be generated by moderate occlusion of the passages (Cass and Hinman, for example showed that the constriction of the bladder neck of female dogs by 50% tended to change losses only slightly e.g. 5% to 15%) it must be assumed that losses in the 90% range are the result of severe proximal obstruction. This is reenforced by the diagnostic information presented in Table 3.

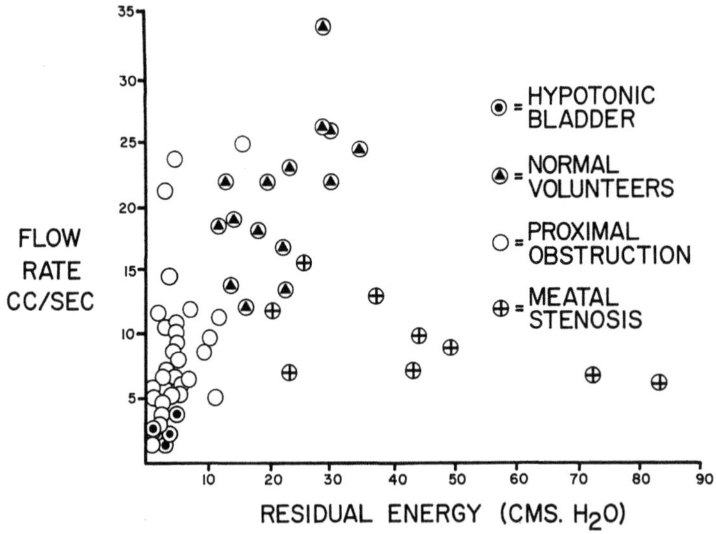

Fig. 10. Flow rate and residual energy can be used to classify male subjects into distinct categories

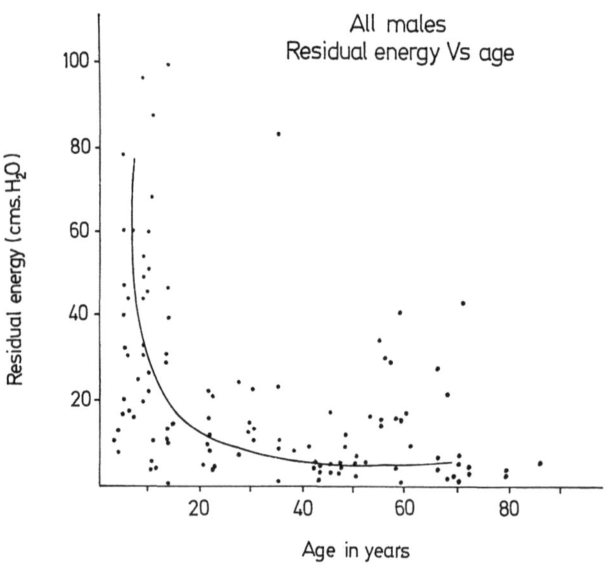

Fig. 11. Residual stream energy in the male plotted against flow rate permits grouping and identification of patients according to their clinical problems. Proximal obstruction presents as low flow/energy: meatal stenosis as low flow/high energy. Seldom do normal present with residual energies below 10 cm of water

Table 4. (Adapted from [WISE *et al.*]). Urethral resistance L^1 in nine normal males between the bladder and a point 4 cm above the meatus. ($E_R{}^1$ = residual energy in urethra)

	p_B	$E_R{}^1$	L^1	L est.
M.R.	34.0	32.5	4.4	21.1
R.B.	65.2	28.0	57.1	59.6
R.B.	41.7	20.0	51.0	56.3
C.S.	74.7	34.0	54.7	59.3
A.W.	48.9	32.2	34.2	45.4
W.B.	66.5	38.8	41.6	51.1
J.S.	42.1	29.0	31.3	41.7
R.H.	49.0	35.9	26.7	36.5
M.N.	49.0	43.7	11.8	24.1

Subjects J.G. and T.R. probably are unobstructed and their percentile loss may be considered normal. This conclusion is reenforced by the results of WISE *et al.* who measured energy four centimeters from the meatus. Losses for most of the urethra (excluding the last four centimeters) were as high as 57.1% so that 60–65% loss is not abnormal in the male (Table 4). We estimated the total loss from the hydrodynamic data contained in WISE *et al.* Computation was carried out by means of the Darcy-Weisbach equation with an assumed friction factor of $f = 0.03$. Even with such a high friction coefficient the distal portion of the pendulous urethra does not contribute too much loss. Obviously J.G. and T.R. (Table 3) cannot be considered to have high loss urethras.

b) A number of observations can be made through examination of individual histories (Figs. 2–9).

α) In most instances the flow rate, residual energy and percentile loss remain steady (time invariant), except for small fluctuations. With the exception of C.L. run number 1 (Fig. 3) which appears to be anomalous it is possible to extract a quasi-steady midflow pattern (Table 3).

This implies that the *internal geometry of the urethra* is fixed during a large portion of the micturion and hence that there exists a characteristic voiding pattern for each subject.

β) *Large, slow* variations in pressure are usually followed by similar changes in flow rate (Fig. 9 at 7 sec, Fig. 8 at 11 seconds). Moderate increases in pressure which occur when bladder pressure is at maximum tend to *reduce* flow rate, whereas the opposite condition of decrease in pressure matched by a temporary increase in flow does not seem to occur (Fig. 3 at 9 and 11 seconds, Fig. 4 at 7 seconds, Fig. 5 at 6 and 10 seconds, Fig. 6 at 4 seconds). The simplest explanation for this phenomenon is that there occurs a temporary closing of the bladder neck or else some other local spasm in the urethra (TANAGHO) while the bladder continues to contract. Theoretically such a local closing or near closing should generate a decrease of residual energy and an increase of loss. It can be seen in Figs. 3, 4, 5 and 6 that the measurements are consistent with theory. (Two possible mechanisms to explain this phenomenon appear in the literature (BOTTACCINI and GLEASON, TANAGHO *et al.*).

γ) Rapid changes in bladder pressure tend to be damped out and not affect the flow rate. Examples of this behavior can be seen in energy study. The most unusual example is seen in Fig. 7. In this instance the bladder pressure fluctuates wildly whereas the loss, flow rate and residual energy tend to remain steady with small fluctuations. In this case the shape of the urethra probably oscillates about a mean, absorbing and returning energy in the stream in the manner of an automobile shock absorber.

c) With the present instrumentation only flow rate, residual energy and area can be computed in the absence of bladder pressure measurements. It is interesting to note that even without pressure some conclusions can be made about a patient's hydrodynamic configuration. A rough correlation has been discovered in a small male population between flow, residual energy and gross hydraulic response. In a larger population, age correlates with residual energy.

α) Residual energy can be plotted versus flow rate and can be compared to diagnosis. (Fig. 10). In the series of males examined, asymptomatic normals cluster in the residual energy range of 12 to 35 cm of water with flow rates above 12 ml/sec. As expected, subjects with only meatal stenosis tend to have moderate-to-low flow rates and normal-to-high residual energy. Patients with proximal obstruction tend to cluster in the residual energy range below 10 cm H_2O with flow rates generally below 13 ml/sec. Four subjects with hypotonic bladders clustered in the lower left hand corner of the figure. The zones appear to be fairly clear for patients with extreme conditions. The "normal" zone, however, is rather obscure. Patients with a hypertonic bladder may appear normal in spite of fairly pronounced obstruction. F.M. (Table 3) would appear in figure 11 as a normal. Urethral dynamometer testing without accompanying bladder pressure measurement appears to be fairly straight forward in the presence of only one resistance generator but is misleading in complex situations involving non-normal detrusors and multiple loss.

β) The residual energy of a series of patients was plotted versus age. Prepubescent boys in the series demonstrated high residual energy, as well as high flow rate; whereas postpubescent males showed moderate to low residual energies (Fig. 11). Such a dramatic shift seems to indicate the possibility of urinary tract maturation in males, such as a postpuberty enlargement or loss of resilience of the prostate. It is not possible to state at this moment whether such a change is true of the general population or whether our results are simply characteristics of the population which tends to visit urological clinics.

6. *Summary*

a) Resistance (K) in hydraulic systems is computed by the expression

$$K = \frac{P_B - E_R}{E_R}$$

Resistance is related to loss (L) where

$$L = \frac{P_B - E_R}{P_B}$$

and

$$K = \frac{L}{1 - L}$$

The parameter

$$R = \frac{P_B - E_R}{Q^2}$$

is meaningless.

b) The loss or resistance appear to be characteristic of a subject and can be related to hydraulic structure.
c) Most subjects generate *nearly constant* bladder pressure, flow rate and resistance through a portion of the micturition. The steady state Bernouilli equation applies. From this we assume that the urethra maintains a fixed geometry exclusive of small fluctuations.
d) Normal females generate losses in the order of 0.50 and normal males somewhat higher, 0.60. Males with prostatic hypertrophy and other severe proximal narrowing can lose as much as 99% of the bladder energy and usually exhibit elevated steady bladder pressures.
e) Flow rate is a function of bladder working rate as well as bladder pressure. The following relationships are to be noted:
α) Slow, large fluctuations (less than one cycle per second) in bladder pressure are matched by similar fluctuations in flow rate.
β) Moderate increase in bladder pressure (1 cycle per second) which occur whenever maximum bladder pressure has been established, tend to decrease flow rate.
γ) Rapid bladder pressure fluctuation tend to be damped out and to have only small effect on flow rate.
f) The fluctuations discussed in β) are consistent with the existence of urethral muscle spasm.
g) Residual energy and flow rate correlate with hydrodynamic structure.
h) There appears to be a drop in residual energy which occurs in males at the age of puberty. It is possible, therefore, that we have discovered a heretofore unsuspected maturation phenomenon.

References

Bottaccini, M. R., Gleason, D. M.: The Effects of Bladder Wall Tension on the Diameter of the Distal Urethral Segment, Hydrodynamics of Micturition, Chapt. 18, Thomas Pub. 1971.

Byrne, J. C., Gleason, D. M., Bottaccini, M. R.: Energy Loss During Micturition: A Measure of Urethral Resistance. Invest. Urol. **10**, 221 (1972).

Cass, A. S., Hinman, F., Jr.: Constant Urethral Flow in Female Dog. II Effects of Constriction of Vesical Neck and External Meatus. J. Urol. **99**, 447 (1968).

Drouin, G., McCurry, E. M.: Catheters for Studies of Urinary Tract Pressure. Invest. Urol. **8**, 195 (1966).

Eyrick, T. B., Many, M., Wise, H. M., Jr.: Analysis of Urethral Resistance. Invest. Urol. **6**, 443 (1969).

Gleason, D. M., Bottaccini, M. R.: The Relation of Urethral Crossectional Area to Flow. Invest. Urol. **5**, 340 (1971).
Gleason, D. M., Bottaccini, M. R., Perling, D., Lattimer, J. K.: A Challenge to Current Urodynamic Thought. Invest. Urol. **97**, 435 (1967).
Gleason, D. M., Bottaccini, M. R.: The Vital Role of the Distal Urethral Segment in the Control of Urinary Flow Rate. J. Urol. **100**, 167 (1968).
Hansen, A. G.: Fluid Mechanics. 418–450. (New York: John Wiley and Sons 1967.)
Melchior, H., Simhan, K. V., Rathert, P., Lutzeyer, W.: Uro-Rheomanometry: Simultaneous Pressure and Flow Monitoring in the Ureter. Symposium of Flow, Paper No. 4-5-228, Pittsburgh, Pennsylvania; May, 1971.
Ritter, R. C., Zinner, N. R., Paquin, A. J., Jr.: Clinical Urodynamics. II Analysis of Pressure Flow Relations in Normal Female Urethra. J. Urol. **91**, 161 (1964).
Tanagho, E. A., Miller, E. R., Lyon, R. P., Fisher, R.: Spastic Striated External Sphincter and Urinary Tract Infection in Girls. Brit. J. Urol. **43**, 69 (1971).
Whitaker, J., Johnston, G. S.: Estimation of Urinary Outflow Resistance in Children: Simultaneous Measurement of Bladder Pressure Flow Rate and Exit Pressure. Invest. Urol. **3**, 379 (1966).
Wise, H. M., Jr., Many, M., Birthwell, W., Eyrick, T., Maguire, M.: Measurement of Urethral Resistance. Invest. Urol. **5**, 539 (1968).

Discussion

Loss of Energy in the Urethra (Internal/External Pressure)

Kinetic energy at the orificium externum urethrae and intravesical pressure must be known for an objective measurement of the energy loss while the urine passes through the urethra. This can be accomplished by simple velocity measurements (Hinman and Gleason). The normal exit velocity of the urine is 170 to 250 cm/sec for men and 200 to 230 cm/sec for women (Zinner).

"External pressure" (Sigel) is a misnomer and is used as a synonym for the kinetic energy remaining in the external urinary stream. There is no pressure in the external stream; only kinetic energy. Accurate measurement of the flow with simultaneous measurement of the force in the external stream, at the exit, or from a stream ejected horizontally and measured at the level of the exit, provides a means of calculating stream velocity. Stream velocity, which relates directly to the kinetic energy of the external stream, can also be measured directly with the spectrometer in a horizontally ejected stream (Zinner).

If the urethra were a smooth rigid uniform channel, the static component of energy would decrease linearly due to uniformly occurring frictional losses. Since the urethra is irregular in contour and consistency, static pressure does not decrease in uniform fashion, but decreases irregularly and can even increase. There is an interaction between static and dynamic elements which always adds to the local value of total pressure.

A measurement of bladder pressure (representing potential energy) and simultaneous exit stream velocity (representing kinetic energy) will provide an indication of loss of energy within the urethra. It will not indicate the location(s) of this energy loss nor will it describe how this energy was lost (Zinner).

Electric-Sensibility of the Bladder

C. FRIMODT-MÖLLER *

It is difficult to determine whether a prolonged cystometrogram with delayed first sensation and augmented capacity is due to a primary sensory lesion or to myogenic decompensation. Research in the field of diabetic bladder diseases has demonstrated the difficulty in distinguishing between an organic lesion and an illfunctioning bladder. Hitherto the sensory loss of the bladder has been established on the basis of the patient's feeling of first sensation and his ability to distinguish between hot and cold fluids. The capacity of the bladder is also an important parameter. But there has been no objective quantitative parameter.
A quantitative and semi-objective test has been developed in the Urological Laboratory, Gentofte Hospital, Copenhagen.

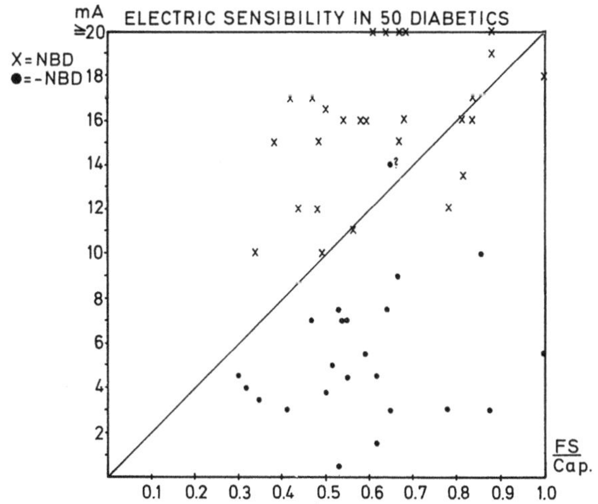

At the end of the urodynamic examination a thin steel or silver wire is passed through the suprapubic catheter to just beyond the end of the catheter. A reference plate-electrode is placed behind the patient's buttock. A universal stimulator delivers constant – current square wave impulses of 1 msec. duration to the patient at intervals of 400 msec. Simultaneous recording of the stimulation current as well as the

* Dept. of Urology, Gentofte Hospital, Copenhagen, Denmark

voltage across the electrode permits calculation of the complex impedance between the stimulation-electrode and the reference plate-electrode.

The results obtained show that normal bladders have an electrical threshold of less than 8–10 mA. Bladders with symptoms of sensory loss show an increased electrical threshold, often exceeding 20 mA. At this level we stop enlarging our impulses because of fairly high output voltage levels which might cause accidents if current from the stimulation electrode should flow to an accidental ground point through the heart.

50 unselected diabetics subjected to urodynamic examination have been used to illustrate the results. On the basis of this the patients were classified into groups of balanced or imbalanced sensory lesions of the bladder, hypoactive-normosensitive bladders, intravesical obstructions of various kinds and finally normal bladders.

The electrical threshold along the ordinate is compared with the ratio of first sensation to bladder capacity, the 2 latter being the parameters used in the documentation ot sensory loss hitherto.

A high ratio would be expected to be equivalent to a sensory loss, and the majority of our cases of sensory bladder lesions with elevated electric threshold are in fact concentrated in the upper right part of the figure. A few black "spots" towards the right in the lower part of the figure represent hypoactive bladders with large volumes of residual urine. Finally, I should mention that patients with infranuclear bladder lesions have raised electric thresholds as should be expected.

Electrostimulation of the Neurogenic Urinary Bladder

TH. BURGHELE*

The present paper reports on the clinical application of electric neurostimulation of the urinary bladder in paraplegic patients with spinal cord injuries above the micturition center.

Since the appearance of our publications on electrical stimulation of the motor (pelvic) nerves of the urinary bladder, other authors have proposed electrical stimulation of the detrusor or the anterior sacral roots, or transrectal electrical stimulation.

We established the optimal parameters of electric stimulus for stimulation of nervous influx to the pelvic nerves without resultant damage by experiment.

An electrical stimulus with optimal parameters induces long and powerful contractions of the urinary bladder, and during a given interval the urodynamic force thus produced overcomes urethral resistance augmented by contraction of the smooth and striate sphincter. Sectioning of the internal pudendal nerves to diminish urethral resistance is not recommended because of the role of these nerves in erection.

Two or three stimulations followed by interruptions of 3–5 minutes result in complete voiding of the bladder in paraplegic patients with injuries to the spinal cord above the micturition center.

In this clinic, electrical stimulation of the pelvic nerves is achieved by means of a receiver implanted in the body. Two platinum electrodes, one for each pelvic nerve, are connected to the receiver by insulated leads.

The receiver, leads and electrodes are covered with plastic well tolerated by the organism, except for the platinum wires.

The receiver is implanted by median subumbilical laparotomy. When the pelvic cavity is free the sites of the second and third anterior sacral foramina are determined by palpation.

The posterior parietal peritoneum is sectioned along the anterior foramina in order to expose the second and third anterior sacral roots. The pelvic nerves appear 1.5 to 2 cm beyond the point where these roots emerge from the foramina and then penetrate into the pelvic plexus. These nerves must be isolated very carefully.

Before the electrodes are definitively fixed on the pelvic nerves they are tested by stimulation of each nerve separately, the effect being appraised by concomitant recording of intravesical pressure by means of an urethral catheter. Pressure induced

* The Clinic of Urology, Panduri Hospital, Bucharest, Rumania

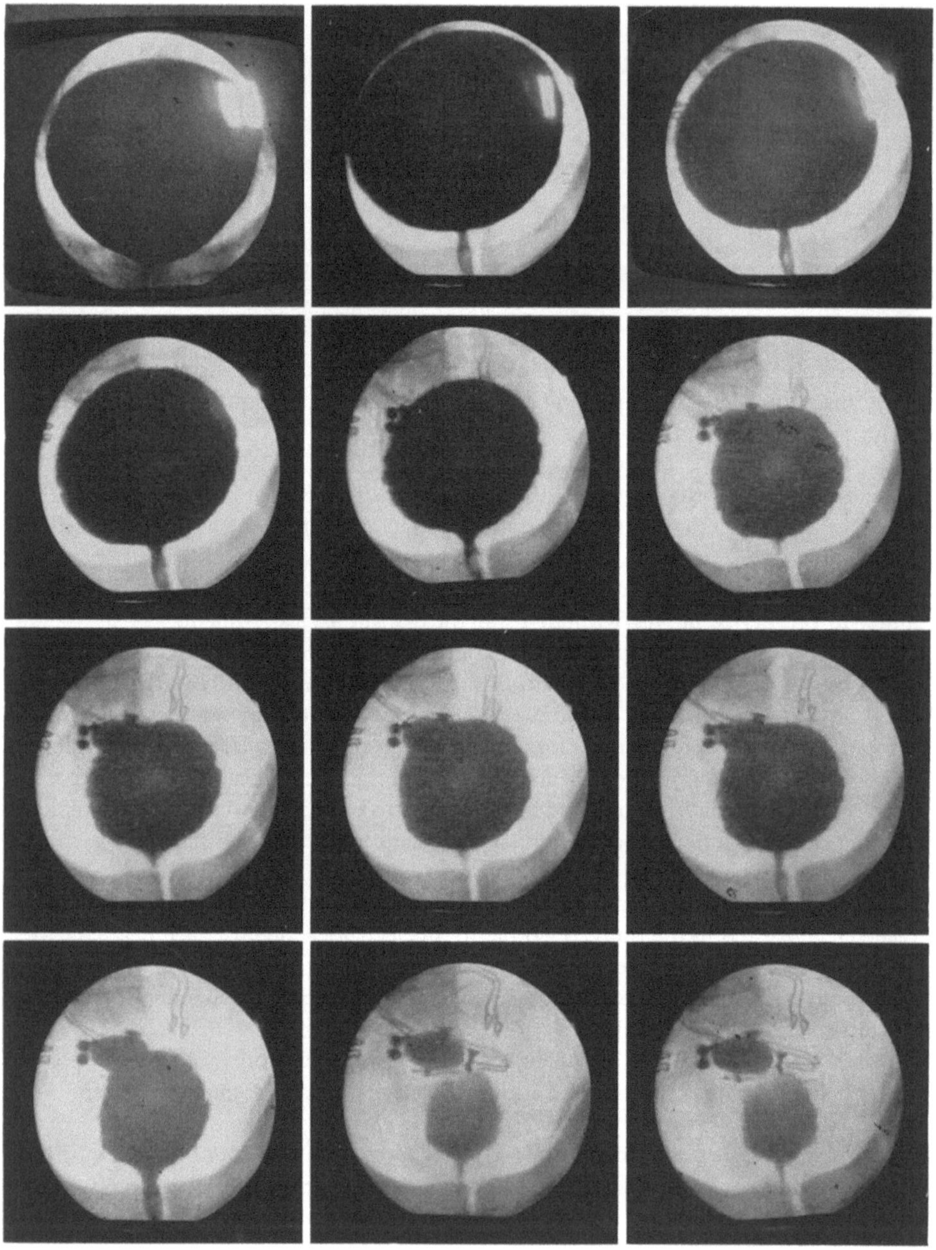

Fig. 1. Male, 31 years, spinal cord injury at Th_9 level, serial voiding cystourethrogram induced by electrical stimulation of the pelvic nerves and followed up in the videorecorder. Two successive stimulations. Note opening of the bladder neck and its closure by contraction of the striate sphincter. The neurostimulator can be seen behind the bladder

by stimulation of one nerve must be 40 cm of water. Bilateral stimulation must produce an intravesical pressure of more than 100 cm water.

Once the electrodes are fixed, the receiver is attached to the anterior aspect of the sacrum retroperitoneally; the peritoneum is then sutured. Electrical stimulation of the nerves is performed on the operating table.

The antenna of an external inducer is brought close to the site of implantation of the receiver before the electrical stimulus is discharged. The bladder is then voided.

In order to appraise the efficiency of electrical neurostimulation, i.e. complete evacuation of the urinary bladder without untoward or harmful side effects tele-recording of electrically induced micturition was performed using an opaque medium (Fig. 1).

These images facilitate several interesting conclusions concerning the physiology of the electrically stimulated bladder. Before stimulation of the pelvic nerves, the base of the bladder describes a curve that may reach the upper edge of the pubic bones. Seven seconds after application of the stimulus, the bladder contracts and becomes perfectly spherical. The base rises over the upper edge of the pubic bones and becomes linear.

When intravesical pressure reaches a certain level, the opaque medium penetrates into the posterior urethra, without any actual breaking of the base plate, as described in physiologic micturition.

This phenomenon might be accounted for by delayed contraction of the trigone-orifice section following irradiation of the electric stimulus.

Careful visualization of the posterior urethra by means of the opaque medium eliminated from the bladder shows a normal appearance with bladder neck, posterior urethra and verumontanum.

Even when electrical stimulation way continues the opaque column gradually disappeared and the urethra appeared to be strangled by contraction of the striate sphincter. Although intravesical pressure either did not vary or even augmented due to occlusion of the evacuation pathway, the opaque medium remained at the level of the bladder neck, which it transformed into a slightly open funnel.

It should be mentioned that in the patient in whom these recordings were taken, the bladder neck was in fact disconnected, i.e. the smooth sphincter was paralyzed.

The question arises as to how partial evacuation of the bladder is possible when there is simultaneous contraction of the detrusor and of the striate sphincter. The explanation lies in the difference between the pressure of the urine produced by contraction of the detrusor and the intra-urethral pressure induced by contraction of the striate sphincter. Initially the urodynamic force exceeds the urethral force; the two forces later become equal and, finally, the urethral force increases above the urodynamic one, and the jet of urine is interrupted.

After an interval of 2–5 minutes, renewed stimulation of the pelvic nerves induces complete evacuation of the urinary bladder contents.

References

Burghele, Th., Ichim, V., Demetrescu, M.: Experimental studies on the physiology of micturition and the empty of the urinary bladder in patients with lesions of the spinal cord. Artificial micturition through distant electromagnetic induction. Rum. med. Rev. **3**, 48 (1959).

Bradley, W. E., Wingrove, R., Chou, S. N.: Use of an implantable R.F. receiver unit for evacuation of the neurogenic bladder. Proc. on Eng. of the 15th annual Conf. in Med. and Biol., Nov. 5–6, 59 (1962).
Dees, J. E.: Contraction of the urinary bladder produced by electric stimulation. Preliminary report. Invest. Urol. **2**, 539 (1965).
Habib, H. H.: Experience and recent contributions in sacral nerve stimulation for voiding in both human and animal. Baltimore: Williams & Wilkins Co. 1967.

Discussion

Range of Indications for Electrical Bladder Stimulation

The range of indications for electrical bladder stimulation includes the treatment of neurogenic bladder dysfunctions if the central nervous system is affected, especially the spinal cord or the encephalon. Direct myotropic electrostimulation of the detrusor vesicae is indicated if the bladder and the center regulating micturition are completely cut off. If the regions above the micturition center are affected, neurogenic bladder stimulation is indicated. While direct myogenic bladder stimulation is generally preferred today, Burghele and Ichim initially used indirect neurogenic stimulation. Because of the selection of patients has so far been unsystematic, it is not possible to say which method promises better results, because the spectrum of diseases of the central nervous system is very broad (Ichim). Traumata and tumors as well as inflammatory changes and myelomeningoceles can cause disturbed micturition or incontinence of urine. Ichim, however, believes that a lower-motor-neuron lesion associated with an atonic bladder without reflexes is a clear indication for electrostimulation. Furthermore, he would use electrostimulation in case of acute trauma of the spinal cord, because this method is more effective for the rehabilitation of the bladder than training of the bladder and intermittent catheterization.

In acute trauma of the spinal cord, it is possible that the normal reflexes of the bladder may regenerate. This generally happens within 2 to 3 months, but even latent periods of 18 months are described (Boyarsky). To wait so long for rehabilitation of the bladder, with the bladder always emptied instrumentally, increases the danger of infection. This can be prevented by an electrostimulation in time (Ichim).

Transurethral Electroresection in Connection with Disturbed Micturition

Neurogenically, disturbed micturition is to be suspected when coordination between detrusor vesicae and sphincter mechanism is upset. Operations on the bladder neck are the major causes of this disorder. Bischoff and Hinman, however, disregard the danger of urine incontinence and remove the bladder neck in its entirety.

Turner-Warwick feels that in a case of disturbed micturition it is important to consider the function of both the detrusor and the sphincter mechanisms, for treatment must be based on fact. Ultimately, he anticipates, it will be possible to render a spastic bladder hypotonic without compromising the bladder sensation or sphincters so that continence can be achieved by adjusting the sphincter mechanism.

Dynamics of the Urinary Bladder Orifices during Electrical Neurostimulation of the Bladder

V. Ichim *

Electrical stimulation of the urinary bladder, and especially of the motor (pelvic) nerves, produces intravesical pressure in excess of 100 cm water.

The urodynamic force induced by electrical stimulation is exerted on the inner surface of the urinary bladder, especially on the trigone with the vesico-urethral and vesico-ureteral orifices, which this force tends to open.

Translucency of the base of the bladder demonstrates the existence of abundant muscular tissue with highly developed vascularization in the region of the vesical orifices. In the posterior portion of the bladder neck this region includes the trigone which surrounds the vesicoureteral orifices while in the anterior portion it forms a ring around the vesicourethral orifice.

It has been demonstrated histologically that the precervical muscular fibers have a fine structure, similar to the muscular fibers of the trigone and, probably, of the same embryologic origin; electrical stimulation of the hypogastric nerve produces contraction of the trigone and smooth sphincter; some authors dispute the anatomic autonomy of the latter formation.

The anatomic absence of a muscular ring between the bladder and urethra does not exclude the function of the sphincter, which is exercised by the precervical muscular fibers arranged in a half ring, curving outward to the front with posterior concavity and continued by the retrocervical elastic connective tissue from the base of the bladder and deep trigone.

Traction applied to the retrocervical elastic-connective tissue by contraction of the precervical muscular fibers draws the fibers of this tissue towards the vesico-ureteral orifice, whereupon the orifice closes.

During filling of the bladder, occlusion of the vesico-urethral orifice, i. e. continence, is assured only by the tonus of the precervical muscular fibers and the elasticity of the retrocervical elastic connective tissue.

When the need to urinate arises and micturition is impossible due to environmental conditions, continence is insured by active contraction of the precervical muscular fibers which induces tension in the elastic connective fibers mentioned.

Continence of the vesico-ureteral orifices is insured during the filling period by pressure of the urine on the intramural ureters and the tonus of the detrusor, and also because the intraureteral pressure is higher than the intravesical pressure. During

* The Clinic of Urology, Panduri Hospital Bucharest, Rumania

micturition, contraction of the detrusor and reflex inhibition of ureteral peristalsis are sufficient to prevent vesico-ureteral reflux.

During physiologic micturition, the vesico-urethral orifice opens due to the intravesical urodynamic force produced by contraction of the detrusor. This force is exerted especially towards the bladder neck, with a fixed insertion.

Contraction of the detrusor also tends to raise the trigone to a vertical position so that the urine pressure is applied to the base of the bladder. The trigone-orifice mucosa, which is perfectly smooth due to its adherence to the deep layers, also facilitates the action of the urodynamic force.

Electrical stimulation of the urinary bladder or of the motor nerves raises the question of the dynamics of the vesical orifices, the efficiency of the stimulation implying opening of the vesicourethral orifice until evacuation of the urine is complete as well as closing of the vesico-ureteral orifices in order to prevent reflux.

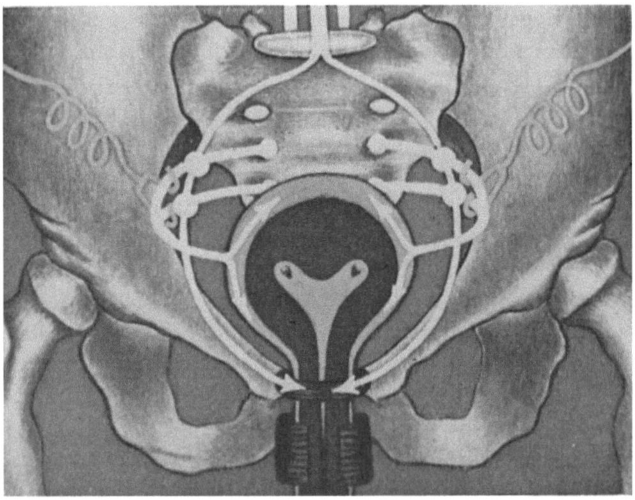

Since the appearance of our publications, several articles on electrical stimulation of the urinary bladder have mentioned an increase in urethral resistance during stimulation and the role of the striate sphincter in this resistance.

Other authors have demonstrated that the striate sphincter incapacitated by severing of the internal pudendal nerves or by insertion of a rigid tube in the urethra surrounded by this sphincter no longer permits complete evacuation of the bladder by electric stimulation.

In order to reach a better understanding of the dynamics of the urinary bladder orifices during electric stimulation of the pelvic nerves it is necessary to elucidate the opening mechanism of the vesico-urethral orifice, the factors responsible for the dynamics of this orifice during stimulation and the closing mechanism of the vesico-ureteral orifices during neurostimulation.

A film of the orifices was therefore taken by means of a cystocope introduced in the bladder during a minor cystostomy to facilitate study of the dynamics of the urinary bladder orifices during experimental electrical stimulation of the pelvic nerves.

As an illustration of the effect of stimulation of the pelvic nerves, Fig. 1 shows the innervation of the detrusor and smooth spincter and the site of stimulation, which is above the point at which the pelvic nerves enter the pelvic plexus.

The vesico-urethral orifice opens a few seconds after the beginning of stimulation, remains open for 20–50 seconds and gradually closes to prevent evacuation of the fluid. If stimulation is repeated after an interval of 3–5 minutes, the same phases succeed one another irrespective of whether stimulation is applied before or after severance of the internal pudendal nerves or sectioning of the urethra above the smooth sphincter.

Opening of the vesico-urethral orifice during electrical stimulation is induced by the urodynamic force of the fluid, which overcomes the resistance of the smooth and striate sphincters.

As already mentioned, the pelvic nerves are stimulated electrically before entering the pelvic plexus. The stimulus courses through the fibers of these nerves and stimulates the hypogastric nerve fibers innervating the smooth sphincter and trigone by irradiation. There is some speculation why contraction of the detrusor and trigone and, implicitly, of the smooth sphincter does not take place simultaneously. Electrical stimulation of the parasympathetic pelvic nerves is followed by an immediate response on the part of the effector organ, whereas stimulation of the sympathetic hypogastric nerve acts with a certain delay upon the effector organ. Consequently, contraction of the detrusor takes place prior to that of the smooth sphincter, so that a certain amount of urine is expelled from the urinary reservoir before contraction of the smooth sphincter.

We eliminated the action of the sympathetic nerves experimentally by administration of sympatholytic drugs, and clinically by severance of the hypogastric nerve and alcoholization of the tissues surrounding the bladder neck. The smooth sphincter no longer contracted and evacuation of the bladder was complete. In contrast, evacuation is interrupted sooner when the hypogastric nerve is stimulated electrically at the same time as or before the pelvic nerves are stimulated.

The dynamics of the vesico-urethral orifice during neurostimulation are therefore induced by the urodynamic force that opens this orifice and by parasite stimulation of the hypogastric nerve that closes this orifice following contraction of the smooth sphincter.

The vesico-ureteral orifices remain closed during neurostimulation of the bladder due to contraction of the detrusor and trigone, the orifices being drawn downwards in the direction of the bladder neck.

References

Bro-Rasmusen, F., Halborg Sorensen, A., Bredahl, E., Kelstrup, A.: The structure and function of the bladder. Urol. int. (Basel) **19**, 280–285 (1965).

Burghele, Th., Ichim, V., Demetrescu, M.: L'éléctroexcitation de la vessie médullaire. Étude expérimentale. J. Urol. Néphrol. **64**, 317 (1958).

Graber, R., Rutishauser, G., Wollf, G.: Experimentelle Untersuchungen über den Öffnungs- und Schließmechanismus der elektrisch stimulierten Blase. Urol. int. (Basel) **21**, 182 (1966).

Hald, T., Rossel, I.: Electrostimulation of the urinary bladder in normal cats. Acta chir. scand. **128**, 414 (1964).

Hald, T., Agrawal, G., Kantrowitz, A.: Studies in stimulation of the bladder and its motor nerves. Surgery **60**, 4, 848 (1966).
Holmquist, B., Olin, T.: Electromicturition in male dogs at pelvic nerve stimulation. An urethro-cystographic study. Scand. J. Urol. Nephrol. **2**, 115 (1968).
Holmquist, B., Staubitz, W. J.: The role of the pudendal nerve in connection with electronic emptying of the neurogenic cord bladder in dogs. J. Urol. (Baltimore) **98**, 198 (1967).

Discussion

Complications of Electrical Bladder Stimulation

The main complication of electrical bladder stimulation is simultaneous irritation of the neighbouring contractile organs. Especially simultaneous irritation of the bladder neck and the striped sphincter externus represents a dangerous complication because it completely cancels out the success of detrusor stimulation.

In neurogenic bladder stimulation the reason for the simultaneous irritation is the closeness of the sympathetic, parasympathetic and somatic nerve fibers in the plexus pelvicus. Upon plexus stimulation, the impulse is also conducted on the efferent fibers. To prevent this interlacing, the stimulator can be connected (insulated) to the parasympathetic nerve fibers, and at the same time the nervus hypogastricus can be cut directly in front of the os sacrum. Desympathization of the bladder neck should be done at the same time to enable separation of the bladder neck from the vesico-ureteral functional entity (Ichim). Such an operation takes time to become effective, because the nerves degenerate over 2 weeks. This surgical treatment reduces the secondary wall contractions caused by simultaneous irritation. However, a side effect on sexual function is unavoidable. For this reason Burghele and Ichim prefer neurogenic stimulation at intervals to obtain micturition.

Energy Balance of Voiding Post-Ileocecocystoplasty*

D. M. GLEASON **, M. R. BOTTACCINI **, J. C. BYRNE ***, R. F. GITTES ****

The initiation of micturition involves opening of the bladder neck and shortening the urethra (LAPIDES *et al.*, LAPIDES). Beyond this, there is little consensus concerning the mechanism. HUTCH summarized 4 possible mechanisms:

1. Contraction of the outer longitudinal muscles of the bladder around a fluid bolus. The contraction of the detrusor pulls open the bladder neck (WOODBURNE).
2. Contraction of the superficial musculature of the trigone which depresses the dorsal lip of the bladder neck, thereby assisting in "breaking the base plate" and funnelling the bladder neck and urethra.
3. Contraction of the internal longitudinal muscular layer of the bladder. The internal longitudinal musculature of the bladder is continuous with the internal longitudinal musculature of the urethra (KRANTZ, CLEGG). It's contraction therefore shortens the urethra and "breaks the base plate" as in Paragraph 2.
4. Contraction of striated musculature found at the bladder neck and around the urethra thereby achieving the same effect as in Paragraph 2. (It should be noted that the striated musculature is not longitudinal, but rather oblique and probably sphincteric (KRANTZ) in the female, and non existant in the male prostatic urethra except at the apex (CLEGG).

WOODBURNE emphasized the importance of radial orientation of smooth muscle fibers in relation to the internal meatus and bladder neck. Bladder filling resulted in more radial disposition of the fibers and thence more effective opening of the bladder neck with detrusor contraction. Perhaps this explains the frequently observed correlation of urinary flow rate with bladder volume and volume voided (VON GARRETS, GLEASON *et al.*).

Our data suggested that the *intact* bladder and it's contained volume were the vital variables in bladder neck opening. The fluid within the bladder acts as a scaffold against which the detrusor fixes itself and contracts generating tension for pulling open the bladder neck. In the patients with ilealcecocystoplasties the cecal pouch vented the bladder allowing fluid mobility, and removing the fixation necessary for development of mural tension. This resulted in micturition with high energy loss

---

* This work made possible through the generous support of Joseph Fishman.
** College of Engineering (Biomedical Engineering) University of Arizona, Tucson, AZ USA.
*** Squier Clinic, College of Medicine Columbia University, New York City, NY USA.
**** College of Medicine (Division of Urology) University of California, San Diego, CA USA

probably secondary to failure of bladder neck opening. Of the four mechanisms summarized by HUTCH, the latter three did not seem pertinent in view of the independence of the hydrodynamic and anatomic variables in this patient group.
The present study measures the flow rate and hydrodynamic efficiency of a group of 6 patients with ileocecocystoplasty from Dr. GITTER's personal series. A portion of the bladder was removed from these patients and an ileocecal pouch was anchored to the bladder stump. Comparisons of these patients to normals allowed some insight into the mechanism of bladder neck opening and the relative importance of some of the variables mentioned previously.

Methods and Materials

Six ileocecocystoplasty patients (Table 1 and 2) included four females and two males. These patients had surgery for causes described in the above tables. All patients had patent urethras. The female subjects had been dilated to 45 French or had undergone urethrotomies. No significant urethral pathology was found on x-ray or cystoscopic study.

Table 1. Clinical histories

Patient	Sex	Age	Original disease	Previous surgery
E. M.	F	40	Interstitial cystitis	TUR bladder neck and bladder 1965
R. B.	F	38	Interstitial cystitis	None
G. L.	M	50	GU tuberculosis contracted bladder, right reflux	Right orchiect-my 1968
Y. S.	F	24	Ectopic left ureter stricture right ureter bilateral hydronephrosis	Ureterostomies 1967 Left nephrectomy 1968
L. B.	F	30	Interstitial cystitis, left reflux	Ruptured bladder following dilation with silver nitrate. Intense abdominal and pelvic scarring 1968
E. T.	M	12	Bilateral reflux, hydro nephrosis	Bilateral loop ureterostomies; Y-V plasty of bladder neck 1964

A minimum of 4 voidings per patient were measured. During voidings, rectal pressure was monitored with a flaccid balloon attached to a Sanborn 267 AC pressure transducer. Following the first voiding the bladder was catheterized with a 5 French infant feeding tube introduced per urethra. After measurement of residual, the bladder was filled with a saline drip to the point of urgency. For bladder pressure measurement the feeding tube was connected to a Sandborn 267 AC pressure transducer. The patient voided around the tube against a vertically mounted disc which measured

Table 2. Surgical procedures

Patient	Date of surgery	Extent of cystectomy	Ileocecocystoplasty data Ureteral reimplantation
E. M.	4/70	Dome only	No
R. B.	6/70	Partial cystectomy cuff 2 cm above posterior edge of trigone	No
G. L.	1/69	Subtotal cystectomy trigone partially excised	To ileum (ileocecal segment)
Y. S.	5/68	Dome only	To ileum (ileocecal segment)
L. B.	10/69	Subtotal cystectomy trigone left.	To ileum (ileocecal segment)
E. T.	1969	Subtotal cystectomy trigone partially excised	To ileum (ileocecal segment)

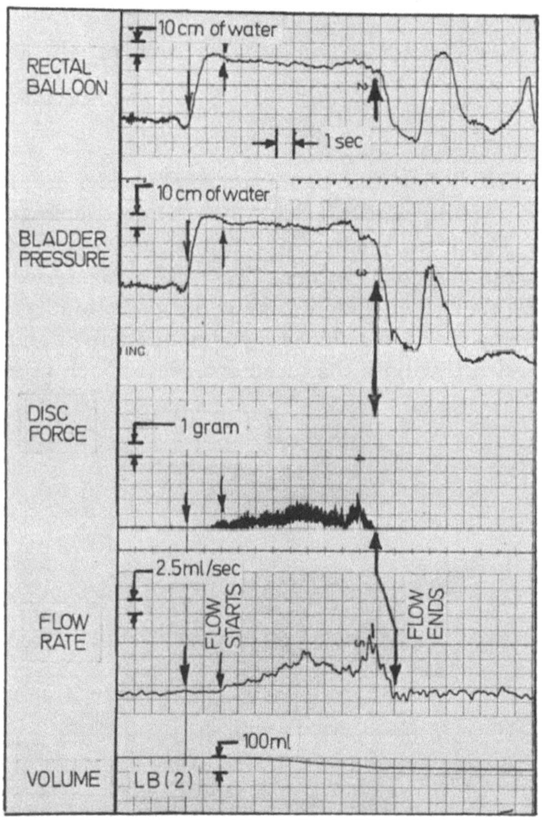

Fig. 1. Five parameters were recorded during most voidings: abdominal pressure; pouch pressure; urinary force (momentum); flow rate; volume voided. Allowing for differences in catheter position or the overlying fluid column, cecal pouch and rectal pressures were similar. Voiding occurred only with abdominal straining; there was no independent generation of pressure by the cecal pouch-bladder remnant (and thus no generation of mural tension)

force. The weight of a beaker which caught the run-off was recorded as volume voided; the data were electronically differentiated to instantaneous flow rate. A total of 5 simultaneous parameters were recorded (Fig. 1). At least two voidings were measured in this manner. A final bladder filling was followed by removal of the urethral catheter. This micturition occurred without any impediments in the urethra. The rectal pressure measurement however, was continued during this final voiding. When the data were calculated, the residual energy per unit of volume of the exit stream (BYRNE *et al.*) was expressed as pressure in centimeters of water. The percentage of energy loss was calculated by the following formulation (BYRNE *et al.*):

$$\% \text{ Loss} = \frac{Pb - Er}{Pb}$$

where

Pb = bladder pressure,
Er = residual energy/unit volume of the voided stream.

Results

The pressures measured both in the rectum and cecal bladder bore a constant relation to each other. The pressures tracked each other during straining, coughing, crede, etc. (Fig. 1 and 2). The primary source of error, damping of transmitted abdominal pressure in the rectum by fecal material, was easily recognized and avoided by replacing or moving the rectal balloon. In only one patient independent contraction of the cecal pouch was noted, but this was not associated with micturition (Fig. 3). The findings suggested that both the rectal and cecal pouch pressures indexed the

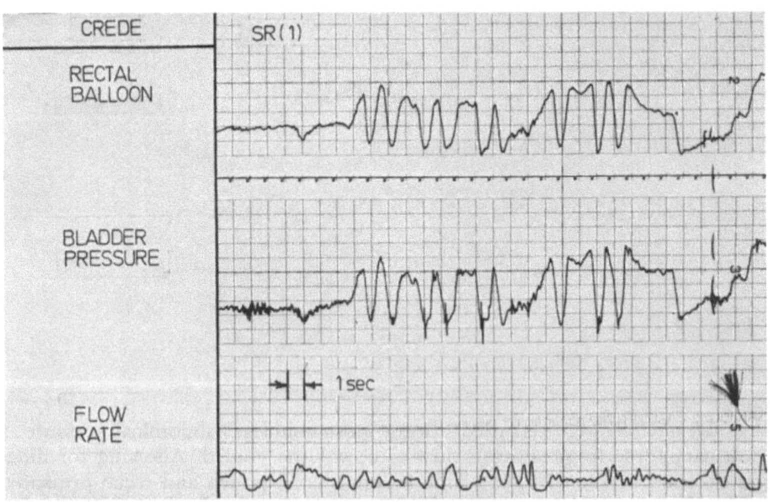

Fig. 2. The duplication of rectal and cecal pouch response to external events is seen in the similarity of their pressure recordings during crede maneuver

Table 3. Hydrodynamic data

Patient	"Bladder" capacity mls	Residual[a] urine mls	Duration of sustained projectile stream	Maximum sustained flow rate (mls/sec)	Force[b] (gms)	residual[b] energy (cmH_2O)	"Bladder"[b] pressure (cmH_2O)	% Energy[b] loss
E. M.	700	180	7	12	1.0	3.4	80	95
R. B.	480	125	21	8	1.4	1.5	37	59[c]
G. L.	600	125	49	18	1.4	2.9	58	95
Y. S.	740	200	8	5	0.5	4.9	56	91
L. B.	840	40	12	25	2.6	4.9	85	94
E. T.	400	75	18	27	2.7	4.5	70	94

[a] The residual volumes were measured on the day of testing. Many of the subjects had demonstrated residuals as low as 5 to 15 mls when tested on other occasions
[b] Taken at the point of most effective sustained flow
[c] Patient voided most effectively by crede, not straining

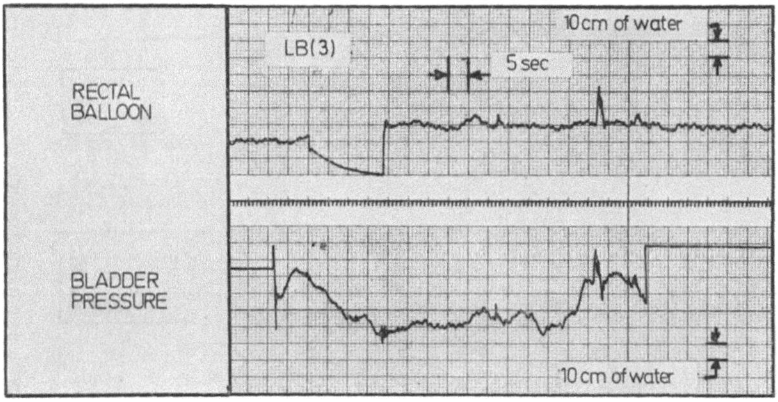

Fig. 3. A single patient developed cecal pouch pressure independent of abdominal straining. The pressure change was not associated with voiding, but was more like a peristaltic contraction

abdominal pressure and that neither significant bladder wall or rectal wall tension was generated. Micturition occurred only with abdominal straining or crede (Fig. 1). In the final voiding the urethral tube was removed and rectal pressures were read as cecal pouch pressures with confidence in all but the one patient mentioned. Here, a correction factor added to rectal pressure compensated for the difference.
Patient hydrodynamic data is summarized in Table 3. Individual cases are presented in Figs 4, 5, 6, 7. The female patients tended to void with a pattern of dribbling interspersed with periods of sustained projectile stream; the male subjects voided a more continuously projectile stream. There was a wide variation in flow rates in the patient group ranging up to 27 mls/sec. Both male and female patients had high energy losses (over 90%) and low residual energies (under 5 cm of water pressure).

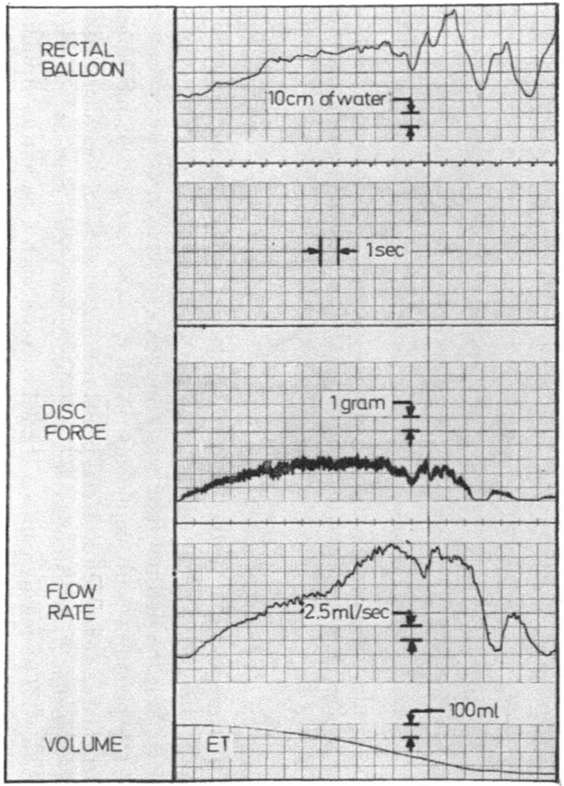

Fig. 4. Patient E. T. a 12 year old boy voided with a sustained pouch pressure and excellent flow rate. The urethral catheter was removed in these and following studies

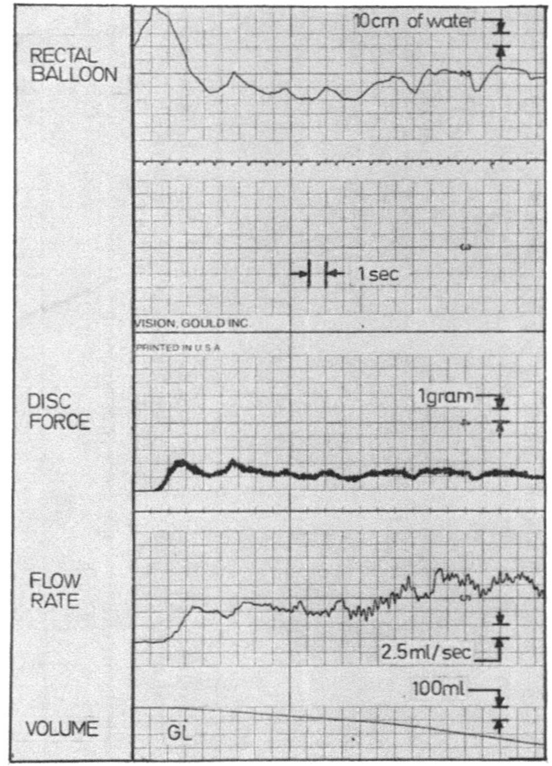

Fig. 5. G. L. demonstrated normal flow rate and a good sustainde flow

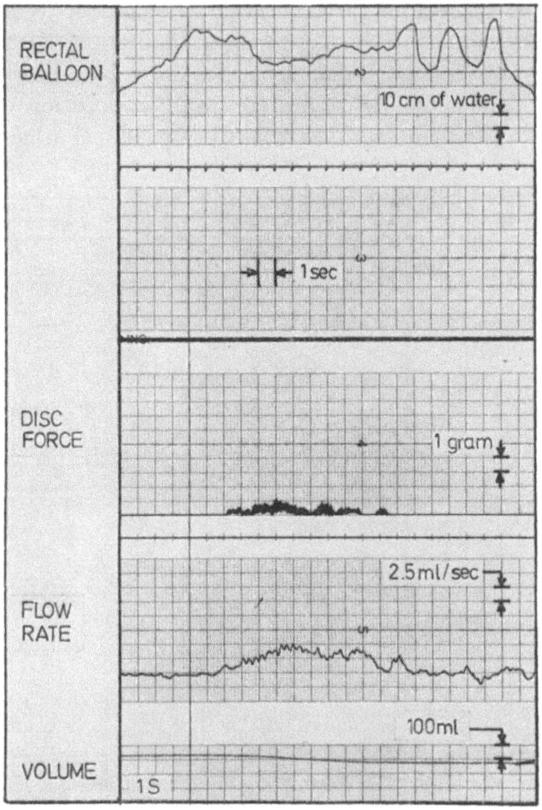

Fig. 6. Y. S. voided with a dribbling stream of long duration interspersed with periods of projectile stream. Note fall of "bladder" pressure preceeding stream

The energy balance patterns resembled patients with proximal obstruction, (such as prostatism, bladder neck contraction, etc. i.e. low to normal flow accompanied by severe energy loss) (GLEASON *et al.*). These excessive energy losses suggested there had been a failure therefore, in the urethral – bladder neck mechanism despite normal flow rates in some patients.

It had been our finding that in normal patients with intact detrusors, female subjects tended to lose only 50% of their initial energy and males about 70% (BYRNE *et al.*). In cecocystoplasty patients the sex difference in urethral geometry and energy loss was overwhelmed by the circumstance of their common post-surgery anatomy; both males and females lost 90% or more of their initial energy. There was no relation between the extent of the bladder resection and energy loss. Two male patients with extensive resection of the trigone voided a better stream (though energy losses were still in the 90% range), than two female patients with only small segments of the dome removed.

Whereas, the normal subject established a steady state of urine flow within 3 seconds after initiation of micturition, and maintained this with stable energy loss persisting through most of the micturition, the ileocecocystoplasty patients demonstrated increasing loss as the micturition proceeded. The ileocecocystoplasty patients apparently did not achieve stable hydrodynamic configuration. Rather, there was an inverse relationship between the instantaneous bladder volume and the energy loss in these patients; as bladder volume decreased during micturition the energy losses increased to the point where micturition stopped and residual urine was left.

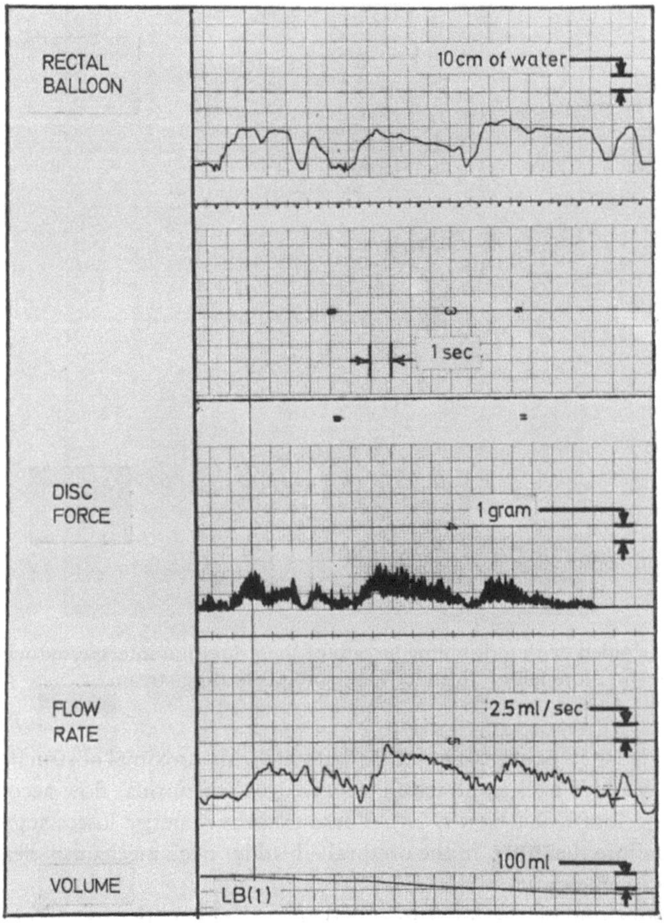

Fig. 7. L. B. voided a dribbling stream which became projectile as abdominal pressure increased. Note the contrast of the abdominal pressure – stream force relationahip in this patient with Y. S., Fig. 6

Discussion

The isolated bladder neck-urethra withoud the *intact* detrusor does not have the capability of becoming a hydrodynamically efficient conduit. The intact bladder can be inferred from these data to play a role in the urethral opening beyond the generation of pressure (supporting previously reported data) (GLEASTON *et al.*). When the bladder is intact, it contracts against it's confined fluid volume, generating mural tension, which forces open the bladder neck. In the cecoplasty patients fluid escapes into the cecal pouch from the bladder segment and mural tension cannot be adequately generated or substained. Intuitively, a balance is suggested between the capability of the detrusor to generate tension necessary to open and mold the proximal urethra, and the compliance of the tissues in this region. Rigid and noncompliant tissues resist opening despite an intact detrusor with good mural tension. Micturition in patients with prostatism and bladder neck contraction is thus characterized by high energy loss. Voiding cystourethrograms in such patients (Fig. 8) demonstrate failure of the bladder outlet to assume an efficient funnel configuration which can be corrected by incising or resecting the rigid bladder neck tissues. Thus, the similarity of voiding

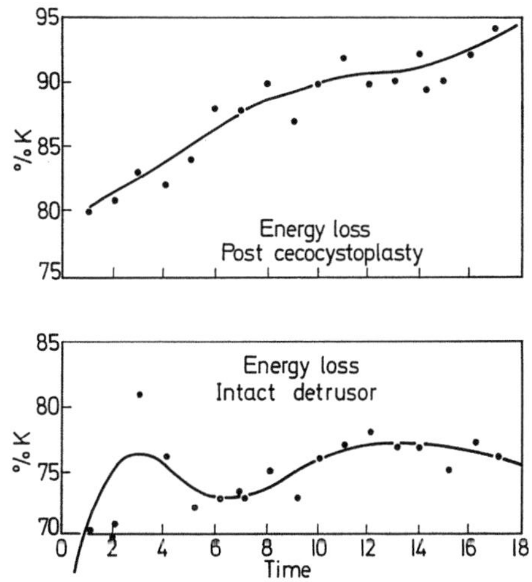

Fig. 8. The level of energy loss is established within the first seconds after voiding and maintains itself fairly constant in patients with intact bladders. In cecoplasty patients energy loss is not stable and changes with time during the micturition. This probably reflects the unstable bladder neck-urethral geometry in these patients

energetics in the ileocecocystoplasty and proximally obstructed patients suggests a common denominator; failure to open the bladder neck and mold the proximal urethra. In the cecocystoplasty patient the problem resides in the failure to develop detrusor mural tension; whereas in the patient with prostatism, or bladder neck contracture, the problem appears to lie in the decreased tissue compliance of the bladder neck prostatic regions.

The present data suggests that satisfactory micturition requires not only the generation of pressure (which can be adequately substituted for by abdominal pressure) but also bladder contraction against its confined fluid thereby generating wall tension. This tension translates into bladder neck opening and funnelling with assumption of an efficient hydrodynamic configuration for voiding. In the absence of the intact detrusor and a fixed contained volume, voiding occurs but it is hydrodynamically inefficient with no evidence that the intrinsic urethral striated muscles, trigonal muscles or longitudinal detrusor fascicles have any significant effect. The micturition is not stable and energy losses are progressive even in the presence of a good flow rate. Extrapolation of these data suggests a mechanism for obstructive disease of the bladder neck including bladder neck contraction and prostatism.

Summary

The hydrodynamic function of the bladder neck-urethral opening mechanism was studied in six ileocecocystoplasty patients. Five simultaneous parameters were measured. Clinically the subjects were able to void effectively, with good functional volumes, and acceptable residual urine. The characteristics which set them apart from normals were: 1. the failure to generate significant "vesical" (cecal pouch) wall tension; 2. very low residual energy in the projected stream, with energy losses in excess of 90% incurred in transporting urine across the urethra; 3. instability of energy dissipation suggesting instability of the bladder neck-urethtal geometry during voiding. The hydrodynamic data resembled that seen in severe proximal urethral obstruction (e.g. prostatism, bladder neck contraction). A unifying mechanism migth relate the compliance of the bladder neck-urethral tissues and the mural tension generated by the detrusor contractor.

References

Byrne, J. C., Gleason, D. M., Bottaccini, M. R.: Energy Loss During Micturition. Inv. Urol. **10**, 3 (1971).

Clegg, E. J.: The Musculature of the Human Prostatic Urethra. J. Anat. **91**, 345–351.

Garrets, B. von: Micturition in the Normal Male. Acta Cher. Scand. **114**, 197 (1958).

Gleason, D. M., Byrne, J. C., Bottaccini, M. R.: The Interrelationship of Some Hydrodynamics Parameters of Voiding. Invest. Urol. **5**, 6 (1968).

Gleason, D. M., Byrne, J. C., Bottaccini, M. R.: The Residual Stream Energy is a Diagnostic Index of Male Urinary Outflow Obstruction. Inv. Urol. **10**, 72 (1972).

Gleason, D. M., Bottaccini, M. R., Perling, D., Lattimer, J. K.: A Challenge to Current Urodynamics Thought. J. Urol. **97**, 935 (1967).

Hutch, J. A.: A New Theory of the Anatomy of the Internal Urinary Sphincter and the Physiology of Micturition. Invest. Urol. **3**, 36 (1965).

Krantz, K. E.: The Anatomy of the Urethra and Anterior Vaginal Wall. Amer. J. Obs. Gyn. **62**, 374 (1951).

Lapides, J.: Structure and Function of the Internal Vesical Sphincter. J. Urol. **80**, 341 (1958).

Lapides, J., Ajemina, E. P., Stewart, B. H., Breahey, B. A., Lichtwardt, J. R.: Further Observations on the Kinetics of the Urethral Vesical Sphincter. J. Urol. **84**, 86 (1960).

Woodburne, R. T.: Stricture and Function of the Urinary Bladder. J. Urol. **84**, 79 (1960).

Woodburne, R. T.: Anatomy of the Bladder, p. 3. The Neurogenic Bladder. Edit. S. Boyarsky, Baltimore: Williams and Wilkins Co. 1967.

Summary and Conclusions of the 1st International Symposium on Urodynamics

W. LUTZEYER *

The foregoing papers and discussions represent our present knowledge of urodynamics. That there is some clash of opinions shows the change from established knowledge based on experiment and clinical experience to hypotheses derived from the results of modern biological investigations and physiological methods of examination. They confirm the topicality of the subject, of the interdependence of morphology and function and their incorporation into the system of regulation by neural and humoral factors.

Urodynamic studies in situ are problematic because of possible technical artefacts. In contrast to the bladder and urethra, there are few effective methods that can be applied to the renal pelvis and the ureter.

Functional diagnosis of the upper urinary tract may be based on urorheography and ureteromanometry, beside the well-established roentgen examination. The lower urinary tract can be studied by simply measuring bladder capacity, and by cystomanometry, sphincteromanometry and electromyography of the pelvic floor. This ensures a sufficient dynamic and static exploration of the urinary tract.

The term "obstruction" was clearly defined by HINMAN and the etiology and clinical manifestations of functional stenosis were discussed. The causes of disturbances of urodynamics, as of infections and stone formation are of special clinical importance and need to be clarified in terms of function.

There is a general tendency to develop urodynamic investigation methods which need no intraluminal instrumentation (HINMAN). It is already possible to draw conclusions from the external urinary stream on the function of the bladder and urethra. The electrical signals are computerized and analysed. The number, size and velocity of the urine drops allow a sufficient analysis of bladder-emptying disorders of different genesis (ZINNER, RITTER). If urinary-stream phonography (STERLING) is used incombination with measurement of the urinary flow, it is possible to identify functional disorders in the urethra.

This kind of a urodynamic exploration requires a mathematical analysis of the results and so demands close cooperation between physician, bioengineer and mathematician.

---

* Abteilung Urologie der Medizinischen Fakultät der Rheinisch-Westfälischen Technischen Hochschule Aachen, Germany

Interdisciplinary cooperation is essential for optimal analysis of biological functions. This involves:

1. Development of physiological investigation methods,
2. development of theoretical models,
3. mathematical abstraction and
4. clinical interpretation.

The morphology of the ureter is characterized by the helical arrangement of the ureteral musculature with distinct climbing angles, despite the controversial results of ALLEN who found different runs in different ureteral segments. The structure of the pyelo-ureteral segment and its definition as a sphincter (MURNAGHAN) exemplity the ureteral architecture. The embryological architecture of certain mesenchymal cells around the ureteral tube will determine the later muscular arrangement of the ureter (TANAGHO). Alterations during the embryonic stage may lead to a longitudinal arrangement of the muscle fibres (ALLEN). This is one possible cause of so-called functional ureteral stenosis.

The nervous control of ureteral peristalsis is difficult to estimate, because such indirect factors as glomerular filtration and sodium excretion are regulated by the sympathetic nerves (BOYARSKY). The dependence on bolus volume, contraction frequency and contraction intensity (MORALES, BOYARSKY) is in balance with the sympathetic nervous control (BOYARSKY, MELCHIOR). KIIL locates the primary site of nervous action in the autoregulation of renal blood flow. Renal blood flow and intrarenal pressure are the parameters which influence diuresis. Nervous control of the ureters cannot be considered separately from the problem of a trigger system or a pacemaker in the ureter: NOTLEY regards the mechanical dilatation of smooth muscle as the impulse generator for contraction, whereas GOLENHOFEN and HANNAPEL advance the hypothesis of a pacemaker system in the renal calyces and the pelvis. Such a system might be demonstrated physiologically (electromyography) but could not be confirmed by a morphological substrate. Control of peristaltic urinary transport could be explained by either or both of these hypotheses. TANAGHO supports the concept of mechanical dilatation as the control system from his clinical and experimental studies. GOLENHOFEN and coworkers base their idea of a pacemaker system on physiological experiments which cannot be altered by dilatation. According to these experiments, the stimulus conduction shows myogenic nature, as demonstrated by an isolated pyelo-ureteral segment in which no alteration of peristalsis was observed after blockade of all nervous elements. Electromyographic studies confirmed that a trigger system or pacemaker is a general cell condition of the renal calyces and pelvis, needing no special morphological arrangements.

The innervation of the ureter was again discussed relative to the demonstration of ganglionic cells and nerve fibres in the ureteral wall: ganglionic cells in the terminal ureter have nothing in common with the well-known periureteral ganglionic cells in the adventitia of the trigone or the base plate.

Electron-microscope studies failed to show any ganglionic cells in the ureter. NOTLEY believes these were misinterpretations of light-microscope pictures.

Electromyography of the ureter is problematic because of the difficulty of interpretation and the still imperfect instrumentation.

The pharmacodynamics of the ureter is difficult to interpret: the spontaneous activity of the ureter as well as its dependence on kidney and bladder function may decisively alter the effect of drugs on the ureter (KILL). For the secondary effects of drugs, a new classification of the general subdivisions of α- and β-receptors is needed (BOYARSKY). The complex processes of ureteral peristalsis can only be expressed by a number of parameters, obtained under constant and reproducible conditions.

The discussions showed clearly that so far no differentiated, exact concept of urodynamics has emerged. There is a variety of examination methods, models and hypotheses, as well as the experimental arrangements and clinical observations employed by the different groups who presented their ideas.

Theories about the structure and function of the ureter, morphological examinations (electron-microscopy, histochemistry) and physiological activity with nervous, humoral, or autonomous regulation represent the basis for further hypotheses.

Our understanding of congenital disturbances of the ureter and bladder dynamics will be advanced by further embryological studies.

Studies relaxed to the concept of pacemakers or trigger systems in the upper urinary tract in connection with pharmacodynamic investigations are time-consuming and difficult. They should certainly yield important results for better understanding.

The term "functional stenosis" has been redefined (ALLEN, MELCHIOR, NOTLEY, RATHERT, TANAGHO, ZIMSKIND). TANAGHO considers responsible an abnormal arrangement of the muscle fibres at the site of the obstruction, while ZIMSKIND regards the obstruction as specifically functional without correlation to morphology. RATHERT and MELCHIOR see the different types as representing stages of one and the same pathological process. The etiology of primary, nonobstructive megaloureter can be explained by abnormal structure of the terminal ureteral segment (GREGOIR).

The different views of the function of the bladder neck are of special interest in relation to continence and incontinence. There is general agreement that a positive pressure difference must exist between bladder and urethra to maintain urinary continence (BOYARSKY, HINMAN, TANAGHO, TURNER-WARWICK). Additional parts in the distal urethra are present as buffers in case of a pressure drop. Small segments of the urethra are sufficient to maintain continence.

The bladder neck, proximal urethra and external sphincter urethrae maintain continence alone or in combination. A model for the dominance of the different factors and the morphological substrate of continence is supplied by radical prostatectomy. Urethral occlusion is effected by sphincter regeneration out of detrusor muscle fibres (TANAGHO), neosphincter (HINMAN), the function of the single external sphincter urethrae muscle (TURNER-WARWICK, GIL VERNET), external sphincter muscle and distal urethral musculature (DONKER), and the combination of such tissue-specific qualities as muscularis elastica and the collagen structures and their morphological and functional interaction.

The clinical indication for electrical bladder stimulation as promoting faster bladder rehabilitation in cases of traumatic lesion of the spinal cord is not convincing (ICHIM, BURGHELE). Such side effects as simultaneous stimulation of adjacent contractile structures, especially the bladder neck, very often make bladder stimulation ineffective. In neurogenous bladder disorders a very subtile and differentiated therapy should be possible. In view of the different functional activities of the sphincter and detrusor, special therapy should be applied instead of simple radical resection of the external sphincter.

Different interpretations are offered of the functional anatomy of the ureteral-vesical junction in cases of vesico-ureteral reflux. The intramural oblique tunnel and the submucosal part of the terminal ureter are seen in combination with maturation during the perinatal period (GIL VERNET), and muscle defects of the terminal ureteral segment (GREGOIR, TANAGHO) are brought into focus by the theory of an embryonal retardation anomaly (TANAGHO): the musculature of the ureter matures from cranical to caudal. This accounts for all stages, from a small reduction of the muscle fibres up to a total muscular defect. This theory can explain the different grades of vesico-ureteral reflux. The importance of other embryological factors for the length of the intramural segment, e.g. the trigone, is stressed by GREGOIR and MELCHIOR.

On the basis of their experimental investigation, GRASSET and GREGOIR do not consider that infection can be a cause of reflux.

Acquired vesico-ureteral reflux of the adult very seldom leads to kidney destruction (BOYARSKY).

The indication for surgical treatment of reflux is generally accepted, especially as early as possible when there are morphological or functional disturbances in the ureter, renal calyces and pelvis, or the kidneys.

Subject Index

MIX
Papier aus verantwortungsvollen Quellen
Paper from responsible sources
FSC® C105338

If you have any concerns about our products,
you can contact us on
ProductSafety@springernature.com

In case Publisher is established outside the EU,
the EU authorized representative is:
Springer Nature Customer Service Center GmbH
Europaplatz 3, 69115 Heidelberg, Germany

Printed by Libri Plureos GmbH
in Hamburg, Germany